Second China Digital Earth Conference

IOP Conference Series: Earth and Environmental Science
Volume 1004

Beijing, China
6 – 8 September 2021

ISBN: 978-1-7138-5446-3
ISSN: 1755-1307

Printed from e-media with permission by:

Curran Associates, Inc.
57 Morehouse Lane
Red Hook, NY 12571

CURRAN ASSOCIATES INC.
proceedings
.com

Some format issues inherent in the e-media version may also appear in this print version.

Institute of Physics
Dirac House, Temple Back
Bristol BS1 6BE UK

Phone: 44 1 17 929 7481
Fax: 44 1 17 920 0979

techtracking@iop.org

Additional copies of this publication are available from:

Curran Associates, Inc.
57 Morehouse Lane
Red Hook, NY 12571 USA
Phone: 845-758-0400
Fax: 845-758-2633
Email: curran@proceedings.com
Web: www.proceedings.com

TABLE OF CONTENTS

The 2nd China Digital Earth Conference

IOP Conf. Series: Earth and Environmental Science 1004 (2022) 011001 doi:10.1088/1755-1315/1004/1/011001

The 2nd China Digital Earth Conference
6-8 September 2021, Beijing, China

Conference Chair

GUO Huodong

Aerospace Information Research Institute, Chinese Academy of Sciences

Editors

CHEN Fulong, NI Wenjian, LIU Zhen, YANG Jiangning, WANG Lei

Aerospace Information Research Institute, Chinese Academy of Sciences

IOP Conf. Series: Earth and Environmental Science 1004 (2022) 011001 doi:10.1088/1755-1315/1004/1/011001

Contents

IOP Conf. Series: Earth and Environmental Science 1004 (2022) 011001 doi:10.1088/1755-1315/1004/1/011001

Digital Earth Technology

Image enhancement method for visual localization and mapping in complex light scenarios

S J Jia, C Liu, H B Wu, C Chen

Active Contour Building Segmentation Model based on Convolution Neural Network

M J Liu, P Liu, B Z Song, Y W Zhang, L Zhang

Digital Earth and Big Data

A Standardized GIS Web Service Oriented Shared-nothing Architecture - Taking Species Distribution Modeling as an Example

X Y Liu, J S D Yu, D D Wei, R J Tong, D Y Li, S Luo

GeoMapViz: a framework for distributed management and geospatial data visualization based on massive spatiotemporal data streams

Q Xu, L G Xiang, H C Wang, X F Guan, H Y Wu

Digital Earth Applications

Real-Time Data Infrastructure for Sea Ice Monitoring in Key Arctic Waters

B Lin

The impacts of trees' canopy occupation on pedestrian thermal environment of township streets

Y M Liu, D Lin, Y B Zhu, C C Chen

Optimal spatial resolution of remote-sensing imagery for monitoring cantaloupe greenhouses

B Cui, W J Huang, H C Ye, Q X Chen, Z C Li, H Y Jiang

Remote-sensing monitoring of rubber plantations using object-oriented characteristics from the vigorous period

B Cui, W J Huang, H C Ye, Q X Chen, Z C Li, H Y Jiang

IOP Conf. Series: Earth and Environmental Science 1004 (2022) 011001 doi:10.1088/1755-1315/1004/1/011001

IOP Conf. Series: Earth and Environmental Science 1004 (2022) 011001 doi:10.1088/1755-1315/1004/1/011001

Preface

The 2nd China Digital Earth Conference

September 6 – 8, 2021

The 2nd China Digital Earth Conference (CDEC) was successfully convened in Beijing, China, from September 6 – 8, 2021. The conference was hosted by the International Society for Digital Earth and the Chinese Academy of Sciences.

With the theme of "Digital Earth promotes sustainable development", CDEC deeply discussed the important role of Digital Earth in achieving the Sustainable Development Goals (SDGs) of the United Nations (UN) through the theoretical and technological innovation, application demonstration and industry development of Digital Earth. It provided a friendly and applicable communication platform for experts and scholars, decision-makers and management officers from various fields and departments to exchange their ideas, experiences and practices.

Since the first International Digital Earth Conference held in 1999 in Beijing, Digital Earth has been developed as a cutting-edge interdisciplinary direction in worldwide, and will increasingly play an important role on the promoting, evaluation and implementation of UN SDG goals, in particular with the advent of the era of Big data and the prosperous development of the digital economy in China.

Editors: CHEN Fulong, NI Wenjian, LIU Zhen, YANG Jiangning, WANG Lei
International Society for Digital Earth
Key Laboratory of Digital Earth Science, Chinese Academy of Sciences

List of Hosts

International Society for Digital Earth
Chinese Academy of Sciences

List of Co-organizers

Chinese National Committee of International Society for Digital Earth
The Big Earth Data Science Engineering Project
The Academy of Digital China (Fujian)
Aerospace Information Research Institute, Chinese Academy of Sciences
Key Laboratory of Digital Earth Science, Chinese Academy of Sciences
State Key Laboratory of Remote Sensing Science
Hainan Key Laboratory of Earth Observation
International Centre on Space Technologies for Natural and Cultural Heritage under the Auspices of UNESCO
Digital Belt and Road Program (DBAR)
CAS-TWAS Centre of Excellence on Space Technology for Disaster Mitigation (SDIM)
Integrated Research on Disaster Risk (IRDR)

IOP Conf. Series: Earth and Environmental Science 1004 (2022) 011002 doi:10.1088/1755-1315/1004/1/011002

Peer Review Statement

All papers published in this volume of IOP Conference Series: Earth and Environmental Science have been peer reviewed through processes administered by the proceedings Editors. Reviews were conducted by expert referees to the professional and scientific standards expected of a proceedings journal published by IOP Publishing. 80% submitted papers have been accepted for publication after the peer-review process.

Characteristics of the Geological Tourism Resources in Hainan Island, China

Q J Wang [1,2*,3], Z C Li[1], Y Chen[1,2]

[1] Key Laboratory of Earth Observation of Hainan Province, Hainan Research Institute, Aerospace Information Research Institute, Chinese Academy of Sciences, 572029, China;
[2] Key Laboratory of Digital Earth Science, Aerospace Information Research Institute, Chinese Academy of Sciences, Beijing 100094, China
[3] University of Chinese Academy of Sciences, Beijing 101408, China

Corresponding author: wangqj@radi.ac.cn

Abstract. In order to deep mine Hainan geological tourism resources, based on remote sensing, geology, geography and field investigation, a Hainan geological tourism resources evaluation method was established, which included data collection, data preprocessing, geological tourism monomers extraction using remote sensing imageries. According to the national standard of classification, investigation and evaluation of tourism resources (GB/T18972-2003), 146 geological tourism monomers in Hainan Island were firstly evaluated. The results showed that there were 13 geological tourism resource types in Hainan Island with the characteristics of large quantity, high quality, long travel time and convenient transportation. Finally, the future development of Hainan geological tourism was sugessted to further strengthen on the geological environment stability to provide references for geological tourism economy development in Hainan Island.

1. Introduction

Geological tourism resources are fossils, rock masses, structural features, ore deposits, geomorphic landscapes and other scenes preserved in the rock stratum due to crustal tectonic changes, magmatic activities, paleogeographic environment evolution, paleontological evolution and other factors in the long evolution process of the earth. They have the value of scientific research, universal education, and attractions to tourists. Geological tourism is a tourism that enables people to understand the knowledge of geology, geography and other related disciplines by visiting and investigating a series of geological heritages and landscapes [1-3].

At present, investigations of tourism resources in Hainan are mostly focused on the analysis on its natural and humanistic characteristics [4-6], investigation and evaluation systems [7-10], and development strategic planning, etc [11-12].

However, there are few studies on the characteristics of the geological tourism resources in Hainan Island. In order to further explore the geological tourism resources, this paper combined remote sensing, geology, geography and field investigations to establish a Hainan geological tourism resources evaluation method. The results are expected to provide a reference for the development of geological tourism economy in such areas as Hainan Island.

2. Study area

Hainan Island is the largest tropical island in China. With the land area of 33920 km^2, its coordinate is 108°37′-111°03′ E and 18°10′-20°10′N and has convenient traffic networks, such as the "three vertical" highway (Haiyu east line, Haiyu middle line and Haiyu west line), Huandao expressway, Eastern Ring Railway, western line railway and Western Ring Railway. In addition, Haikou Meilan International Airport, Sanya Phoenix International Airport, Haikou Port, Sanya port, Basuo port, Qinglan port and Yangpu Port are connected at home and abroad.

Hainan Island has a tropical monsoon marine climate. The annual average temperature is 23.8℃. In which, January is the coldest, with an average temperature of about 18.0℃. Frost can occur in the central mountainous areas, and the extreme minimum temperature is 0.1℃; July is the hottest month, with an average temperature of 28.2 ℃ and an extreme maximum temperature of 39℃. The rainfall of Hainan Island varies greatly from place to place, and is controlled by seasonality. The rainy season lasts from May to October, accounting for 75.5% ~ 90.0% of the annual rainfall; The dry season is from November to April, and the rainfall accounts for only 10.0% ~ 24.5% [13].

Terrain of Hainan Island is high in the middle and low in around. Landforms of the study area are composed of mountains, platforms and plains. In which, with an elevation of 500-800m, mountains are main landforms and distributed in the center of the island, which account for 38.7% of total area.

There are 154 rivers in the study area, most of which originate from the central mountainous and 38 rivers have a catchment area of more than 100 km^2. Nandu River, Changhua River and Wanquan River are the three major rivers in Hainan Island, with a catchment area of more than 3000 km^2, and their drainage areas account for 47% of the whole island. It is these rivers that form beautiful waterfalls in mountains and become an important part of geological tourism resources.

With oceans, mountains, rivers, lakes and coastal zones, Hainan island has formed more than 100 famous scenic spots. Among them, there are 1 World Geopark, 2 National Geoparks, 6 scenic spots with 5A grade, 21 scenic spots with 4A grade, 10 national protected areas and 23 provincial protected areas.

3. Methodology

Shown in the Figure 1, the method mainly includes four steps: data collection, data preprocessing, extraction of geological tourism monomers using remote sensing imageries and evaluation on geological tourism resources based on field investigations [14].

(1) Data collection

Remote sensing images, geological, topographic and other basic geographic information data of Hainan Island were collected to form a priori background knowledge database.

(2) Data preprocessing

After collecting data, they are registed using a unified projection to make an target on different maps in a same location. Then, they are put together to form a database.

(3) Geological tourism monomers extraction using remote sensing imageries.

Various geological tourism monomers in Hainan island, such as coasts, islands, valleys, rivers, caves, etc., are extracted from remote sensing imageries. Their characteristics on true colored remote sensing images were shown in table 1. From which, we can extract different geological tourism monomers respectively.

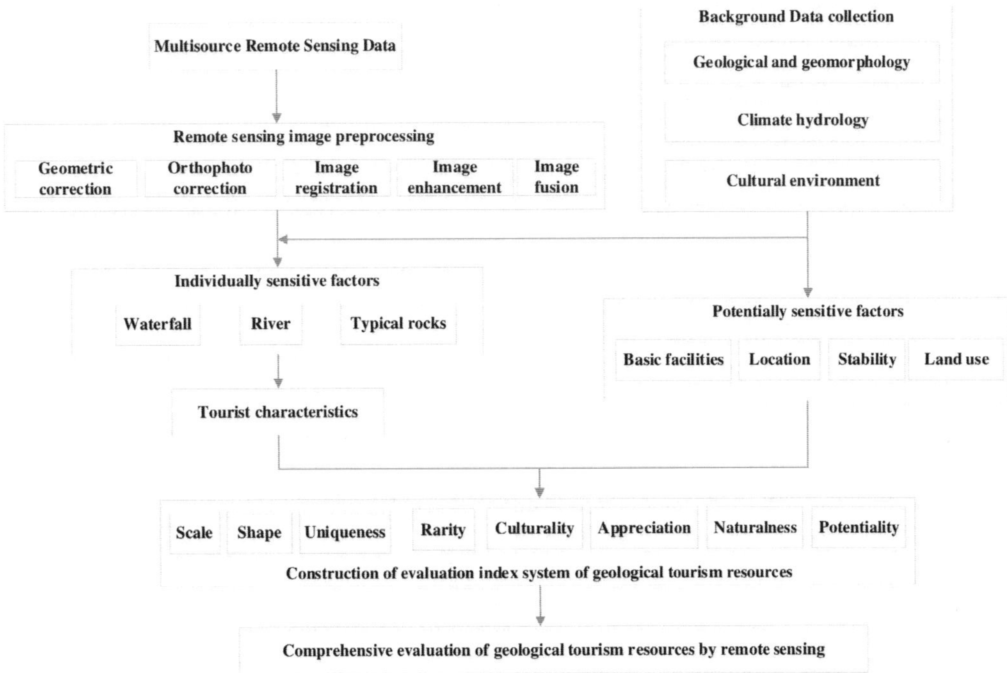

Figure 1. Flow chart of the geological tourism resources evaluation

Table 1. Characteristics of geological tourism monomers on true colored remote sensing images

Geological tourism monomers	Characteristics on true colored remote sensing images
Coasts	One side is land with white or green colors, the other is sea with blue or black colors.
Islands	Small land with white or green colors in the sea whose color is blue or black.
Rivers	Curved shapes with black or blue colors.
Caves	Round shapes with black colors.
Beaches	In land with belt shapes and white colors near to sea.
Hills	Higher altitude to backgrounds with green vegetation.
Volcano and lavas	Round shapes with green vegetation.
Hot springs	Higher temperature to backgrounds with poisonous vegetation whose color is slightly yellow.
Valleys	Long and narrow shapes with black or blue colors.
Waterfalls	Higher elevation difference along rivers.
Lakes	Round shapes filled by water with black or blue colors .
Ore accumulations	Round shapes caused by artificial activities with white, blue or red colors.
Meteorite craters	Round shapes caused by meteorite activities with green vegetation.

(4) Evaluation on geological tourism resources based on field investigations

According to the standard of classification, investigation and evaluation of tourism resources (GB / t18972-2003), from the aspects of geographical location, structure, composition, genetic mechanism and evolution process, scale and volume, environmental backgrounds, regional access conditions, protection and development status, etc., field investigations were carried out on each Hainan geological tourism monomer in terms of ornamental and recreational value, historical and cultural value, scientific value, rarity or strangeness, scale and abundance, integrity, popularity and influence, scope of use, pollution status, etc.

According to the evaluation criteria of geological tourism resources (Table 2) [15], the expert scoring method is used. It mainly includes the individual quality and structure evaluation, development conditions and comprehensive evaluation, and the final evaluation score is obtained. According to the total score, monomers are divided into five grades: grade 5 tourism resources score is more than 90; grade 4 tourism resources score is between 75 and 89; grade 3 tourism resources score is between 60 and 74; grade 2 tourism resources score is between 45 and 59, and grade 1 tourism resources score is between 30 to 44.

Table 2. Evaluation criteria of geological tourism resources

Items	Criteria	
The ornamental value, recreation value or use value provided by the monomer for tourists	All or one of them has the highest ornamental value, recreational value and use value.	30-22
	All or one of them has the higher ornamental value, recreational value and use value.	21-13
	All or one of them has high ornamental value, recreational value and use value.	12-6
	All or one of them has general ornamental value, recreational value and use value.	5-1
The historical value, or cultural value, or scientific value, or artistic value contained in the monomer	At the same time or one of them has historical value, cultural value, scientific value and artistic value of world significance	25-20
	At the same time or one of them has national historical value, cultural value, scientific value and artistic value	19-13
	At the same time or one of them has provincial historical value, cultural value, scientific value and artistic value	12-6
	Historical value, or cultural value, or scientific value, or artistic value has regional significance	5-1
Rarity and singularity	There are a large number of rare species, or the landscape is very strange, or such phenomena are rare in other areas	15-13
	There are many rare species, or the landscape is strange, or such phenomena are rare in other areas	12-9
	There are a few rare species, or the landscape is prominent, or such phenomena are rare in other areas	8-4
	There are some rare species, or the landscape is more prominent, or such phenomena are more common in other areas	3-1
Individual size, group structure and density	The scale and volume of independent monomer are huge; Tourism resources have perfect single structure and excellent density; Natural scenes and human activities occur periodically or frequently	10-8
	Independent monomer has large scale and volume; The single structure of combined tourism resources is very harmonious and the density is good; Natural scenes and human activities occur periodically or frequently.	7-5
	Independent monomer with medium scale and volume; The individual structure of combined tourism resources is harmonious and the density is good; Natural scenes and human activities occur periodically or frequently	4-3
	Independent monomer has small scale and volume; The	2-1

	single structure of combined tourism resources is harmonious and the density is general; Natural scenes and human activities occur periodically or less frequently	
Is it disturbed and damaged by nature or man-made, or is the preservation completed?	Keep the original shape and structure	5-4
	The morphology and structure changed a little, but not obvious	3
	There are obvious changes in morphology and structure	2
	Significant changes in morphology and structure	1
Popularity	Well-known in the world or constitutes a famous brand recognized by the world	10-8
	Well-known nationwide or constitutes a national famous brand	7-5
	Well-known in the province or constitutes a famous brand in the province	4-3
	Well-known in the region or constitutes a famous brand in the region	2-1
Travel time	Suitable sightseeing days exceed 300 days a year, or suitable for all tourists to use and participate	5-4
	Suitable sightseeing days are more than 250 days a year, or it is suitable for about 80% of tourists to use and participate	3
	Suitable for sightseeing days are more than 150 days a year, or it is suitable for about 60% of tourists to use and participate	2
	Suitable for sightseeing days are more than 100 days a year, or it is suitable for about 40% of tourists to use and participate	1
Environmental safety	It has been seriously polluted or has serious potential safety hazards.	-5
	It has been moderately polluted or has obvious potential safety hazards	-4
	It has been slightly polluted or has certain potential safety hazards	-3
	Engineering protection measures have been taken to ensure environmental safety.	3

4. Results

According to the method, 146 geological tourism monomers in Hainan Island were firstly evaluated combining with the evaluation results of Hainan Geological Survey Institute [1]. In which, 5 field investigations in November 2016, July 2017, March 2018, July 2018 and November 2018 were carried out.

The results showed that Hainan Island was rich in geological tourism resources. There were 85 excellent geological tourism monomers, including 13 basic geological tourism resources: islands, rocks, beaches, hills, volcano and lavas, rock caves, hot springs, valleys, waterfalls, lakes, seismic relics, ore accumulations and meteorite craters. Among these types, beaches accounted for 32.9%, followed by hills and islands, with a proportion of 15.3% respectively; Next was the volcano and lavas, with a proportion of 10.6%.

From Figure 2, we can see that the coastal areas are mainly islands and beaches, the northern volcanic eruption hilly areas are mainly volcanic lava, the low mountain areas in the Eastern are mainly waterfalls and hot springs, the high mountain areas in the middle are mainly valleys and waterfalls, and the low sedimentary mountain areas in the Western and Southern are mainly rocks, caves and waters.

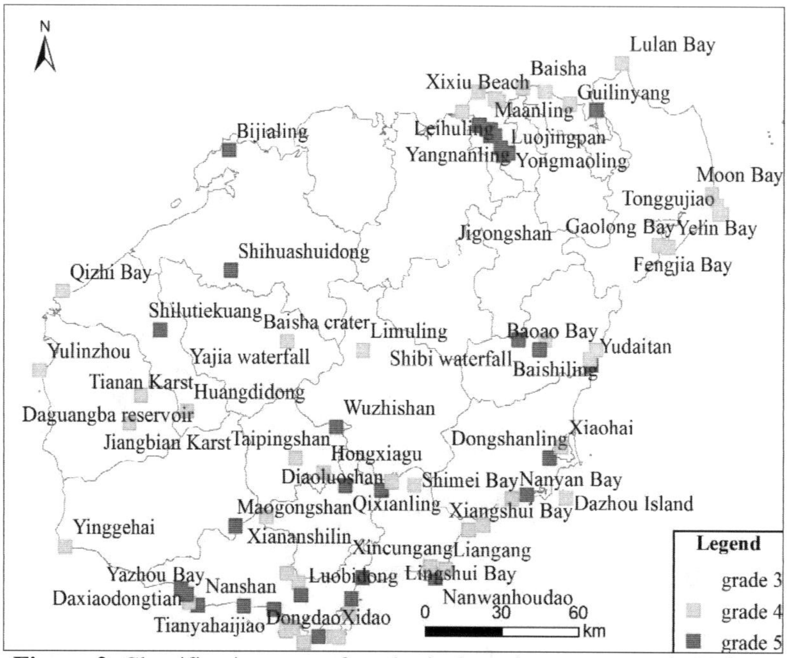

Figure 2. Classification map of geological tourism resources in Hainan

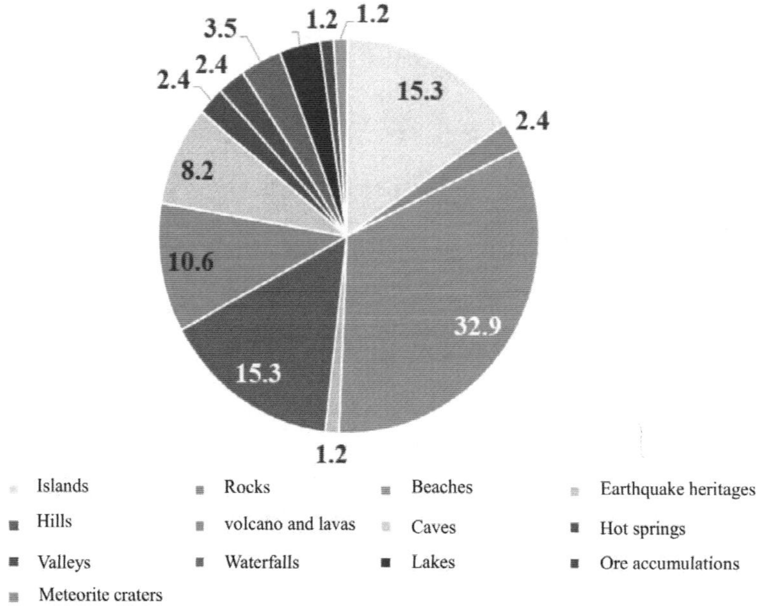

Figure 3. Proportion chart of different geological tourism resources in Hainan

5. Conclusions

In this paper, 146 geological tourism monomers in Hainan Island were firstly evaluated. Main conclusions can be drawn as follows.

(1) Hainan Island is rich in geological tourism resources

Most geological tourism monomers in Hainan Island are excellent geological tourism monomers, which include islands, rocks, beaches, hills, volcano and lavas, caves, hot springs, valleys, waterfalls, lakes, earthquake heritages, ore accumulations and meteorite craters.

(2) Hainan Island geological tourism resources have the characteristics of wide suitable population and long tourism time.

The geological tourism resources in Hainan Island can be available for a long time. In winter, Sanya in the southern is an ideal place for overwintering; When the temperature is high in summer, the central mountains are summer resorts; In spring and autumn, most of the scenic spots in Hainan Island are suiltable for whole family to travel mountains, rivers and leisure sightseeings.

(3) A Hainan Island geological tourism resources evaluation method was established.

Combing remote sensing, geology, geomorphology and field investigations, a Hainan geological tourism resources evaluation method was established, which provided a reference for geological tourism resources evaluation in similar scenic spots.

Most scenic spots in Hainan Island are relatively mature, but the traffic conditions of some scenic spots, especially the geological tourism spots that have not been commercialized, are needed to be further improved in geological tourism infrastructures and popularities.

In future, it is necessary to strengthen on the transformation of geological tourism infrastructures, evaluation on geological environment stability, detailed investigations on geological tourism resources etc in Hainan Island.

References

[1] Xue G C 2011 Demonstration report on Tourism Geological Survey and Evaluation in Hainan Hainan Geological Survey Institute

[2] Dong H M, Liu H F 2017 Analysis on the Concept and Characteristics of Geological Tourism *Tourism overview.* **10** pp 65-67

[3] He X Q, Liu Y, Xiong G B 2015 Current Situation and Prospects of Studies on Geotourism *Tropical Geography* vol 35 **1** pp 130-138

[4] Wang X J 1996 Structural Characteristics and Development Evaluation of Tourism Resources in Hainan *Tropical Geography* vol 16 **2** pp175-182

[5] Li W T 2020 SWOT Analysis of Tourism Development in Hainan Island under the Background of Global Tourism.*Technology and economy guide* vol 28 **17** pp 87-88

[6] Wei X D 2020 Analysis on the Current Situation of Global Tourism Development in Hainan Province *Economic Research Guide* **15** pp 148-150+156.

[7] Fu G J 2010 Investigation, Classification and Evaluation of Natural Tourism Resources in Hainan *Natural Science Journal of Hainan University* 28 **1** pp 52-58

[8] Yang X J, Wei X D, Liang Y et al 2020 Evaluation of Tourism Ecological Efficiency based on SBM-DEA model -- a case study of Hainan Province *China Economic and Trade Guide* **10** pp 41-44

[9] Wang M Z, Luo Z Y, Han F 2020 Investigation and Study on Tourism Geological Resources of Paracel Islands *Rural Economy and Technology* vol 31 **17** pp 55-57+64

[10] Chen Y X, Xie R J 2020 Classification and Evaluation of Rural Tourism Resources in Yazhou District, Sanya *Technology and Industry* vol 20 **7** pp 71-77.

[11] Fu Q J, Lin C, Xu X X 2015 Preliminary Study on the Development of Karst Geological Relic Landscape in Tian'an, Hainan Province *West-China Exploration Engineering* **6** pp 117-119

[12] Zhao H X 2020 Study on the Coordinated Development of Regional Tourism Industry and Economy under the Construction of Hainan Free Trade Port *Financial Circles* **20** pp 21-22

[13] Office of Hainan Local Chronicles Compilation Committee 1992 Hainan chronicles series "summary of Hainan county situation" *Nanhai Press*

[14] Wang Q J, Chen Y, Zhou H Y 2017 Investigation and Evaluation of Sanya Geological Tourism Resources. *Journal of Hainan Normal University(Natural Science)* vol 30 **4** pp 443-449

[15] General Administration of Quality Supervision Inspection and Quarantine of the People's Republic of China 2003. Classification, Investigation and Evaluation of Tourism Resources (GB/T18972-2003)

Acknowledgments
This work was financially supported by Hainan Provincial Department of Science and Technology under Grant No. ZDKJ2019006, and Hundred person project of Hainan province under Grant No. [2018]31.

Soil erosion dynamic monitoring in the Northeast Inner Mongolia of China from 1978 to 2018

Wei Jiang[1,2,3], Zhiguo Pang[1,2], Xiangdong Qin[1,2], Hongrun Ju[4], Ding Xiaohui[5], Lin Li [1,2], June Fu[1,2] Hao Li[1,2], Jingxuan Lu[1,2], and Tianshi Feng[1,2]

[1]China Institute of Water Resources and Hydropower Research, Beijing 100038, China;

[2]Research Center of Flood and Drought Disaster Reduction of the Ministry of Water Resources, Beijing 100038, China;

[3]Postdoctoral Workstation of China Reinsurance (Group) Corporation, Beijing 100033, China;

[4]School of Tourism and Geography Science, Qingdao University, Qingdao 266071, China;

[5]Guangzhou Institute of Geography, Guangdong Academy of Sciences, Guangzhou 510070, China

E-mail: pangzg@iwhr.com

Abstract. The black soil area in Northeast Inner Mongolia is an important ecological barrier and food production base in China and the soil erosion is a key risk factor in regional ecological security and sustainable development. Multi-source satellite remote sensing and auxiliary data, including meteorology, land use, vegetation coverage and digital elevation were collected to establish a distributed model of soil erosion. Then, the quantitative analysis of soil erosion in the Northeast Inner Mongolia from 1978 to 2018 was carried out. The results show the following: (1) the areas of soil erosion in 1978 and 2018 were 203,538 km^2 and 102,886 km^2 respectively, and the area of soil erosion was decreased by 100,653 km^2. The decreased of soil erosion was mainly located in the Songliao Plain Sand Prevention Farmland Protection Zone; (2) wind erosion is the main erosion type in the Northeast Inner Mongolia. Wind erosion area decreased from 188,982 km^2 in 1978 to 88,466 km^2 in 2018, and water erosion area decreased from 14,556 km^2 in 1978 to 14,420 km^2 in 2018.(3) the intensity of soil erosion is mainly mild. The area change rates of mild, moderate, strong, intensive and severe soil erosion intensity are -48.91%, -41.41%, 32.63%, -91.15% and -91.14% respectively. This study demonstrated how to use multi-source remote sensing technology to explore the long term series soil erosion dynamic monitoring and provided support for decision-making and management of soil erosion in Northeast Inner Mongolia under the background of sustainable development.

IOP Conf. Series: Earth and Environmental Science 1004 (2022) 012002 doi:10.1088/1755-1315/1004/1/012002

1. Introduction

Water and soil resources are the basic conditions and prerequisites for human existence and play a vital role in the sustainable development of social and economic [1]. With the impact of climate change and human activities, the issue of soil erosion has attracted widespread attention [1-2]. Generally, soil erosion leads to land degradation, sedimentation and ecological deterioration, which inevitably threats to ecological security and food production. Therefore, long-term series soil erosion investigation will be helpful to understand the evolution pattern of soil erosion scientifically, objectively evaluate the control effects of soil and water conservation measures and promote sustainable development of the natural environment and economic production [3-4].

Currently, site observation and remote sensing technology are the mainly means to conduct soil erosion dynamic monitoring [5]. Site observation can achieve long-term series monitoring in local area by installing a series of instruments such as sand accumulators and rain gauges. Compared with site observation, remote sensing has the advantages of wide observation range, short observation interval and low cost, which provides potential for long-term soil erosion dynamic monitoring using multi-source satellite imagery. Various soil erosion parameters, including digital elevation model, vegetation cover and land use, can be extracted from multi-source satellite imagery to establish the distributed soil erosion model [6-7]. At present, there are many soil erosion models across worldwide and the universal soil loss equation is the widely used soil erosion model to assess the soil erosion at different spatial scale [8-9].

To investigate soil erosion dynamic monitoring in the Northeast Inner Mongolia over the past four decades, multi-source satellite imageries at five key time nodes (1978, 1985, 1995, 2005 and 2018) were collected to extract the soil erosion parameters. Then, the distributed model of soil erosion combined with various parameters was established to calculate the intensity of water erosion, wind erosion and soil erosion respectively. Finally, the dynamic changes of soil erosion were analysed to explore temporal and spatial pattern. This research will provide decision-making basis for soil and water conservation planning and management in the Northeast Inner Mongolia.

2. Study area

Northeast Inner Mongolia is situated in the northeast China with a land area of 387,488 km^2. The location and altitude of the study area are showed in figure 1 and it is one of the three largest black soil belts in the world, namely the northeast black soil area. The land is mainly covered by forest, meadow grassland and farmland and the climate is mid-temperate continental monsoon. Currently, the Northeast Inner Mongolia is the important base of livestock production, grain production and ecological security. Due to its unique climate and soil condition, coupled with the high intensity of human activities, the soil erosion has become the risk factor for sustainable development of the ecological environment and economic production.

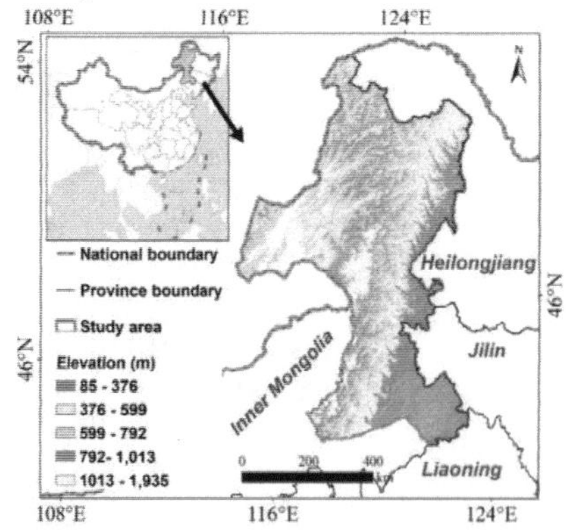

Figure 1. The location and altitude of Northeast Inner Mongolia

IOP Conf. Series: Earth and Environmental Science 1004 (2022) 012002 doi:10.1088/1755-1315/1004/1/012002

3. Methodology

According to the requirements of the provincial soil erosion dynamic monitoring planning, the technical flow chart is shown in figure 2 and the soil erosion type types were defined as wind erosion and water erosion. Firstly, the remote sensing data, meteorological data, topography data and field survey data were collected and pre-processed, and then, the soil erosion factors such as slope and length, wind speed, land use and vegetation coverage were extracted and resampled at 30 m spatial resolution. Secondly, according to soil erosion classification standards, the slope, length and vegetation coverage indicators were used to construct a distributed model of water erosion and the water erosion was reclassified into six levels (slight, mild, moderate, strong, intensive and severe). Moreover, the land use, wind speed and vegetation coverage were employed to calculate the distributed model of wind erosion by referring universal soil loss equation, and the wind erosion was also reclassified into six levels. The soil erosion (S) for each pixel was calculated used the followed formula:

$$S = \mathrm{Max}(S_{water}, S_{wind}) \quad (1)$$

where S_{water} represents the level of water erosion and the S_{wind} represents the level of wind erosion. If $S_{water} = S_{wind}$, then, the erosion type of the pixel is the wind erosion.

Finally, the dynamics of soil erosion in the Northeast Inner Mongolia from 1978 to 2018 were analysed from the perspective of spatial and temporal.

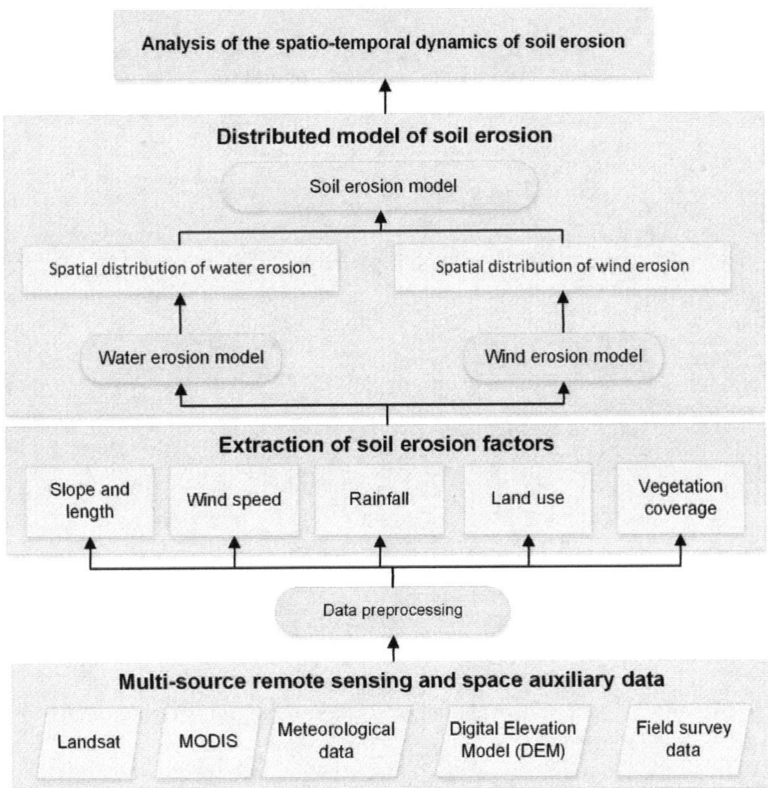

Figure 2. The flow of soil erosion dynamic monitoring

IOP Conf. Series: Earth and Environmental Science 1004 (2022) 012002 doi:10.1088/1755-1315/1004/1/012002

4. Results and discussion

4.1. Spatial Characteristics of Soil Erosion Dynamic Monitoring

Based on the flow of soil erosion dynamic monitoring, the soil erosion dynamic changes in the Northeast Inner Mongolia from 1978 to 2018 are demonstrated in figure 3. The spatial characteristics of soil erosion dynamic monitoring can be summarized as followed: (1) the intensity of soil erosion in the study area has shown a downward trend as a whole, and the area of soil erosion in the Northwest and Southwest regions has decreased, which indicates that the soil and water conservation measures have been significantly improved over the past four decades. (2)the erosion type in this area is dominated by wind erosion, and the decline areas are located in Hulunbel and Tongliao regions, which are highly related to the changes of vegetation coverage.

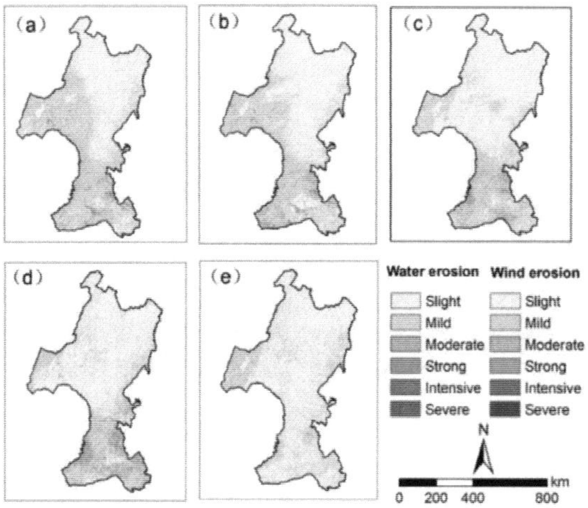

Figure 3. Soil erosion dynamic changes in the Northeast Inner Mongolia. (a) 1978, (b) 1985, (c) 1995, (d) 2005, (e) 2018.

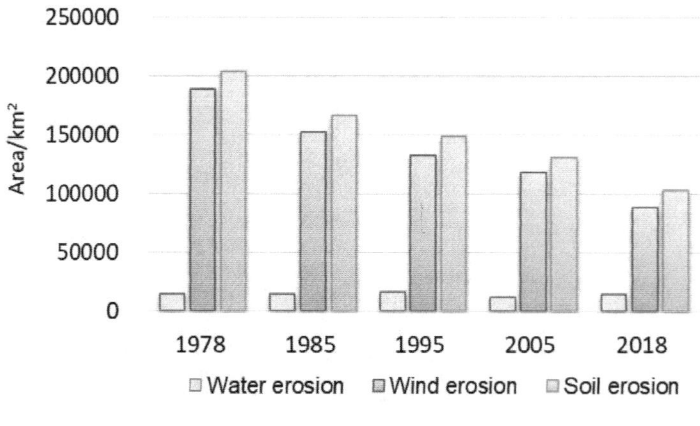

Figure 4. Soil erosion area change in the Northeast Inner Mongolia from 1978 to 2018.

4.2. Temporal Characteristics of Soil Erosion Dynamic Monitoring

The areas of soil erosion in the Northeast Inner Mongolia from 1978 to 2018 are shown in figure 4. The total area of soil erosion in the Northeast Inner Mongolia has been decreasing in the past 40 years. In 1978, the area of soil erosion was 203,538 km^2, while in 2018, the area of soil erosion was 102,884

IOP Conf. Series: Earth and Environmental Science 1004 (2022) 012002 doi:10.1088/1755-1315/1004/1/012002

km^2, a decrease of 100,654 km^2during the past four decades. In the process of decreasing soil erosion area, the wind erosion area decreased most significantly, with an area of 100,517 km^2. Combined with the spatial distribution map of soil erosion, the reduction areas of soil erosion are mainly located in the Songliao Plain sand-proof farmland protection area. This region is an important part of the fragile ecological environment in northeast China. The sharply soil erosion area decreased may be related to the sharply increase of the vegetation coverage of forest and grass after the establishment of wind-proof solidification sarin system.

4.3. Intensity Characteristics of Soil Erosion Dynamic Monitoring

The intensity of soil erosion in different periods is summarized in the table 1. From 1978 to 2018, the mild soil erosion accounted for the largest proportion in each period and the change rate of mild soil erosion was -48.91%. Moreover, the area of intensive and severe soil erosion decreased by more than 91%, followed by the mild and moderate soil erosion, while the area of strong soil erosion increased by 268 km^2. The results indicate that the decrease rate of high intensity soil erosion is sharply and decrease area of low intensity soil erosion is larger. Thus, the soil and water conservation measures such as restoring vegetation, controlling slope (channel) erosion and preventing grassland degradation have achieved great performance over the past four decades.

Table 1. Soil erosion intensity for different periods

Soil erosion intensity/ km^2	1978	1985	1995	2005	2018	Change rate (1978-2018)/%
Mild	187419	148008	124509	78775	95743	-48.91
Moderate	9448	10797	20102	42346	5535	-41.41
Strong	820	1852	2648	7949	1088	32.63
Intensive	4103	4281	1244	1349	363	-91.15
Severe	1748	1593	354	149	155	-91.14

5. Conclusions

This study introduced a method for soil erosion monitoring based on multi-source remote sensing data, and explored the temporal and spatial dynamic changes of soil erosion in Northeast Inner Mongolia from 1978 to 2018. The conclusions were as follows:

(1) In 1978 and 2018, the areas of soil erosion were 203,538 km^2 and 102,884 km^2 respectively, and the area of soil erosion decreased by 100,654 km^2, which were mainly located in the Songliao Plain Sand Prevention Farmland Protection Zone. The ecological environment management, especially the increase of vegetation coverage, made a great contribution to the decline of soil erosion.

(2) The type of soil erosion in Northeast Inner Mongolia is dominated by wind erosion. The area of wind erosion decreased from 188,982 km^2 in 1978 to 88,466 km^2 in 2018, and the area of water erosion decreased from 14,556 km^2 in 1978 to 14,420 km^2 in 2018.It was highly related to the wind-proof solidification sarin system.

(3) The intensity of soil erosion is mainly mild, and the area change rates of mild, moderate, strong, intensive and severe soil erosion intensity are -48.91%, -41.41%, 32.63%, -91.15%, -91.14%, respectively, indicating that the implementation of water and soil conservation project effectively curbed soil degradation.

Multi-source remote sensing technology plays a unique role in the long-term series dynamic monitoring of soil erosion, which is helpful to explore the pattern of soil erosion evolution and provide a decision-making basis for sustainable development of soil and water conservation planning. In the future, more meteorological data will be collected to conduct the freeze erosion assessment in the study area.

References

[1] Wang X, Zhao X, Zhang Z, et al. Assessment of soil erosion change and its relationships with land use/cover change in China from the end of the 1980s to 2010[J]. Catena, 2016, 137: 256-268.

[2] Guo B, Yang G, Zhang F, et al. Dynamic monitoring of soil erosion in the upper Minjiang catchment using an improved soil loss equation based on remote sensing and geographic information system [J]. Land Degradation & Development, 2018, 29(3): 521-533.

[3] Li X, Wu B, Zhang L. Dynamic monitoring of soil erosion for upper stream of Miyun Reservoir in the last 30 years[J]. Journal of mountain science, 2013, 10(5): 801-811.

[4] Litvin L F, Kiryukhina Z P, Krasnov S F, et al. Dynamics of agricultural soil erosion in European Russia [J]. Eurasian soil science, 2017, 50(11): 1344-1353.

[5] Biswas S S, Pani P. Estimation of soil erosion using RUSLE and GIS techniques: a case study of Barakar River basin, Jharkhand, India [J]. Modeling Earth Systems and Environment, 2015, 1(4): 1-13.

[6] Wang H, Zhao H. Dynamic changes of soil erosion in the Taohe River Basin using the RUSLE Model and Google Earth Engine [J]. Water, 2020, 12(5): 1293.

[7] Singh G, Panda R K. Grid-cell based assessment of soil erosion potential for identification of critical erosion prone areas using USLE, GIS and remote sensing: A case study in the Kapgari watershed, India [J]. International Soil and Water Conservation Research, 2017, 5(3): 202-211.

[8] Alexakis D D, Tapoglou E, Vozinaki A E K, et al. Integrated use of satellite remote sensing, artificial neural networks, field spectroscopy, and GIS in estimating crucial soil parameters in terms of soil erosion[J]. Remote Sensing, 2019, 11(9): 1106.

[9] Phinzi K, Ngetar N S. The assessment of water-borne erosion at catchment level using GIS-based RUSLE and remote sensing: A review [J]. International Soil and Water Conservation Research, 2019, 7(1): 27-46.

Acknowledgments

This research was funded by the National Key Research and Development Program of China (2017YFA0605003), the National Natural Science Foundation of China (51779269) and the Young Innovative Talents Promotion Program Ministry of Research Center of Flood and Drought Disaster Reduction of the Ministry of Water Resources.

IOP Conf. Series: Earth and Environmental Science 1004 (2022) 012003 doi:10.1088/1755-1315/1004/1/012003

Spatiotemporal variation patterns and aggregation of crop residue burning at county scale in Northeast China

Yi Wang[1], **Xiaoyi Guo**[1, 2]

[1] School of Geographical Sciences, Northeast Normal University, Changchun 130024, China
[2] Key Laboratory of Geographical Processes and Ecological Security in Changbai Mountains, Ministry of Education

Corresponding author: guoxy914@nenu.edu.cn

Abstract. Crop residue burning (CRB) has been a major environmental concern in key agricultural regions of China. To reveal spatiotemporal patterns of CRB events, a county-level assessment in Northeast China was conducted using Moderate Resolution Imaging Spectroradiometer (MODIS) thermal anomalies and land cover data. The results revealed that cropland fire occurrence increased significantly from 2003 to 2019, and annual CRB occurrence from 2014 to 2019 was significantly higher than that from 2003 to 2013. The majority of detected CRB events occurred in spring and autumn, accounting for 58.50% and 28.98% of the total cropland fires, respectively. The time scan statistics showed that CRB events lasted for approximately 40 days each year and were more severe in spring than autumn. The number of cropland fires monitored by Terra and Aqua satellites was relatively more balanced, accounting for 49.74% and 50.26% of the total, respectively, indicating that farmers may intentionally burn crop residue at night. The counties with higher CRB intensity were mainly in the Songnen Plain, Sanjiang Plain, and along the main stream of the Liaohe River. The Moran's I value of 0.054 to 0.452for CRB intensity demonstrated a strong spatial autocorrelation at the county level. The local autocorrelation Getis-Ord G_i^* showed 86 counties with significant hot spots and 91 counties with significant cold spots. The county-level temporal and spatial patterns of CRB events may be the results of policies and socio-economic conditions. This study could provide support for identifying the influencing factors of CRB and offer suggestions for crop residue management practices in this agriculture-dominant region.

1. Introduction

Burning in fields is a common land management practice worldwide to remove crop residue [1-4]. Globally, crop residue burning (CRB) plays an important role in biomass destruction. CRB is a human activity to prepare fields for cultivation [5], particularly in the post-harvesting and pre-planting periods [6]. CRB is a primary contributor to greenhouse gases and particulate emissions, such as CO_2, CH_4, black carbon, PM_{10}, $PM_{2.5}$, which substantially and adversely affect air quality and human health at local and regional scales [7-9].

In recent years, CRB has become a major growing environmental concern in China, the world's largest crop residue producer[10]. China produces about 600 to 800 Tg of crop residues each year [11-14] and accounts for approximately 20% of global agricultural waste burning emissions [15]. Chinese farmers have adopted various crop residue management practices. Field burning is considered a more labor- and cost-effective residue management practice than removal and incorporation into the soil [16]. The Chinese government has implemented a series of policies to prevent CRB, including encouraging the use of crop residue in various agricultural-based applications and industrial processing, such as energy, fertilizer, fodder, and raw materials [17]. Still, approximately 20 to 30% of crop residues are burned in open fields.

CRB has temporal and spatial heterogeneity due to the different climatic conditions and farming methods in China [18]. Although CRB occurs throughout the country for a short period after harvest, it is impractical to monitor using traditional surveys such as point-by-point manual inspection at a broad scale [4; 9]. Satellite remote sensing and dataset derivatives can supply information to characterize fires in terms of intensity, extent, and spatiotemporal variations [19-20], making them effective approaches to monitor CRB conditions in large spatial areas over extended periods [4; 21-23]. In particular, Moderate Resolution Imaging Spectroradiometer (MODIS) data products are designed for observing actively burning fires twice per day under relatively cloud-free conditions. Compared with previous sensors, MODIS sensors provide an improved combination of spectral, temporal, and spatial resolution for fire detection [21]. The current fire monitoring algorithm is mainly the context algorithm [24].

Korontzi et al. first used MODIS Terra thermal anomaly products to explore the temporal and spatial characteristics of agricultural fires in croplands and found that from 2001 to 2003, agricultural fires accounted for 8 to 11% of global fire events, and the contribution rate was significantly higher on a regional basis [14]. However, satellite-based burned area methodologies specifically calibrated for CRB are limited. McCarty et al. [25] adopted a satellite observation-based hybrid approach to estimate the amount of burned crop residues, which combined MODIS 8-day differencing of Normalized Burn Ratio (dNBR) burned area mapping with MODIS active fires calibrated into the area. Based on the high-resolution Advanced Spaceborne Thermal Emission and Reflection Radiometer (ASTER) data, the effectiveness of the hybrid approach was verified. The strong correspondence with ASTER (mean $R^2 = 0.92$) suggested that the hybrid approach could be used to monitor and quantify fire activities in cropland areas. Yang et al. estimated the open biomass burning emission by using the Visible Infrared Imaging Radiometer Suite (VIIRS) sensors capable of detecting more fires, particularly in relatively smaller areas [26]. Spectral characteristics of the ground surface change before and after CRB. With two Landsat 8 images in the Songnen Plain, the normalized burned ratio (NBR), the normalized burned ratio incorporating the thermal band (NBRT), and the burned area index (BAI) were used to classify the areas with and without crop residue burned, which reached overall classification accuracies of 91.9%, 92.3%, and 87.8%, respectively [27]. Moreover, Landsat 8, GF-1, and HJ-1A/B data were used comprehensively to improve the remote sensing spatial resolution, time series analysis was performed with a revisit period of about 1.6 days, and the object-oriented image analysis methods were adopted to extract the straw burning areas in Taikang County [28].

The Northeast region is an important commodity grain base of China. Compared to other regions, CRB events in the Northeast accounted for a major proportion of the total CRB spots in China and had the strongest inter-annual variation [4]. In October and November 2015, the number of CRB in Heilongjiang, Jilin, and Liaoning Provinces accounted for 69.02% and 93.13% of all CRB events in China, respectively [29]. With respect to spatial distribution, CRB events mainly occurred in west

Liaoning, Jilin, and northeast Heilongjiang. CRB events in the vicinity of airports and railways were also very frequent in Northeast China [30]. Furthermore, MODIS fire activity data in cropland, statistical socio-economic data, and resource and environment data indicated that biophysical factors, agricultural input-output, economic development, and crop types were significant determinants of CRB intensity in Northeast China. The spatiotemporal distribution of CRB events has received considerable attention since it can be used to estimate the emissions inventory [3-4; 18]. In autumn and winter, the change in CRB sites significantly altered PM2.5 levels in Northeast China [18]. Moreover, the seasonal spatial distributions of the Air Quality Index in 41 cities in Northeast China correspond to the seasonal variation of fire spots, and pollutants appear in the peak periods of fire [9].

CRB decreases the availability of nutrients in the soil, including nitrogen, potash, and Sulphur [5; 16], which may lead to decrease grain yield in the long run. Nevertheless, CRB in the fields has not been studied in great detail or characterized at the county level. Less emphasis has been placed on understanding the spatial dependence and heterogeneity of CRB events distribution. Therefore, more effective measures should be exercised to monitor and manage CRB in Northeast China. The objectives of this study are to investigate the temporal distribution of CRB events in Northeast China and examine spatial patterns of CRB events at the county level.

2. Methods

2.1. Study area

This study focuses on the area in Northeast China situated between 38° 42'-53° 35' N, 115° 32'-135° 9' E (Figure 1). The study area covers Heilongjiang, Jilin, Liaoning, and eastern Inner Mongolia, including the Hulun Buir City, Hinggan League, Tongliao City, and Chifeng City. The study area was selected because it is the main grain production base in China. Croplands occupy approximately 3.12×10^5 km^2 [31], which is over 25% of the regional land. Northeast China has been the largest crop residue contributor, producing approximately 20% of the total crop residue in China [32]. This is a single-cropping agricultural region due to the climatic conditions. The waste products of rice and maize are major crop residues. It was estimated that 1.59×10^9 t of crop residues was produced in Northeast China in 2015, and the amount has been increasing by 6.1% per year [33].

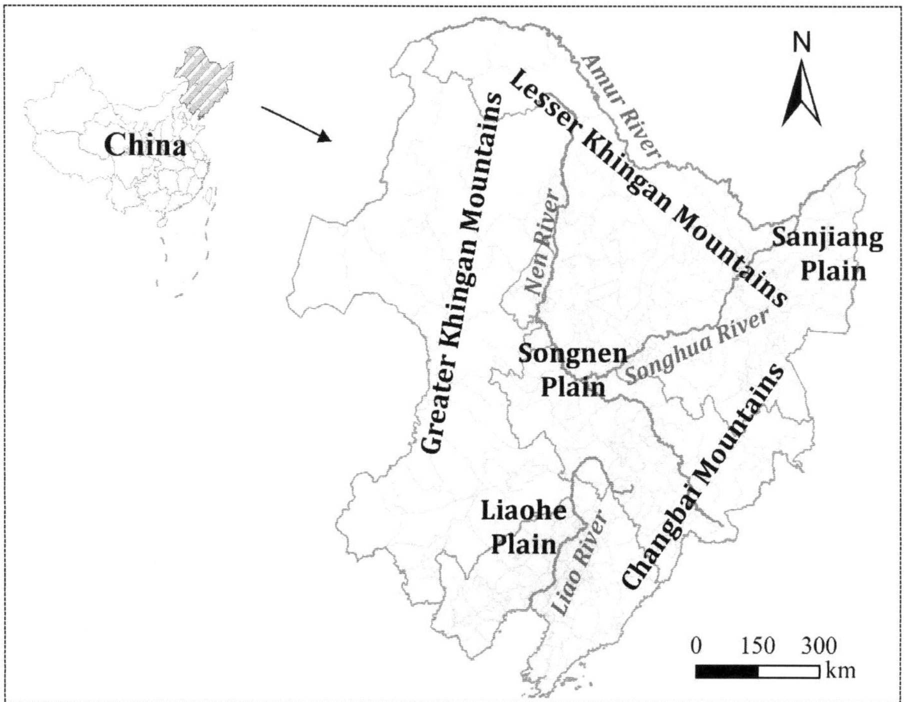

Figure 1. Location of the study area and distribution of cropland in 2019.

2.2. MODIS datasets

In this study, the MODIS data was used to extract the active fire within croplands, including the 1 km MODIS active fire product and 500 m MODIS Land Cover Product from 2003 to 2019. The datasets are openly accessible from https://search.earthdata.nasa.gov/.

The data product was generated with the latest and improved collection 6 MODIS Active Fire Detection Algorithm [34]. Due to the difference in the local passing times, two daily products, MOD14A1 and MYD14A1, were used in combination in the study. High, medium, and low confidence fire pixels were retained for the following study [35]. Moreover, the raster layer of the fire dataset was converted to vector points. Due to limitations in the pixel size, overpass of satellites and cloud cover, the number of active fires is not representative of the actual burned area. The number of fires within croplands enables a relative comparison of the proportion of pixels affected by fire within the various counties and is useful to broadly characterize the intra- and interannual spatiotemporal variability of CRB.

To derive the distribution of CRB events, active fire observation was combined with the annual MODIS land cover dataset (MCD12Q1). The International Geosphere-Biosphere Programme classification scheme was used to extract croplands [36], which reached an overall accuracy of 75% [37].

The number of active fires within cropland cannot reflect the magnitude of CRB among various counties because of the different cropland areas. Therefore, the CRB intensity was defined as follows:

$$y_i = \frac{N_i}{S_i} \tag{1}$$

where y is CRB intensity in the ith county, N is the number of fires within croplands in the ith county, and S is the area of croplands in the ith county.

2.3. Traditional statistical analysis

A two paired sample t-test was applied to test whether there was a difference in the CRB intensity of different periods at the 95% level of significance. The temporal variations in CRB intensity were analyzed using least-squares linear regression for each county. The year was defined as the independent variable, and CRB intensity was defined as the dependent variable. The temporal dynamics exhibited a positive trend if the slope was greater than 0 and a negative trend if the slope was less than 0. The results were considered significant at a p-value less than 0.05.

2.4. Time Scan Statistics

Time scan statistic was used to analyze the CRB aggregation period of each year. Overlapping candidate time windows of a particular year were defined, which varied from 1 to the maximum specified size. The maximum scanning window was set to 50% of the days in each year. The number of fire pixels in each window is examined, and the statistic log-likelihood ratio (LLR) is calculated for each window [38]. The LLR function of Poisson distribution can be calculated as follows:

$$LLR = \left[\frac{k}{E(k)}\right]^{c}\left[\frac{K-k}{C-E(K)}\right]^{K-k} I() \tag{2}$$

where K is the total number of CRB events in a certain year, k is the CRB events observed in the window, and $E(k)$ is the expected value of each window. $I()$ is an indicative function. $I()=1$ if the actual CRB events in the window are higher than expected; otherwise, $I()=0$. The Monte Carlo hypothesis test method was used to calculate the p-value, and $p<0.01$ was set to make it a cluster. The window with the highest LLR was selected, corresponding to the aggregation period each year.

2.5. Spatial autocorrelation analysis

To investigate whether CRB intensity distribution has spatial autocorrelation, the global and local measures of spatial association were calculated using global Moran's I statistics [39] and local Getis-Ord G_i^* statistics [40].

The Moran's I can be calculated as follows:

$$I = \frac{n\sum_{i=1}^{n}\sum_{j=1}^{n} w_{ij}\left(y_i - \bar{y}\right)(y_j - \bar{y})}{\left(\sum_{i=1}^{n}\sum_{j=1}^{n} w_{ij}\right)\sum_{i=1}^{n}(y_i - \bar{y})^2} \tag{3}$$

where n is the number of counties, w_{ij} is the spatial weighting factor between counties i and j, y_i and y_j are the CRB intensities of counties i and j, respectively, and $\bar{y}$ is the average CRB intensity of all counties. To characterize the spatial relation between counties, we conducted a binary spatial weights matrix based on queen contiguity. The Moran's I falls between -1 and 1. A Moran's I greater than 0 presents a clustered pattern, a Moran's I less than 0 presents a dispersed pattern and a Moran's I equal to 0 presents a random pattern.

The Getis-Ord G_i^* statistic can detect local 'hot' and 'cold' areas of CRB intensity. The calculation formula of G_i^* is as follows:

$$Z(G_i^*) = \sum_{j=1}^{n} w_{ij} y_j - \frac{\bar{y} \sum_{i \neq j}^{n} w_{ij}}{s \sqrt{n \sum_{j=1}^{n} w_{ij}^2 - \frac{(\sum_{i \neq j}^{n} w_{ij})^2}{n-1}}} \tag{4}$$

$$s = \left(\frac{1}{n} \sum_{j=1}^{n} (y_j - \bar{y})^2 \right)^{1/2} \tag{5}$$

where $Z(G_i^*)$ is a z-score, and s is the standard deviation of CRB intensity for all counties. A high positive $Z(G_i^*)$ for a feature indicates a spatial clustering of high values, suggesting a hot spot area. A low negative $Z(G_i^*)$ for a feature indicates a spatial clustering of low values, suggesting a cold spot. $|Z(G_i^*)|$ greater than 1.65, 1.96, and 2.58 represent clustering patterns at the 90%, 95%, and 99% level of significance.

3. Results and discussions

3.1. The temporal dynamics of CRB

3.1.1. Annual dynamics of CRB. Annual fires within croplands increased from 2,898 to 100,930 according to MODIS active fire and land cover data from 2003 to 2019 (Figure 2). The lowest and highest annual fire events were in 2004 and 2017, respectively. The firs in croplands exhibited an increasing trend starting from 2003. The overall CRB trend was similar to that of China [4]. In addition, CRB emissions in Northeast China showed a fluctuating upward trend from 2003 to 2016 [41]. Two reasons led to the large amount of redundant crop residue remaining in open fields. On the one hand, crop residue is an important domestic fuel source for rural households. However, it has been replaced by commodity energy, such as coal and electricity, recently [28]. On the other hand, as an important domestic fuel source for rural households, crop residue has recently been replaced by commodity energy, such as coal and electricity [42]. On the other hand, crop residue production is directly related to grain yield. Northeast China has a high crop residue density [14]. The total grain yield in 2019 was almost double that in 2003. Hence, the crop residue increased rapidly with the increasing grain yield.

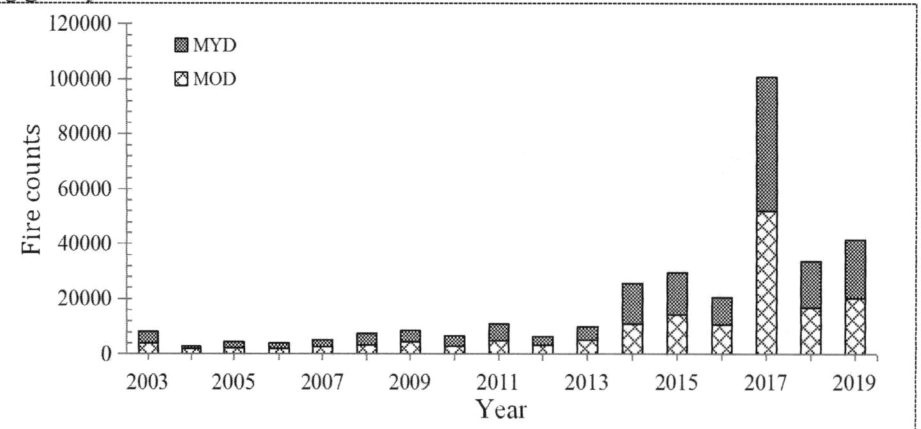

Figure 2. Annual distribution of MODIS fires within croplands from 2003 to 2019 in Northeast China.

A notable 258% interannual increase in the total number of fires within croplands was observed between 2013 and 2014. Accordingly, the temporal trajectory of CRB was divided into two sections during the study period. CRB events appeared to increase from 2003 to 2013 and remained high in numbers from 2014 to 2019. The two paired sample *t*-test showed that the mean number of fires was significantly different between the two periods ($p < 0.01$).

By searching the related policies, we noticed that the Heilongjiang Provincial Government issued the Notice of Implementation Plan for Banning Straw Burning to Improve Air Environment Quality in October 2016, which established more explicit requirements for straw burning prohibition. With the accumulation of previous policies and strict control measures, the total amount of straw burning in Heilongjiang Province in October 2016 was 40.9% lower than that in the same period in 2015 [43], and the effect of banning straw burning was significant. However, the comprehensive straw utilization measures in Heilongjiang Province have not been widely implemented and promoted. As a result, a large number of stalks produced during the harvesting period were not effectively treated, which in turn caused a large-scale rebound in straw burning in April 2017, reaching the highest level in the same period in 5 years.

For the study period, MODIS data from Terra and Aqua satellites showed that 49.74% and 50.26% of the fires occurred within croplands, respectively, reflecting that farmers' choice of burning time was fair because of the different local times of the two satellites. Farmers assumed that they might avoid detection by satellite and people if burning crop residue at night.

3.1.2. Monthly dynamics of CRB. The monthly distribution of total fire events within croplands showed considerable variation (Figure 3). The monthly total fire events showed two different peaks. The first peak was observed in April, with the highest number of fires. The second peak occurred in October. This is similar to the two seasonal peaks of agricultural fire activity in the Mississippi Valley, while the opposite is that the first, smaller peak, occurring in June during the spring harvesting of wheat, and the second, bigger peak, in October during the fall harvesting of rice and soy [44]. For monthly variations of CRB in different regions [4], detected fire spots in June were mainly concentrated in Central China (CC), East China (EC), and North China (NC). There are obvious differences in the month of CRB in Northeast China and other regions due to the different mechanisms of planting crops caused by climate distribution. CRB has been shown to be a seasonal activity, which is consistent with the crop management practices in the region [22]. The April peak is attributed to the practice of clearing cropland to prepare for spring planting in Northeast China, while the October peak directly corresponds to the rice and maize harvests. About 58.50% of the fires occurred within croplands in spring (from March to May) during the study period, followed by autumn (from September to November), contributing 28.98%. Fire events were the lowest in summer (from June to August). Similarly, high emissions in the Northeast Plain were mainly attributed to in-field CRB in the March to April period (sowing period) and October to November period (harvest period) from 2003 to 2014 [45]. The situation of CRB in spring is more severe than that in autumn.

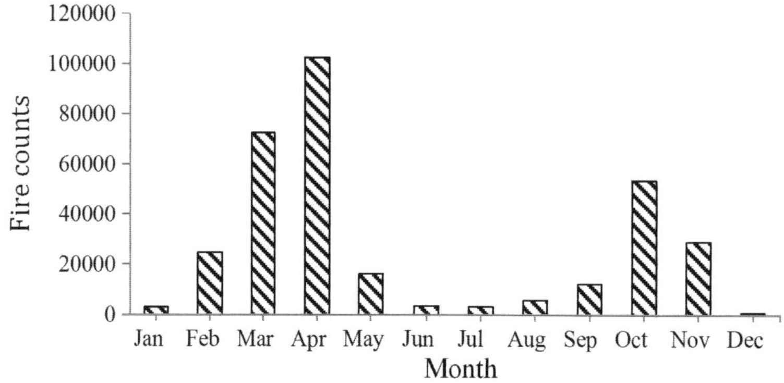

Figure 3. Monthly distribution of MODIS fire events within croplands from 2003 to 2019 in Northeast China.

The monthly distribution patterns of cropland fires are difficult defined the start of date (SOD), the end of date (EOD) and the length of seasons (LOS) of CRB for each year. Based on time scan statistic, Figure 4 shows the annual SOD, EOD, and LOS of CRB from 2003 to 2019. The results showed that CRB had significant clusters for all years (p = 0.001). The entire annual CRB aggregation period appeared in spring and autumn, which was consistent with the monthly distribution patterns of MODIS fire events. There were 12 and 5 seasons of CRB in spring and autumn, respectively. By burning crop residue in the spring, farmers could receive specific benefits in the short term, such as fertilizing the soil and controlling weeds, disease, and pests. In autumn, the government strictly prohibits CRB because the combination of CRB and coal-fired heating generates serious air pollution in Northeast China.

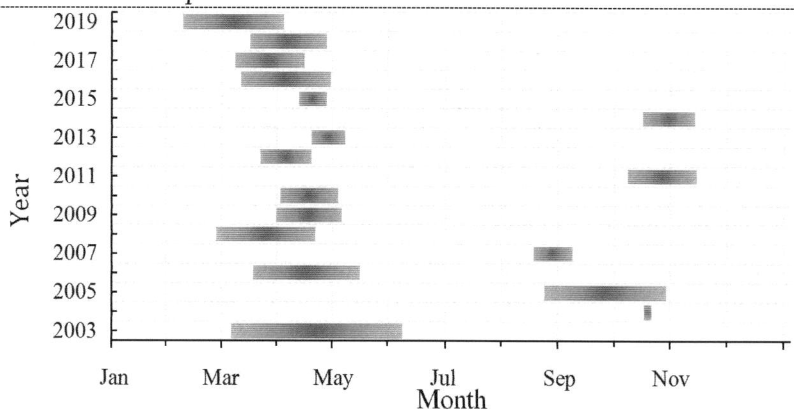

Figure 4. Temporal clustering of MODIS fire events within cropland for each year during the study period.

The SOD, EOD, and LOS of CRB were clearly different in the various years. The longest LOS of CRB appeared in the spring of 2003. Meanwhile, the shortest LOS appeared in the autumn of 2004. The average LOS of CRB was approximately 40 days, indicating that the average annual duration of aggregated crop residual burning in Northeast China was over one month. The average SOD and EOD are March 19 and April 30 in spring, respectively. The average SOD and EOD are September 19 and October 20 in autumn, respectively.

3.2. The spatial patterns of CRB intensity

3.2.1. The CRB intensity at county level. Figure 5 shows the spatial distribution of CRB intensity. Since CRB was affected by both natural and anthropogenic factors, it was not evenly distributed across the study area at the county level (Figure 5a). The counties with relatively high CRB intensity are mainly in the northern Songnen and Sanjiang Plains and along the main stream of the Liao River, all of which are over 0.1. Conversely, counties in the southern Greater Khingan and Changbai Mountains showed lower CRB intensities. The occurrence frequency of cropland fires in Northeast China from 2003 to 2013 was generally low (Figure 5b), while cropland fires frequently occurred from 2014 to 2019 (Figure 5c).

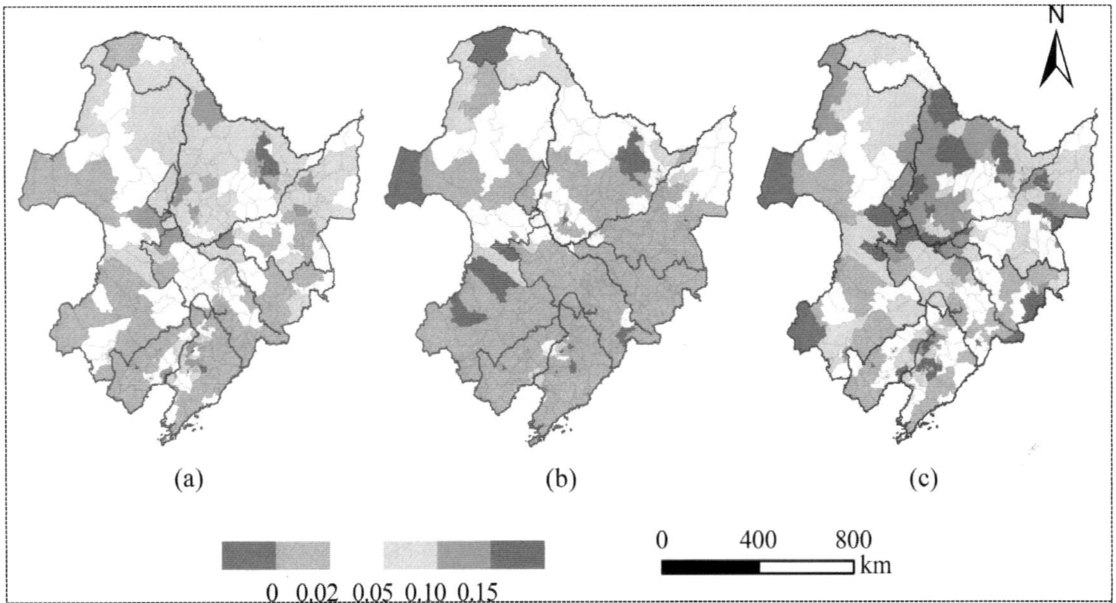

Figure 5. Spatial distribution of the average annual CRB intensity (a) 2003-2019, (b) 2003-2013 and (c) 2014-2019 in Northeast China.

The spatial distribution of CRB intensity indicated that approximately 66.77% of the total counties exhibited significant increasing trends from 2003 to 2019 (Figure 6). The relatively high slope was commonly distributed in the Songnen Plain, the Sanjiang Plain, and along the Liaohe River.

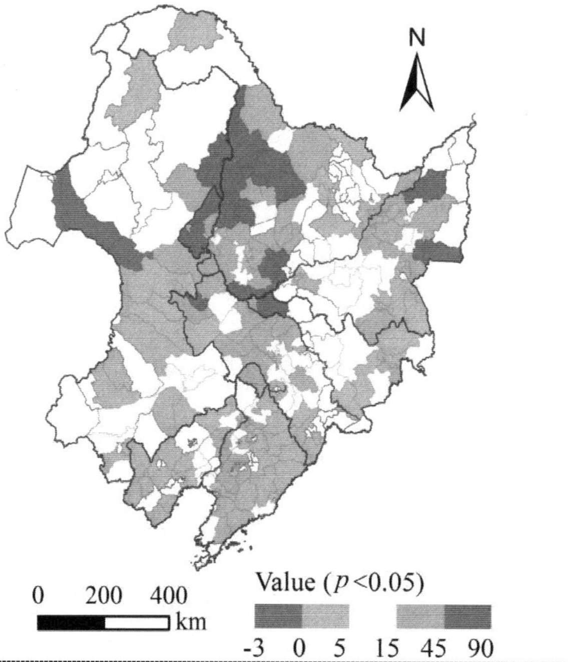

Figure 6. The trend of CRB intensity in counties from 2003 to 2019 in Northeast China.

3.2.2. Spatial autocorrelation of CRB intensity. To further explain the spatial pattern evolution characteristics of the average CRB intensity, we conducted the global spatial autocorrelation test of the average burning intensity of each year. Figure 7 exhibits the trend of the spatial process of the average CRB intensity. The Moran's I of CRB intensity ranged from 0.054 to 0.452 from 2003 to 2019. Therefore, the spatial patterns of CRB in Northeast China show a significant spatial autocorrelation for each year. Moran's I was the highest and lowest in 2017 and 2012, respectively. The differences in the spatial aggregation of the average CRB intensity may be the results of the CRB ban policy for various years. Heilongjiang Province was greatly affected by the policy in northeast China. Due to the large crop residue production and low comprehensive utilization rate in the Northeast Plain, CRB events in Heilongjiang Province rebounded to the highest level in 2017 [43]. Smaller Moran's I indicated weak spatial CRB aggregation with loose banning policy.

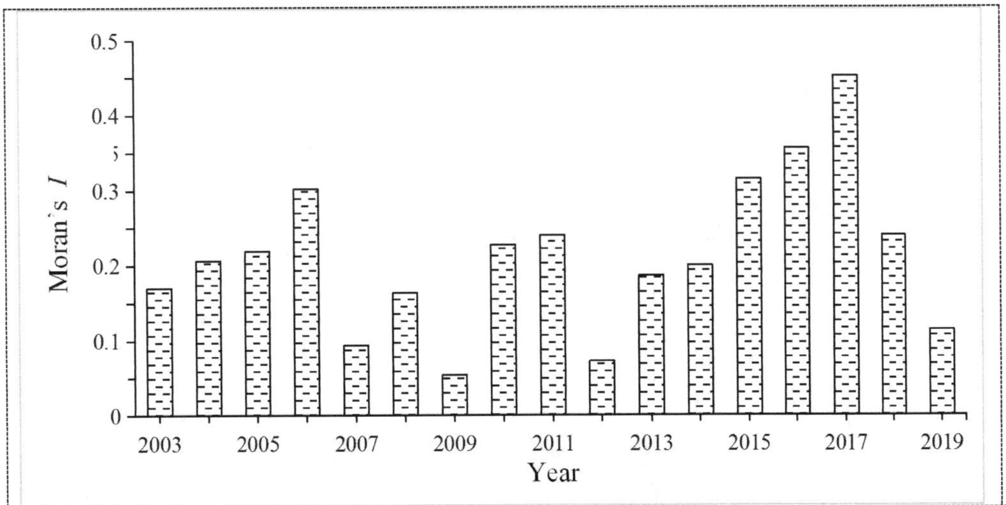

Figure 7. Annual Moran's I of CRB intensity from 2003 to 2019 in Northeast China.

The average annual CRB intensity of the G_i^* statistics showed 86 and 91 significant hot spots and cold spots, respectively (Figure 8). The hot spots are mainly along the upper reaches of the Nen River, on the Sanjiang Plain, and the Songnen Plain. The cold spots are in the middle and southern parts of the Changbai Mountain regions, the western part of the Liaohe Plain, and the southern part of the Greater Khingan Mountains.

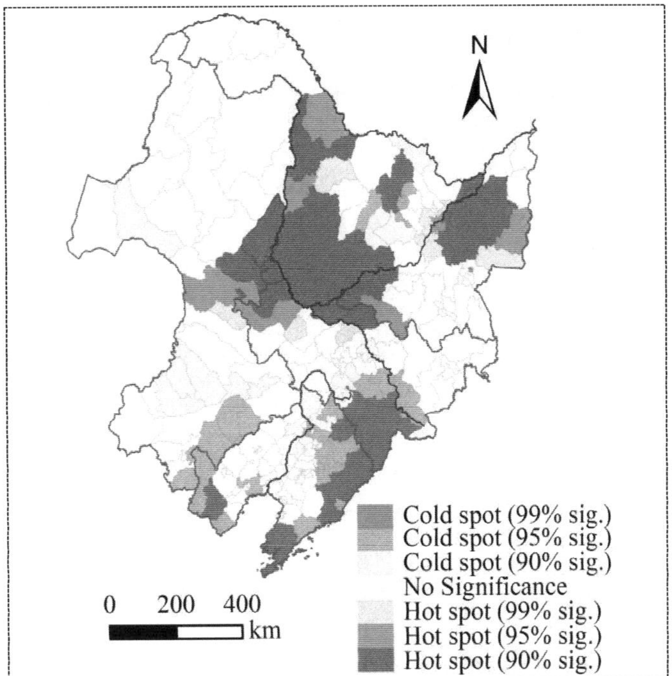

Figure 8. Spatial distribution of cold and hot spots of CRB intensity in Northeast China.

4. Conclusion

CRB is an important issue affecting regional sustainable development. In this study, the spatiotemporal patterns of CRB were explored by active fire and land cover data products of MODIS from 2003 to 2019 in Northeast China.

Over the study period, CRB showed a general increase throughout the study area. However, CRB increased greatly during the recent period from 2014 to 2019. Distinct differences in the seasonality of CRB were observed. CRB was the most severe in spring and autumn. More CRB events appeared in spring than autumn. These differences were attributed to the crop calendar. Generally, CRB lasted approximately more than one month. Terra and Aqua satellites detected similar accounts of CRB, indicating that farmers' choice of burning time was fair. Regarding the spatial patterns, satellite observations indicated that CRB was common in Northeast China. Regions with a high CRB intensity included the Sanjiang Plain, the area along the Nen River, and the main stream of the Liao River. Annual Moran's I of the CRB intensity indicated a statistically significant spatially positive correlation during the study period. At the county level, 86 and 91 significant hotspots and coldspots of average annual CRB intensity were identified, respectively.

CRB has raised many concerns worldwide, particularly in developing countries. This study provided useful information for further exploring the relationships between CRB and socio-economic factors at the county level, which could help governments implement decisions toward green and sustainability agriculture in different regions.

References

[1] Bhuvaneshwari S, Hettiarachchi H and Meegoda N J 2019 *Int. J. Environ. Res. Public. Health* Crop Residue Burning in India: Policy Challenges and Potential Solutions **16** 832

[2] Junpen A, Pansuk J, Kamnoet O, Cheewaphongphan P and Garivait S 2018 *Atmosphere* Emission of Air Pollutants from Rice Residue Open Burning in Thailand, 2018 **9** 449

[3] McCarty J L, Korontzi S, Justice C O and Loboda T 2009 *Sci. Total Environ.* The spatial and temporal distribution of crop residue burning in the contiguous United States **407** 5701–12

[4] Zhuang Y, Li R, Yang H, Chen D, Chen Z, Gao B and He B 2018 *Remote Sens.* Understanding Temporal and Spatial Distribution of Crop Residue Burning in China from 2003 to 2017 Using MODIS Data **10** 390

[5] Haider M Z 2013 *J. Environ. Manage.* Determinants of rice residue burning in the field **128** 15–21

[6] Brye K R, Longer D E and Gbur E E 2006 *Soil Sci. Soc. Am. J.* Impact of Tillage and Residue Burning on Carbon Dioxide Flux in a Wheat–Soybean Production System **70** 1145–54

[7] Crutzen P J and Andreae M O 1990 *Science* Biomass Burning in the Tropics: Impact on Atmospheric Chemistry and Biogeochemical Cycles **250** 1669–78

[8] Li Y, Liu J, Han H, Zhao T, Zhang X, Zhuang B, Wang T, Chen H, Wu Y and Li M 2019 *Atmos. Environ.* Collective impacts of biomass burning and synoptic weather on surface PM2.5 and CO in Northeast China **213** 64–80

[9] Wang Z Z, Zhao J J, Xu J W, Jia M R, Li H and Wang S J 2019 *Int. J. Environ. Res. Public. Health* Influence of Straw Burning on Urban Air Pollutant Concentrations in Northeast China **16** 1379

[10] Chen J, Gong Y, Wang S, Guan B, Balkovic J and Kraxner F 2019 *Sci. Total Environ.* To burn or retain crop residues on croplands? An integrated analysis of crop residue management in China **662** 141–150

[11] Cao G L, Zhang X Y, Wang Y Q and Zheng F C 2008 *Chin. Sci. Bull.* Estimation of emissions from field burning of crop straw in China **53** 784–790

[12] Liu H, Jiang G M, Zhuang H Y and Wang K J 2008 *Renew. Sust. Energ. Rev.* Distribution, utilization structure and potential of biomass resources in rural China: with special references of crop residues **12** 1402–18

[13] Cai Y Q, Qiu H G and Xu Z G 2011 *J. Nat. Resour.* Evaluation on potentials of energy utilization of crop residual resources in different regions of China **26** 1637–46

[14] Jiang D, Zhuang D, Fu J, Huang Y and Wen K 2012 *Renew. Sust. Energ. Rev.* Bioenergy potential from crop residues in China: Availability and distribution **16** 1377–82

[15] Bi Y Y, Wang Y J and Gao C Y 2010 *Agri. Mech. Res.* Straw resource quantity and its regional distribution in China **32** 1–7

[16] Ahmed T, Ahamad B and Ahamad W 2015 *Land Use Policy* Why do farmers burn rice residue? Examining farmers' choices in Punjab, Pakistan **47** 448–458

[17] Huang X, Li M, Li J and Song Y 2012 *Atmos. Environ.* A high-resolution emission inventory of crop burning in fields in China based on MODIS Thermal Anomalies/Fire products **50** 9–15

[18] Yin S, Wang X F, Xiao Y, Tani H, Zhong G S and Sun Z Y 2017 *Environ. Pollut.* Study on spatial distribution of crop residue burning and PM2.5 change in China **220** 204–221

[19] Lentile L B, Holden Z A, Smith A M S, Falkowski M J, Hudak A T, Morgan P, Lewis S A, Gessler P E and Benson N C 2006 *Int. J. Wildland Fire* Remote sensing techniques to assess active fire characteristics and post-fire effects **15** 319–345

[20] McCarty J L 2011 *J. Air Waste Manag. Assoc.* Remote Sensing-Based Estimates of Annual and Seasonal Emissions from Crop Residue Burning in the Contiguous United States **61** 22–34

[21] Korontzi S, McCarty J, Loboda T, Kumar S and Justice C O 2006 *Glob. Biogeochem. Cycles.* Global distribution of agricultural fires in croplands from 3 years of Moderate Resolution Imaging Spectroradiometer (MODIS) data **20** GB2021

[22] Verma S, Dar J A, Malasiya D, Khare P K, Dayanandan S and Khan M L 2019 *Ecol. Indic.* A MODIS-based spatiotemporal assessment of agricultural residue burning in Madhya Pradesh, India **105** 496–504

[23] Xie H, Du L, Liu S, Chen L, Gao S, Liu S, Pan H and Tong X 2016 *ISPRS Int. J. Geo-Inf.* Dynamic Monitoring of Agricultural Fires in China from 2010 to 2014 Using MODIS and GlobeLand30 Data **5** 172

[24] Giglio L, Descloitres J, Justice C O and Kaufman Y J 2003 *Remote Sens. Environ.* An enhanced contextual fire detection algorithm for MODIS **87** 273–282

[25] McCarty J L, Loboda T and Trigg S 2008 *Appl. Eng. Agric.* A Hybrid Remote Sensing Approach to Quantifying Crop Residue Burning in the United States **24** 515–527

[26] Yang G Y, Zhao H M, Tong D Q, Xiu A J, Zhang X L and Gao C 2020 *Sci. Total Environ.* Impacts of post-harvest open biomass burning and burning ban policy on severe haze in the Northeastern China **716** 136517

[27] Ma J H, Song K S, Wen Z D, Shao T T, Li B N and Qi C 2015 *Chin. J. Appl. Ecol.* Quantification of crop residue burned areas based on burning indices using Landsat 8 image **26** 3451–56

[28] Wu X H, Liu T, Cheng Y Z, Wang L G, Guo Y, Zhang Y and He J 2017 *Tran. Chin. Soc. Agric. Eng.* Dynamic monitoring of straw burned area using multi-source satellite remote sensing data **33** 153–159

[29] Yin S, Wang X F, Zhang X R, Zhang Z X, Xiao Y, Tani H and Sun Z Y 2018 *Atmos. Environ.* Exploring the effects of crop residue burning on local haze pollution in Northeast China using ground and satellite data **199** 189–201

[30] Tu X M, Yang X H, Zhang J Y, Luan X C and Ning K X 2020 *Geogr. Res.* Analyzing geographical characteristics of straw burning in China during 2014-2019 **39** 2379–90

[31] Ning J, et al 2018 *J. Geogr. Sci.* Spatiotemporal patterns and characteristics of land-use change in China during 2010–2015 **28** 547–562

[32] Li H L, Wang C, Sun H T, Yan X L and Liang Q 2017 *J. Agric. Mech. Res.* Comprehensive utilization and sustainable development of agriculture straw **39** 256–262
[33] Wang J W, T H and Wang J F 2017 *Trans. Chin. Soc. Agric. Mach.* Comprehensive utilization status and development analysis of crop straw resource in Northeast China **48** 1–21
[34] Giglio L, Schroeder W and Justice C O 2016 *Remote Sens. Environ.* The collection 6 MODIS active fire detection algorithm and fire products **178** 31–41
[35] Roy D P, Boschetti L, Justice C O and Ju J 2008 *Remote Sens. Environ.* The collection 5 MODIS burned area product-Global evaluation by comparison with the MODIS active fire product **112** 3690–3707
[36] Loveland T R, Reed B C, Brown J F, Ohlen D O, Zhu Z, Yang L and Merchant J W 2000 *Int. J. Remote Sense.* Development of a global land cover characteristics database and IGBP DISCover from 1 km AVHRR data **21** 1303–30
[37] Liang D, Zuo Y, Huang L S, Zhao J L, Teng L and Yang F 2015 *ISPRS Int. J. Geo-Inf.* Evaluation of the Consistency of MODIS Land Cover Product (MCD12Q1) Based on Chinese 30 m GlobeLand30 Datasets: A Case Study in Anhui Province, China **4** 2519–41
[38] Kulldorff M, Athas W F, Feurer E J, Miller B A and Key C R 1998 *Am. J. Public. Health* Evaluating cluster alarms: a space-time scan statistic and brain cancer in Los Alamos, New Mexico **88** 1377–80
[39] Anselin L 2010 *Geogr. Anal.* Local indicators of spatial autocorrelations-LISA **27** 93–115
[40] Ord J K and Getis A 1995 *Geogr. Anal.* Local Spatial Autocorrelation Statistics: Distributional Issues and an Application **27** 286–306
[41] Xu Y Q, Huang Z J, Jia G L, Fan M, Cheng L X, Chen L F, Shao M and Zheng J Y 2019 *Sci. Total Environ.* Regional discrepancies in spatiotemporal variations and driving forces of open crop residue burning emissions in China **671** 536–547
[42] Zhang L, Liu Y and Hao L 2016 *Environ. Res. Lett.* Contributions of open crop straw burning emissions to PM2.5 concentrations in China **11** 014014
[43] Xu B B, Fan M, Chen L F, Jiang T, Tao J H, Chen L X, Ji X Y and Wu W L 2020 *J. Rem. Sens. Chin.* Analysis of temporal and spatial characteristics and influencing factors of crop residue burning in major agricultural areas from 2013 to 2017 **24** 1221–32
[44] Korontzi S, McCarty J and Justice C 2008 *J. Air & Waste Manage. Assoc.* Monitoring agricultural burning in the Mississippi River Valley Region from the moderate resolution imaging spectroradiometer (MODIS) **58** 1235–39
[45] Wu J, et al. 2020 *Env. Int.* The moving of high emission for biomass burning in China: View from multi-year emission estimation and human-driven forces **142** 105812

Acknowledgments

This work is supported by the National Natural Science Foundation of China (Grant No. 42071359), Science and Technology Development Plan of Jilin Province (Grant No. 20190103151JH), Foundation of the Education Department of Jilin Province (Grant No. JJKH20211291KJ), and the Fundamental Research Funds for the Central Universities (Grant No. 2412019FZ002).

IOP Conf. Series: Earth and Environmental Science 1004 (2022) 012004 doi:10.1088/1755-1315/1004/1/012004

Dynamic Monitoring and Analysis of Ecological Environment in the Coastal Tourist Destinations of Sanya City, Hainan

Zhichao Li[1*], Jingjing Xie[2, 3] , Jingyi Yang[2, 3] and Dingkun Chang[2, 3]

1. Key Laboratory of Earth Observation of Hainan Province, Hainan Research Institute, Aerospace Information Research Institute, Chinese Academy of Sciences, Hainan, China, 572029
2. Key Laboratory of Digital Earth Science, Aerospace Information Research Institute, Chinese Academy of Sciences (CAS), CAS, Beijing, China, 100094
3. University of Chinese Academy of Sciences, Yanqi Lake Campus of Chinese Academy of Sciences University, Beijing, China, 101408

*Corresponding author: Zhichao Li, 1006320484@qq.com

Abstract. Taking the Sanya Bay as the study area, based on the remote sensing images of Gaofen-2 in 2015, 2017, 2019 and 2020, this paper extracted 7 land use types of Sanya Bay. The main objectives of this paper includes: (1) Applying the theory of landscape ecology to calculate the landscape fragmentation and disturbance intensity in order to indicate the ecological changes in the study area; (2) Analysing the impact of landscape changes on the surroundings areas driven by the diversity of land use and perform the "Dynamic Monitoring and Analysis Technology for the Ecological Environment of Coastal Tourism Regions". In which to monitor the dynamics of the ecological environment in the area of research. The results showed that the quality of the ecological environment in the study area changed dynamically with the construction of tourism sites: from year 2015 to 2017, due to a large number of construction, the ecological environment deteriorated to a certain extent; From 2017 to 2020, with the suspension of construction and environment restoration work, the quality of the ecological environment in the study area improved. Technology established in this paper provides a reference for the dynamic monitoring of the ecological environment in the coastal tourist area.

Keywords: remote sensing image; ecology; dynamic monitoring

Introduction

Over the past few years, with the rapid development of the world's economy, people's living standards have gradually improved and therefore seeking after more ways for leisure and entertainment. Generally, tourism has become an accessible lifestyle that average residents can engaged in. This phenomenon subsequently aids in the accelerated growth of tourism, especially in developing coastal destinations where tourism act as a vital economic driver. Correspondingly, tourism has become a pillar industry in some coastal cities, promoting and enhancing the socioeconomic benefits. Besides, the attention paid to the integrated development of "tourism + ocean" is increasing, making clear that there are a lot of opportunities and challenges in developing the coastal regions [1]. However, the excessive development of coastal tourism has led to the deterioration of the ecological environment and should not be ignored. Therefore, this paper will discuss the timely monitoring and analysis of the changes on coastal ecological environment which in turn may provide scientific theoretical support for the development of future coastal tourist attractions.

With the development of remote sensing technology and the continuous accumulation of high-resolution remote sensing images, more and more experts and scholars adopted remote sensing methods to conduct large-scale, accurate and dynamic monitoring of the ecological environment [2-10]. For example, Guan Xiaoying et al [11] discussed the application and analysis of remote sensing monitoring technology in the ecological environment protection of the Qilian Mountains; Ye Huping [12] performed remote sensing monitoring and analysis on the ecological environment of the coastal oceans of Sri Lanka; Ren Qiang [13] implemented a 20 year follow-up investigation on the dynamic monitoring of the changes in the ecological environment of the main stream of the Tarim River; Liu Shuqin et al [14] used remote sensing to monitor land use changes in Yanchi County and analyzed the ecological environmental effects. According to the aforementioned analysis, it is feasible to use remote sensing technology to dynamically monitor the coastal tourist area and identify the changes of the ecological environment in combination with relevant landscape ecology knowledge. Based on the current status of remote sensing instruments, these high spatial resolution imageries can provide relevant scientific theoretical support for future development of the coastal destinations.

Sanya City, located in the southern part of Hainan Island, is famous for its unique geographical location, good weather conditions, and abundant tourism resources. Since the introduction of National Strategies, tourism has become the main driving factors in promoting this international coastal destination, strategies such as the development of an international tourist island, the construction of the One Belt and One Road Initiative, and the construction of a free trade port, new requirements have been implemented for the development of Sanya. As a coastal tourist city, Sanya has been developing coastal tourism vigorously in recent years. The development of many coastal tourist attractions has provided strong evidence for the economic growth of Sanya. Nevertheless, in the early development of the Sanya Coastal

Tourism Area, due to the problem of over-development, the ecological environment of the field of study was destroyed, this affected the subsequent development of the relevant area [15-17]. Therefore, in order to prevent the same incident from happening, it is particularly important to use remote sensing technology to conduct a multi-period dynamic and scientific ecological environmental monitoring for the coastal tourist area of Sanya City. At the same time, by adopting the knowledge of ecological landscape, this paper aims to analyze the impact of scenic spots construction on the ecological environment changes, thus providing guidance for the rational development and sustainable development of tourist areas.

1. Study area

The investigation area is the Sanya Bay area, located in the southern part of Sanya City (Figure 1), with convenient public transportation and a dense road network. This region belongs to the tropical maritime monsoon climate and has abundant coastal tourism resources, such as the Sanya Bay, Phoenix Island, Luhuitou Scenic Area. It covers an area of about 1.9 square kilometres and consists of 5 ecological landscapes: vegetation, beaches, water bodies, artificial beaches, and buildings.

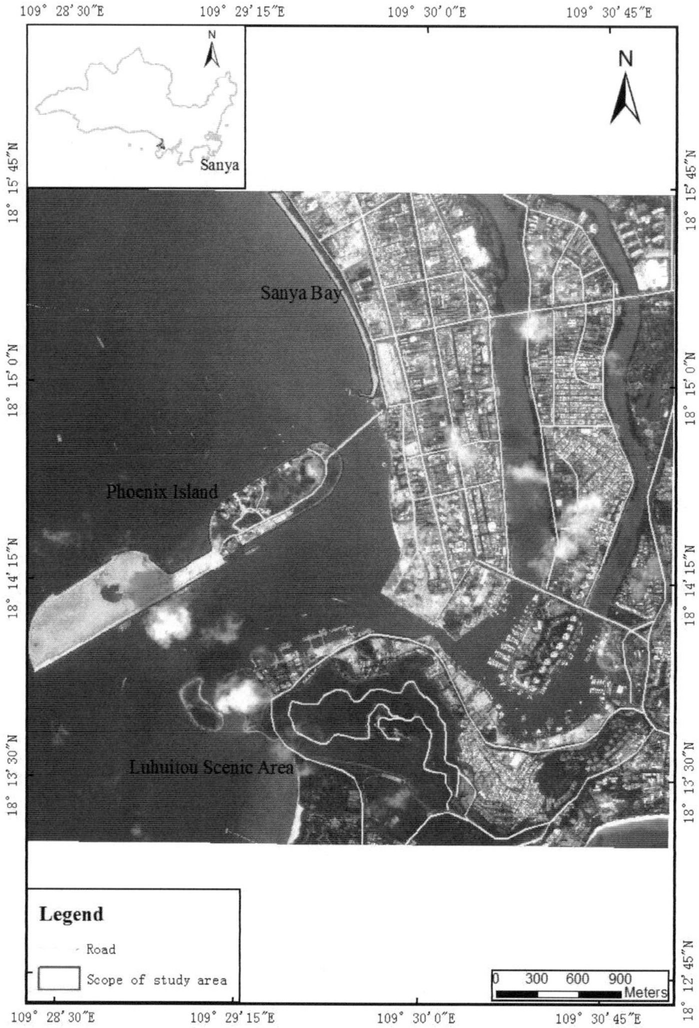

Figure 1. Location of study area.

2. Data and method

2.1. Data

In this paper, the High-resolution Earth Observation System Hainan Data and Application Center (http://gfdc.hainan.gov.cn/) was selected to show four phases of Gaofen-2 data in 2015, 2017, 2019, and 2020 (with a resolution of 0.8 m). (Table 1).

Table 1. Gaofen-2 data used in this paper

Sensors	Resolutions	Date	Serial numbers
PMS2	0.8m	2015.10.17	1689106
PMS1	0.8m	2017.02.16	3339700
PMS2	0.8m	2017.04.11	3542231

PMS2	0.8m	2019.11.23	7119110
PMS1	0.8m	2020.03.20	7526740

First, the images are pre-processed, including radiometric calibration, atmospheric correction, orthorectification, image fusion, image mosaic, etc. Secondly, the NDVI index is used to extract vegetation information. In reference to the processed images, combined with the urban and rural land layout map of Sanya City, the land use types in the study area are then divided into seven categories: vegetation, beaches, artificial beaches, water bodies, buildings, corridors, and open spaces (Table 2). Next, four phases of SHP data are generated for correlation analysis.

Table 2. Land use classification

Land use classification	Description
vegetation	Contains ground covered by vegetation, such as various trees, shrubs, lawns, etc.
beaches	Mainly refers to the beach of Sanya Bay
artificial beaches	Mainly refers to the man-made island covered by sand, i.e. Phoenix Island
water bodies	Including pool, sea surface
buildings	Including construction buildings, such as workers' sheds, storage sheds and buildings
corridors	Including roads, railways, dams, ditches
open spaces	Open spaces mainly refers to all unused or undeveloped land other than the above six types of land use

2.2. Method

After the initial investigation and data collection, this paper utilized remote sensing technology along with ecological landscape thesis to carry out research on the study area [18]. The ecological landscape pattern index was used to obtain the change of the ecological landscape pattern of the area studied. After collecting the data, we then observe the impact of the change of the ecological landscape pattern on the environment. Next, we analyse the data using the responses of land use diversity and forest coverage (Refer to the technical flowchart as shown in Figure 2).

IOP Conf. Series: Earth and Environmental Science 1004 (2022) 012004 doi:10.1088/1755-1315/1004/1/012004

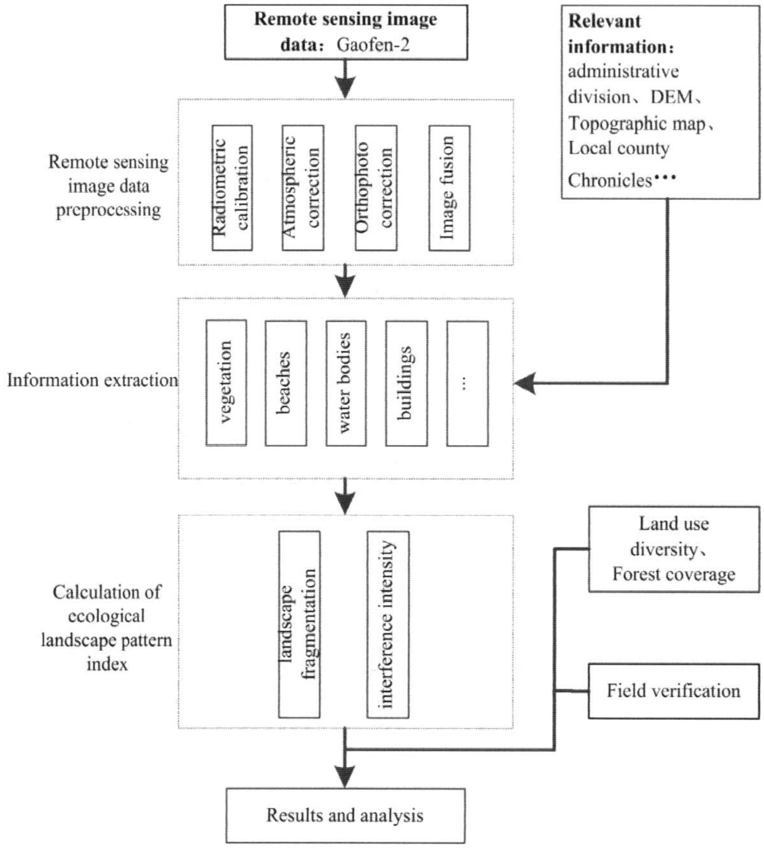

Figure 2. Technical flowchart.

2.2.1. Ecological landscape pattern index

Common indexes used to analyse landscape pattern changes include landscape fragmentation, landscape separation, interference intensity and naturalness, and landscape diversity index. Based on the classification and monitoring results of land use of the study area and the opinions of scholars, this paper selects two types of indices, landscape fragmentation and disturbance intensity to investigate the changes in the ecological landscape pattern of the interest area.

(1) Calculation of landscape fragmentation

Landscape fragmentation represents the degree of fragmentation of the landscape, which reflects the complexity of the spatial structure of the landscape. It also reflects the degree of human interference to the landscape to. It can be calculated as follows:

$$C_i = N_i / A_i \quad (1)$$

In the formula, C_i is the fragmentation degree of landscape *I*; N_i is the number of patches of landscape *i*, and A_i is the total area of landscape *i*.

(2) Calculation of interference intensity

The interference intensity is the interference effect of human beings on the ecological environment, formula as shown below:

$$W_i = L_i \ / \ S_i \quad (2)$$

In which, W_i represents the intensity of interference; L_i refers to the total length of corridors (roads, railways, dams, ditches) in the ith ecosystem, and S_i refers to the total area of the ith ecosystem.

2.2.2. Land use diversity

Land use diversity is an index to measure the stability of land use, which reflects the stability of land use in the development process through the changes of land use types and area proportions in the study area. Its formula is as follows:

$$H = -\sum_{k=1}^{n} p_k \ln p_k \quad (3)$$

In which, H is the land use diversity index; p_k is the area of the kth second land use type; n is the number of land use types.

2.2.3. Forest coverage

Forest coverage refers to the ratio of forest area to the total land area. It is an important indicator reflecting the actual level of forest resources and forest land occupation in the study area. It is generally expressed as follows:

$$F = (S / L) * 100\% \quad (4)$$

In which, F is the forest coverage rate; S is the total forest area; L is the total area of the region.

3. Results and analysis

3.1. Changes in the ecological landscape pattern

Based on the data of the ecological landscapes obtained by remote sensing image interpretation, combined with the landscape fragmentation index and the interference intensity index, the analysis of the ecological landscape pattern changes in the study area was formulated.

3.1.1. Change and analysis of landscape fragmentation

There are five types of ecological landscape elements in the study area: vegetation, beaches, artificial beaches, water bodies, and buildings. Based on the interpretation results of multiple periods of remote sensing, the landscape fragmentation of the study area was calculated. According to the monitoring results (Table 3, Figure 3 and Figure 4), it can be seen that from 2015 to 2020, there were 4 types of ecological landscape elements in 2015 and 5 types after 2017. Subsequently, from 2015 to 2020, there were 4 types of ecological landscape elements in 2015 and 5 types after 2017. The artificial beach is newly added as a landscape ecological element. The most obvious change in landscape fragmentation is water, followed by vegetation and buildings. The fragmentation degree of water landscape increased significantly from a lower value of 0.00001993 in 2015 to a higher value of 0.002616 in 2020. The fragmentation of the architectural landscape increased significantly from 2015 to 2017, and then remained constant and declined slightly. The fragmentation of vegetation landscape

IOP Conf. Series: Earth and Environmental Science 1004 (2022) 012004 doi:10.1088/1755-1315/1004/1/012004

increased slightly from year 2015 to year 2017, and decreased significantly after 2 years (from 2017 to 2019), and then remained static.

According to the research of the changes of the landscape fragmentation index, the complexity of the landscape spatial structure within the study area increases at the beginning and then decreases. In which, the water body changed obviously. According to the monitoring, it was caused by the replacement of seawater by artificial beaches, which resulted in the index to change sharply. Corresponding to the analysis of the natural environment of the area studied, the main reason for the changes in landscape fragmentation is human interference. The early development and construction of scenic spots led to the increase of the landscape fragmentation index in the study area; the cessation of later development and construction and the follow-up of restoration work resulted in the decrement of the landscape fragmentation index in the study area.

Table 3. Landscape fragmentations in the study area

Year	Landscape types	Number of plaques	Area (m^2)	Landscape fragmentations
2015	vegetation	111	1061779.54	0.0001045
	beaches	1	67677.50	0.00001478
	water body	9	451672.00	0.00001993
	buildings	34	40303.73	0.0008436
2017	vegetation	120	1001549.98	0.0001198
	beaches	1	50440.24	0.00001983
	water body	7	6938.33	0.001009
	artificial beaches	1	337219.8	0.000002965
	buildings	41	42255.99	0.0009703
2019	vegetation	70	1173092.75	0.00005967
	sandy beach	1	48160.91	0.00002076
	water body	7	5425.56	0.001290
	artificial beaches	1	339131.49	0.000002949
	buildings	33	32987.83	0.0010
2020	vegetation	62	1131766.67	0.00005478
	beaches	1	51815.10	0.0000193
	water body	6	2293.30	0.002616
	artificial beaches	1	343520.16	0.000002911
	buildings	37	37345.42	0.0009908

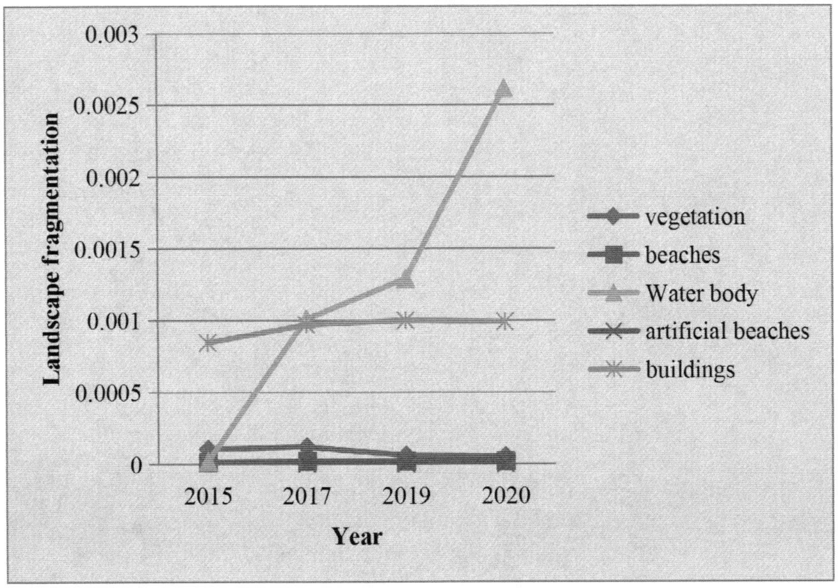

Figure 3. Diagram of landscape fragmentations in the study area (2015-2020).

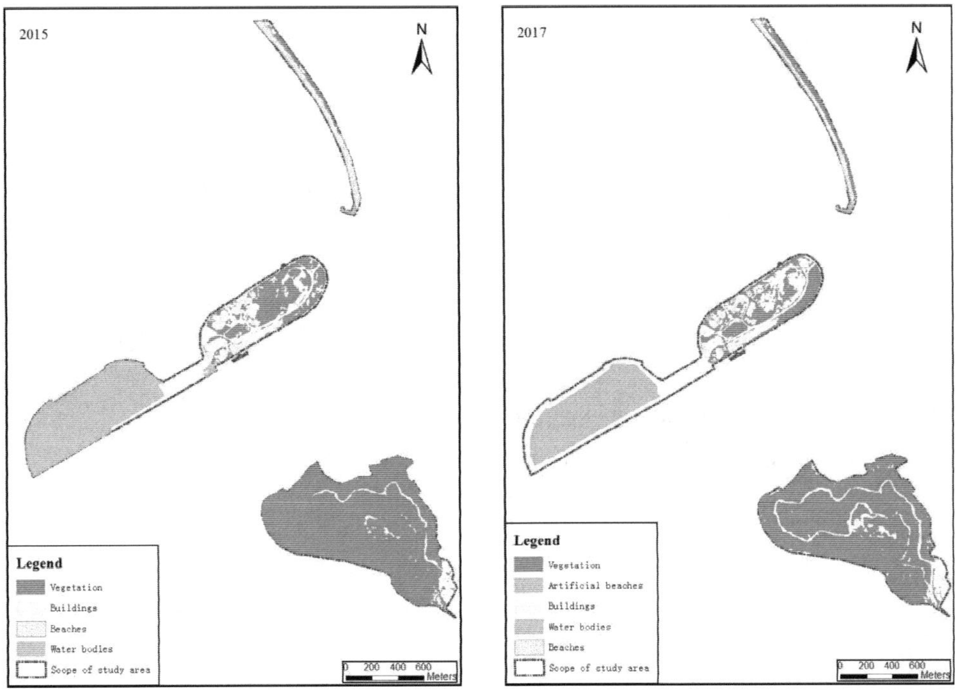

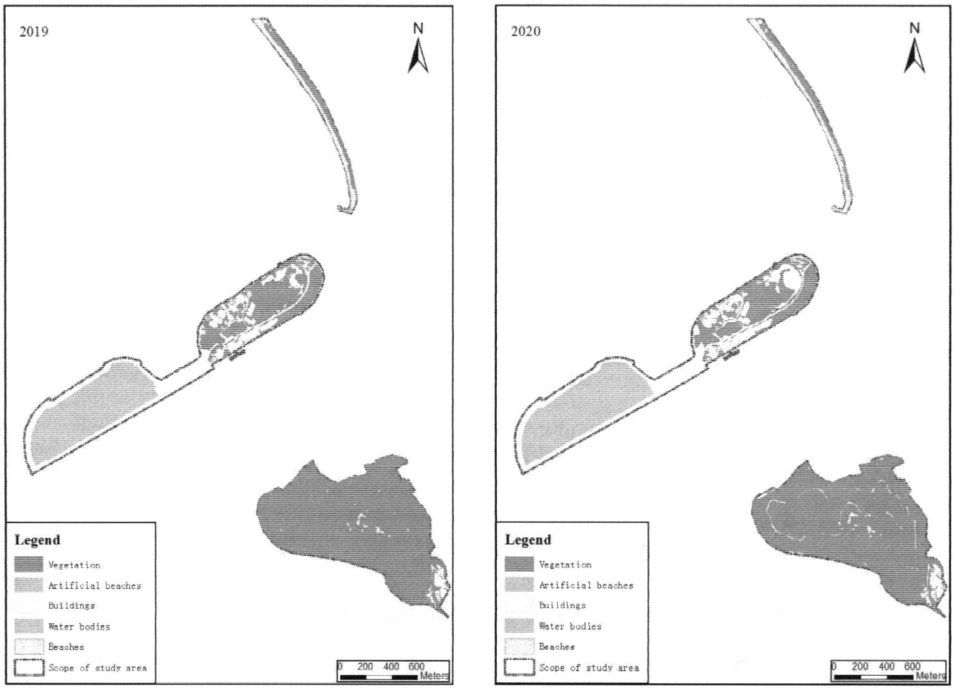

Figure 4. Maps of ecological landscape types in the study area (2015-2020).

3.1.2. Changes and analysis of interference intensity

In compliance with the results (Table 4, Figure 5 and Figure 6), it is noticeable that the length of the corridor increased from 2017 to 2019. In which, the interference intensity increased obviously. While from 2019 to 2020, it was unchanged.

The interference intensity represents the human interference effection. Analysis of the interference intensity index changes in the study area shows that the human interference has been increasing from 2015 to 2019, and then it has reached a relatively flat stage. The main reason for the changes in interference intensity in the surveyed area from 2015 to 2017 was the increment in the length of the dams caused by reclamation on Phoenix Island; the main reason for the change from 2017 to 2019 was the completion of the construction of the landscape road in the Luhuitou Scenic Area.

Table 4. Interference intensity in the study area

Years	Length of corridors (m)	Interference intensity
2015	7209.25	0.003760
2017	9747.01	0.005084
2019	13424.35	0.007002
2020	13590.71	0.007089

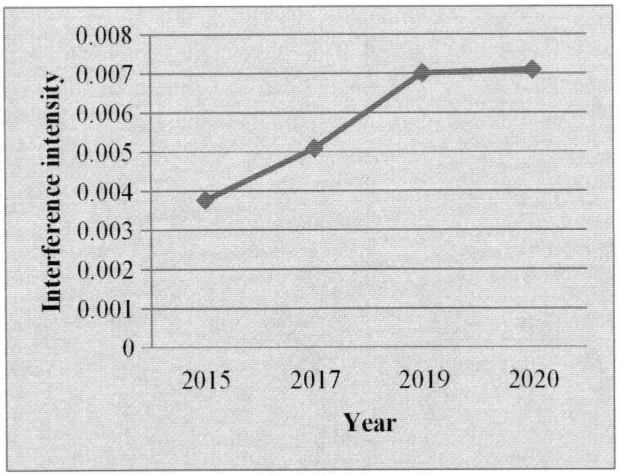

Figure 5. Diagram of interference intensity in the study area (2015-2020).

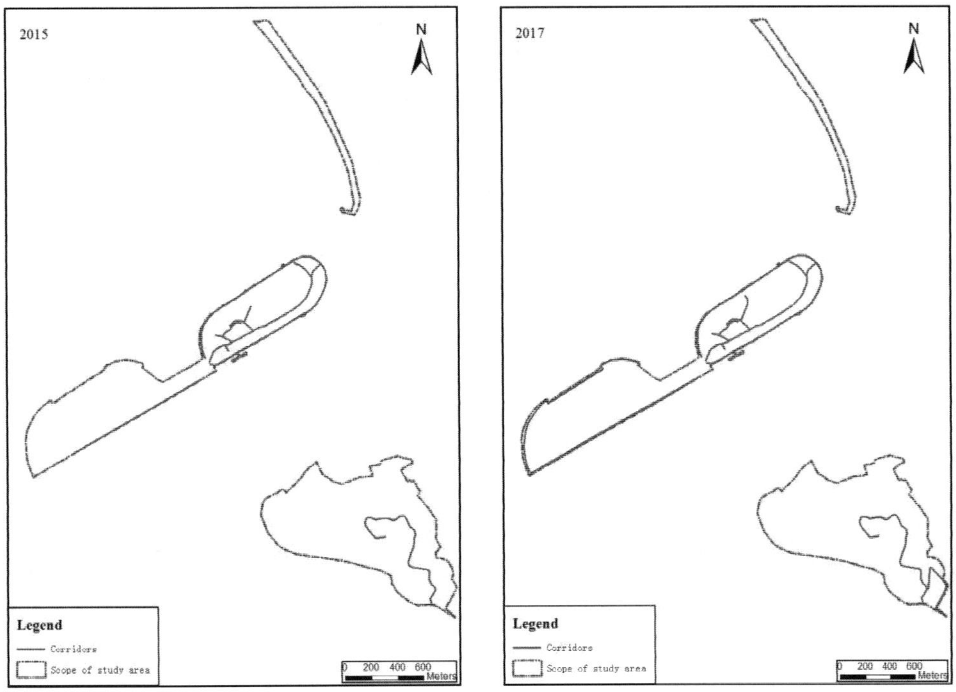

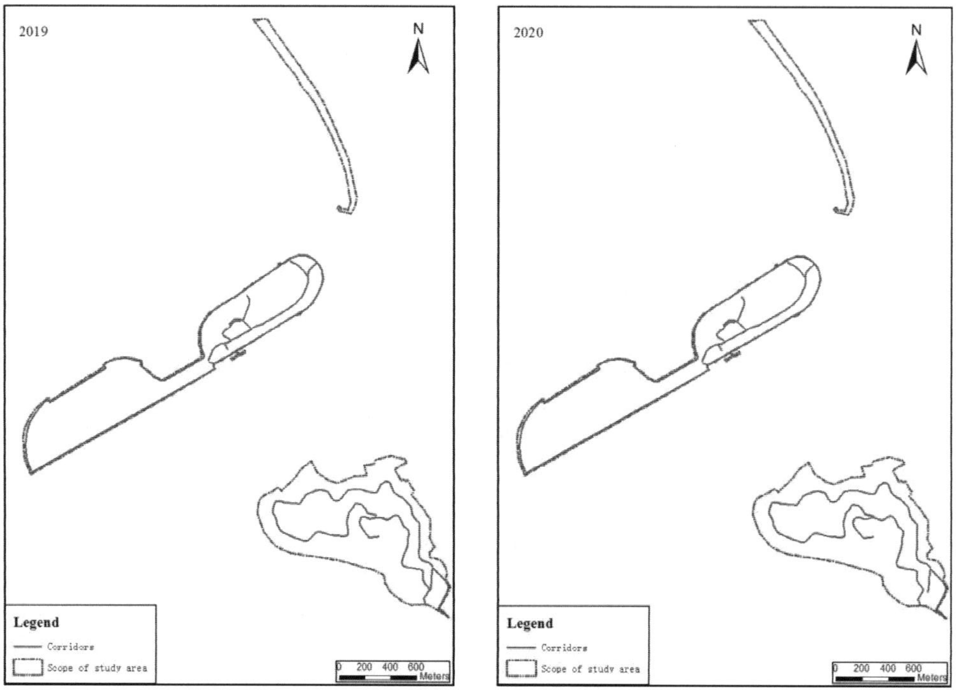

Figure 6. Maps of interference intensity in the study area (2015-2020).

3.2. Changes and analysis of land use diversity

Using remote sensing images to conduct dynamic monitoring of land use types in the study area (Table 5, Figure 7 and Figure 8), it can be seen that the land use types in 2015 were vegetation, beaches, water bodies, buildings, corridors, and open spaces. Following that, in year 2017, artificial beaches were added, as a result there are 7 types of land use; Apart from that, vegetation has always been the largest land use type in all four periods. From the perspective of the land-use diversity index, from 2015 to 2017, the land-use diversity index rose significantly, and the index from 2017 to 2019 declined significantly, and then the index rose more significantly from 2019-2020.

The changes of the land use diversity index reflect the changes of the land use type in the study area from 2015 to 2020 with the development and construction of scenic sites. During the period of 2015-2017, with the implementation of the reclamation project on Phoenix Island, the types of land use zones increased, which causes the index to increase. Within this period (year 2017 to year 2020), alongside the cessation of construction and ecological restoration work, the index dropped and then rise again.

Table 5. Diversity of land use types in the study area

Years	Land use types	Area (m^2)	Land use diversity Indexes
2015	vegetation	1061779.54	1.2116
	beaches	67677.50	

	water bodies	451672.00	
	buildings	40303.73	
	corridors	36046.25	
	open spaces	259653.60	
	vegetation	1001549.98	
	beaches	50440.24	
	water bodies	6938.33	
2017	artificial beaches	337219.8	1.2736
	buildings	42255.99	
	corridors	48735.05	
	open spaces	429993.23	
	vegetation	1173092.75	
	beaches	48160.91	
	water bodies	5425.56	
2019	artificial beaches	339131.49	1.1696
	buildings	32987.83	
	corridors	67121.75	
	open spaces	251212.33	
	vegetation	1131766.67	
	beaches	51815.1	
	water bodies	2293.3	
2020	artificial beaches	343520.16	1.2021
	buildings	37345.42	
	corridors	67953.55	
	open spaces	282438.42	

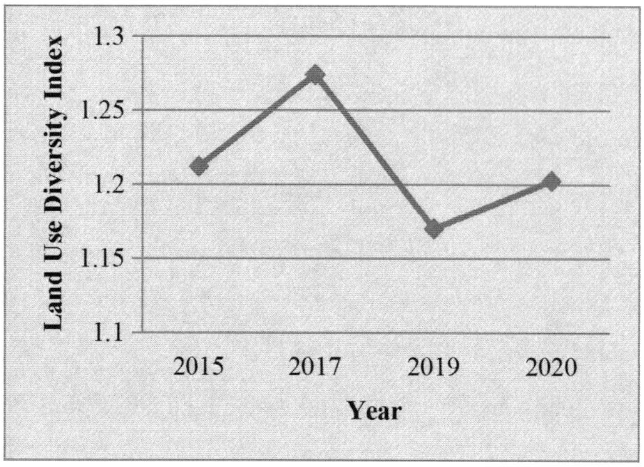

Figure 7. Diagram of land use diversity index in the study area (2015-2020).

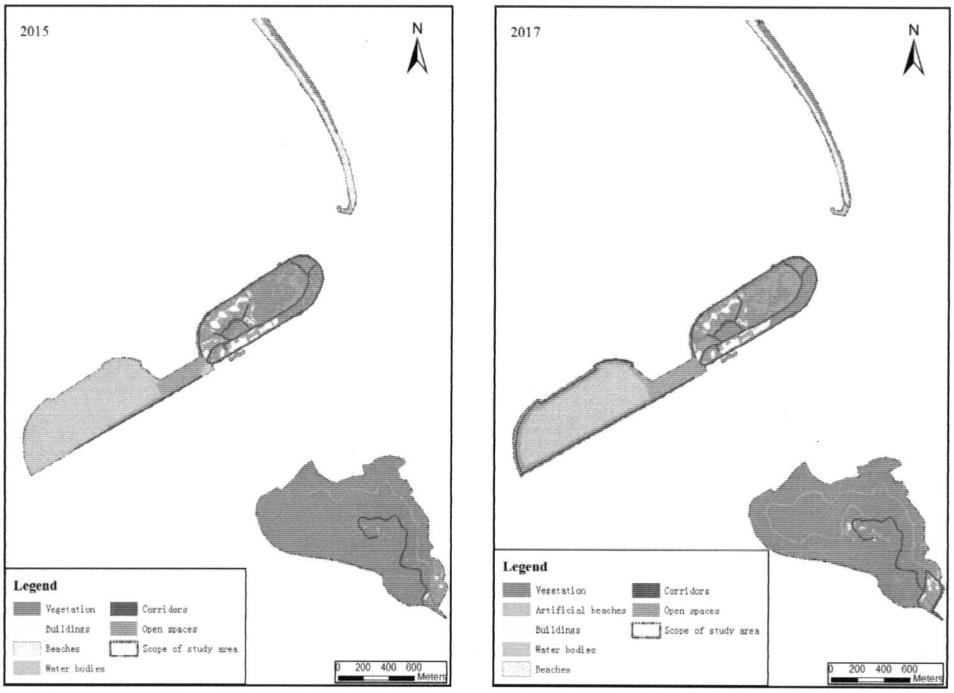

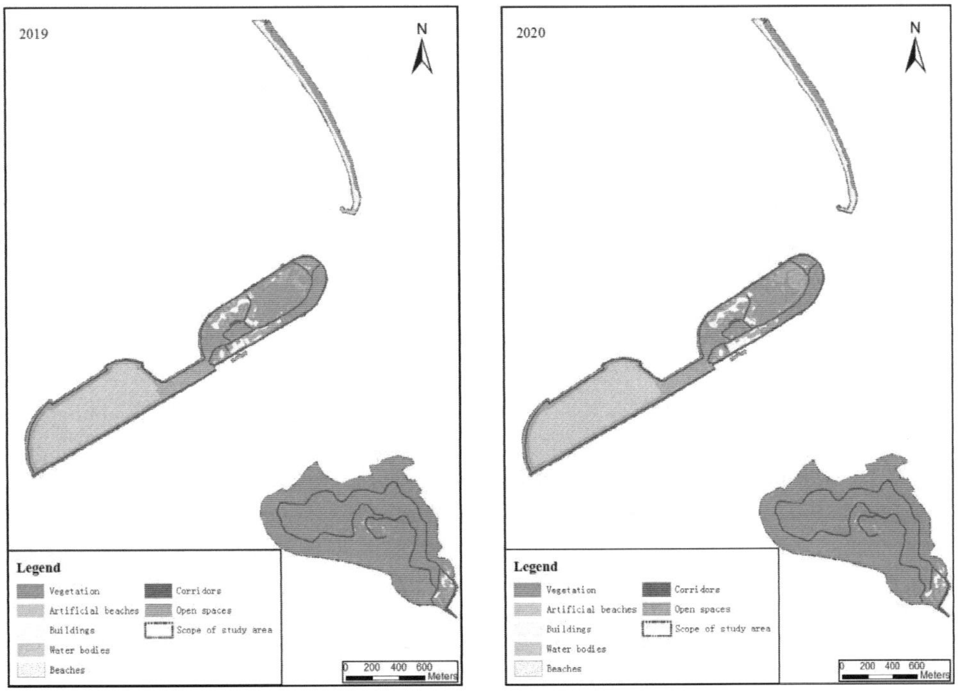

Figure 8. Maps of land use diversity in the study area (2015-2020).

3.3. Changes and analysis of forest coverage

Table 6, Figure 9 and Figure 10 illustrate the forest coverage of the area of study. As can be seen from the chart, the forest region decreased in size significantly in year 2017, and the forest coverage rate dropped to 41.39%. In 2019, the forest area rebounded, with a coverage rate of 46.43%, which was the highest in all 4 periods. However, there was a slight decline in year 2020, and the coverage rate dropped again to 45.33%.

Changes in the ecological environment will be reflected in the forest coverage rate. From the above changes in forest coverage, we can see that the ecological environment in the study area was damaged to a certain extent and the forest coverage rate decreased during 2015-2017. While during 2017-2020, the ecology was restored and the forest coverage rate rose up again.

Table 6. Forest coverage in the study area

Years	Forest areas (m2)	Forest coverage
2015	862959.09	45.01%
2017	793527.67	41.39%
2019	890141.83	46.43%
2020	869035.93	45.33%

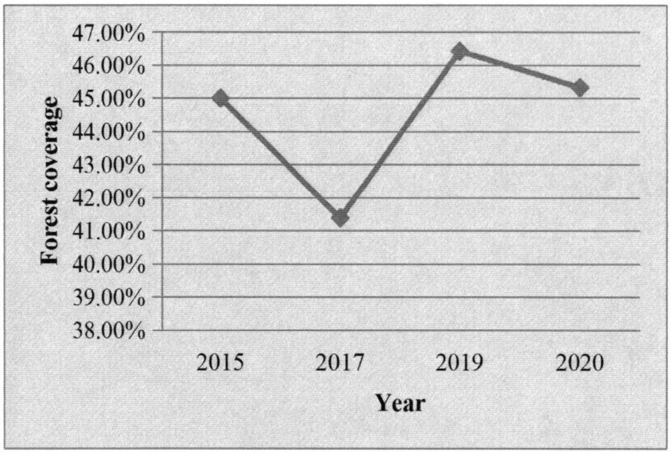

Figure 9. Diagram of forest coverage in the study area (2015-2020).

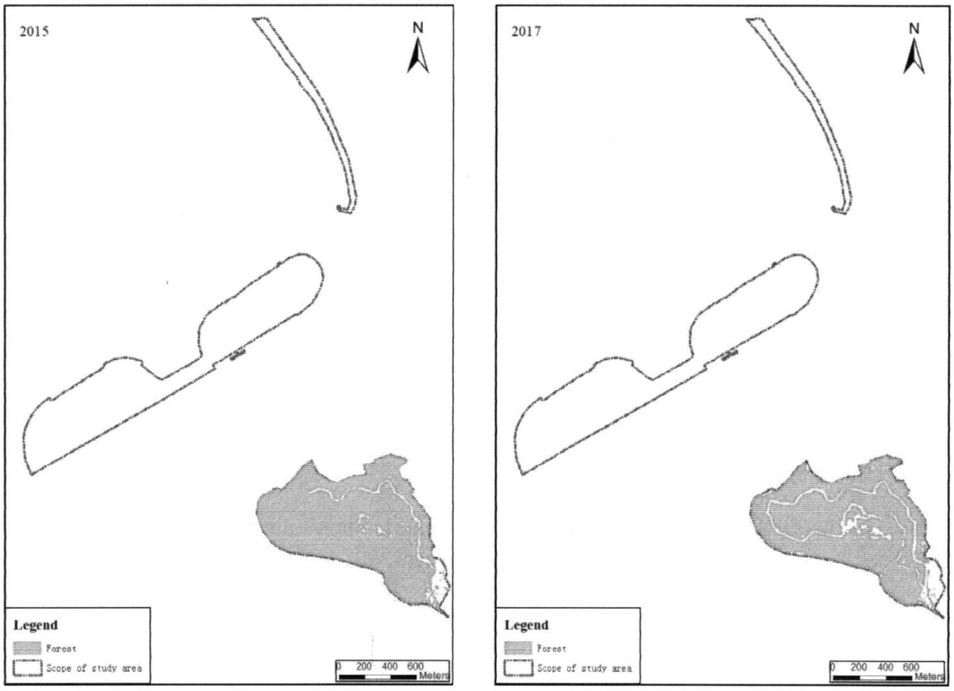

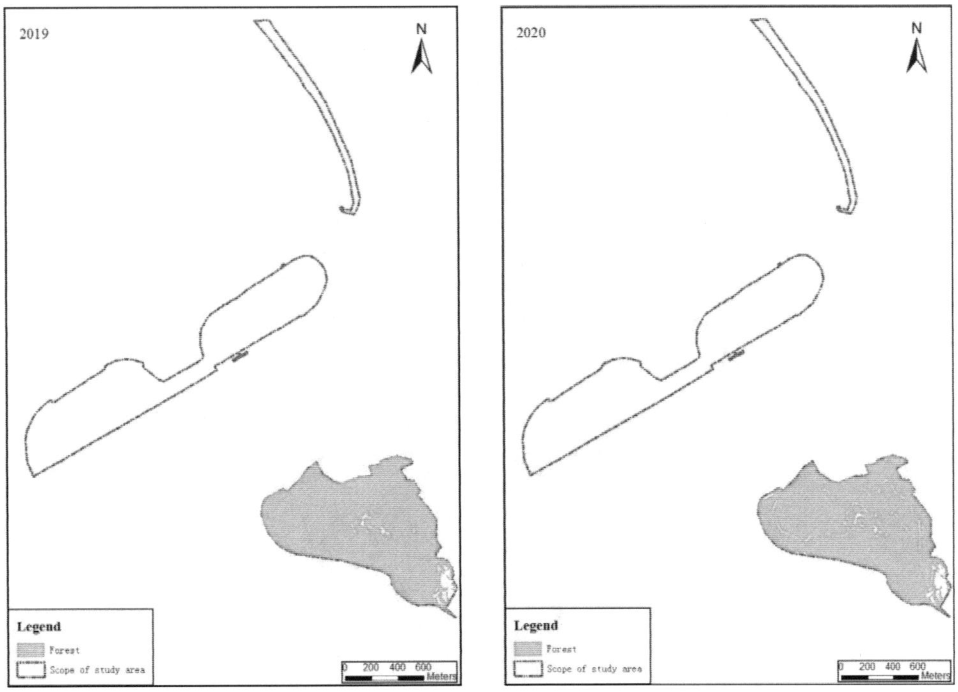

Figure 10. Forest coverage in the study area (2015-2020).

3.4. Field investigation and analysis

At the end of October 2020, we conducted a field investigation and found that there were clear muddy beach phenomenon at Sanya Bay (Figure 11). The beach had a thick mud layer, while the sandbars remain the same. At this point, the reclamation of Phoenix Island scenic spot and the construction on the island has been completed, therefore there is no active construction taking place; The reconstruction of Luhuitou Scenic Area has been completed, and there is no construction on site.

Figure 11. Field investigation on the study area

The ecological environment in the study area fluctuates to a certain extent, from the deterioration in the early stage to the improvement in the later stage. Changes of vegetation in the landscape fragmentation is basically consistent with the changes of forest coverage. The increment in landscape fragmentation from 2015 to 2017 was also reflected in the diversity of land use. The changes in interference intensity illustrates the development of scenic spots in the study area to a certain extent. During the period of 2017 to 2019, all construction work has basically stopped and certain repair measures have been taken throughout the time, which is in line with the actual situation verified in the area.

Based on all the above data and field verification, it can be seen that the changes in the ecological environment reflected by two types of ecological landscape pattern indexes have a good correlation with the diversity of land use and forest coverage. That is, the changes in the ecological landscape pattern indexes in three stages of 2015-2017, 2017-2019, and 2019-2020 were consistent with the changes of two indexes.

4. Discussion and conclusions

4.1. Discussion

The ecological environment of the research area is dynamically changing between 2015 and 2020, as evidenced by changes in ecological landscape pattern, land use variety, and forest coverage. During the period from 2015 to 2017, the ecological environment declined and suffered a certain amount of damages. In the following year, the ecological environment began to recover and the damages were somewhat rectified. The reclamation of the sea and the construction of scenic spots are the main causes leading to changes in the ecological environment. Based on the actual field research, the construction period in the study area is basically the same as the fluctuation period as shown in the monitoring.

This article employs multi-phase high-resolution remote sensing images to extract various landscape ecological landscapes in the scenic area in different periods, and then uses the ecological landscape pattern indexes to indicate the various changes in the ecological environment. The impact of these changes on the environment is reflected in the diversity of land use and forest coverage. Along with field investigation, this paper conducts dynamic monitoring and analysis of the ecological environment of the study area. The technical route of this article has good feasibility for the dynamic monitoring of the ecological environment of coastal tourist areas. In future, the technical route can be expanded with more indexes in the future to make it more ubiquitous, thus increase its universality.

4.2. Conclusions

According to an examination of dynamic monitoring findings, the ecological environment of the coastal tourist area around Sanya Bay has fluctuated significantly over the six-year period 2015-2020. The following are the conclusions:

(1) From 2015 to 2017, the natural environment deteriorated substantially, which was clearly linked to the building of scenic spots in the tourist region, such as the second-phase land reclamation project of Phoenix Island and the upgrade of Luhuitou scenic spot. The ecological environment recovered to some extent between 2017 and 2020, which was related to the suspension of construction of scenic spots. For example, the second phase of the land reclamation project on Phoenix Island was halted in 2018, whereas the Luhuitou Scenic Area was upgraded completely in 2017.

(2) The ecological impact of artificial beaches in the Phoenix Island scenic includes muddy beaches and loss of the beaches of Sanya Bay, and will not be eliminated by 2020. The upgrading and renovation of Luhuitou Scenic spot also resulted in a severe environmental degradation in a short period of time. However, the ecological environment of the study area was improved after the restoration, as evidenced by an increase in vegetation landscape fragmentation in 2017. When construction was totally completed in 2020, the vegetation fragmentation declined and the forest coverage rate rose gradually since then.

(3) The use of remote sensing images to monitor the ecological environment should take into account the local climatic conditions during the time the images were acquired. The forest coverage rate declined slightly from 2019 to 2020. Compared to the forest coverage rate of the 4th period, the forest coverage rate in year 2020 was not much different from year 2015, there's only a slight increase. In contrast to the sharp decline in 2017, the ecological environment of the study area has been recovered. In addition, the local climate and environmental conditions were analysed, date of image acquisition in 2019 was November, the vegetation grew well during rainy season. Date of image acquisition in 2020 is in March,

IOP Conf. Series: Earth and Environmental Science 1004 (2022) 012004 doi:10.1088/1755-1315/1004/1/012004

in which, Sanya has already entered into the dry season, therefore there is no rain for a long period of time, so the vegetation growth declined.

(4) The land use diversity index can better reflect the changes in land use types in the study area. The increment of indexes from 2015 to 2017 is a response to the major construction situation of land reclamation. The first decrement and then the increasement after 2017 is a reflection of the changes in land use types caused by the suspension of construction in the scenic area and the restoration of the ecological environment.

(5) In the scientific and reasonable development of the coastal tourism area, dynamic monitoring of the ecological environment is substantial, it can be used for reference by the relevant departments when the development has a greater impact on the ecological environment. For example, when the damage caused by the upgrading and reconstruction of Luhuitou Scenic spot has been repaired, there is no need to interfere too much; On the other hand, during the second phase of the Phoenix Island reclamation project, it appears to be an irreversible impact on the environment, in that case it should be handled right away.

References

[1] Department of Marine Strategic Planning and Economics M o N R 2020 2019 China Marine Economic Statistics Bulletin

[2] Yuan W and Dong X 2020 Research on Application of Ecological Environment Monitoring and Protection Based on Remote Sensing Technology *China Science and Technology Information* p 89-90

[3] Qiao M, Zhang Y, and Yu L 2020 Dynamic Monitoring of Beijing's Ecological Environment Quality Based on Remote Sensing Ecological Index *Journal of North China University of Technology (Natural Science Edition)* **42** p 17-23

[4] Sawuti R, Abuli A, Li H, Kasimu N, and Li X 2020 Monitoring and Evaluation of Dynamic Changes of Ecological Environment in Fukang City Based on Remote Sensing Ecological Index Model *Soil and Water Conservation Research* **27** p 283-289+297

[5] Li Q 2020 Environmental protection and dynamic monitoring based on remote sensing ecological index model *Resource conservation and environmental protection* p 26-27

[6] Su Y, Teng M, Xiao W, Huang Z, Lei J, Zeng L, and Wang P 2015 Construction of Small Watershed Ecosystem Health Evaluation System Based on Fishnet Grid *Journal of Huazhong Agricultural University* **34** p 65-71

[7] Wei L, Wan D, Bi J, and Xie J 2020 Remote Sensing Dynamic Monitoring and Evaluation of Ecological Environment Quality in Xikuangshan Antimony Mining Area *Environmental Science and Technology* **43** p 230-236

[8] Liu Y, Qi J, Lu H, Wang G, Gao X, Wang J, and Zhang T 2020 Remote Sensing Monitoring and Analysis of Ecological Environment Change in Qingdao City *Surveying and Mapping Bulletin* p 60-65

[9] Zhu D, Chen T, Niu R, and Zhen N Monitoring and Evaluation of Ecological Environment in Mining Areas Based on Remote Sensing Ecological Index of Moving Window *Journal of Wuhan University (Information Science Edition)* p 1-8

[10] Jie W, Zhang C, Wang B, Zhang E, Niu H, Li J, and Dong S 2020 Remote sensing dynamic monitoring and ecological environment changes of Bashang Plateau wetland from 2000 to 2018 *Mineral exploration* **11** p 2720-2728

[11] Guan X and Pan X 2020 Application Analysis of Remote Sensing Monitoring Technology in Ecological Environment Protection of Qilian Mountain National Nature Reserve *Southern Agriculture* **14** p 165-167

[12] Ye H, Liao X, He X, and HuanyinYue 2020 Remote Sensing Monitoring and Analysis of the Changes of the Marine Ecological Environment in the Offshore of Sri Lanka *Journal of Geo-information Science* **22** p 1463-1475

[13] Ren Q, Long A, Yang Y, and Liu J Remote Sensing Monitoring and Analysis of the Ecological Environment Changes in the Mainstream of the Tarim River in the Past 20 Years *Water conservancy and hydropower technology* p 1-11

[14] Liu S, Bai C, Bian Z, An T, Xia C, Hou R, Chen J, Zhang M, and Zhang K 2020 Land Use Change and Its Eco-environmental Effects in the Fragile Area on the Edge of the Desert: A Case Study of Yanchi County on the Southern Edge of the Mu Us Desert *Grassland Science* **37** p 2175-2184

[15] He S 2019 R&F destroys mangroves, Sunac illegally reclaimed the sea... Real estate companies in Hainan "inflate their pockets and ruin the ecology", how to solve? *China Economic Weekly* p 95-97

[16] Zhang L, Liao J, Yuan X, Mu X, Song Q, and Bi J 2020 Remote Sensing Analysis on the Characteristics of Coastline Change of Hainan Island from 1987 to 2017 *Tropical geography* **40** p 659-674

[17] Ren J 2005 Research on Land Use Change and Environmental Impact Assessment Supported by GIS *Liaoning University of Engineering and Technology*(Liaoning)

[18] Zhao Y, Yang J, Yin Y, Zhao H, He J, and Zhang H 2019 Research on Remote Sensing Monitoring of Development Status of Coastal Zirconium-Titanium Placer Deposits in Eastern Hainan Island and Countermeasures for Ecological Restoration and Governance *Remote Sensing of Land and Resources* **31** p 143-150

Acknowledgement: This work was sponsored by the Hainan Provincial Department of Science and Technology, China under Grant No. ZDKJ2019006 and the Hundred Person project of Hainan province, China under Grant No. [2018]31.

A High-Performance Cloud-based Remote Sensing Data Reprojection Method

Yi Dong[1,2], **Chen Xu*** [3,4]

1. Aerophotogrammerty & Remote Sensing Bureau of China Coal (**ARSC**)
2. Shaanxi Engineering Technology Research Center for Geospatial Information
3. Key Laboratory of Digital Earth Science, Aerospace Information Research Institute, Chinese Academy of Sciences
4. University of Chinese Academy of Sciences

E-mail: xuchen20@mails.ucas.ac.cn

Abstract. Remote sensing (RS) data are the cornerstone of the digital earth. At present, the productions from Earth observation satellites are updated frequently with refined data quality. However, traditional methods are implemented mostly based on a single machine. When processing large-scale data in batches, there are limitations in computing power and storage, and the expansion is relatively cumbersome. This demands a more efficient and complex computation platform, such as cloud computation. Specifically, the reprojection procedure, unlike other data processing procedures, is both computationally intense and I/O-intense. This paper proposes a high-performance cloud-based RS data reprojection method called OCRM (optimised cloud-based reprojection method). First, the data process flow of reprojection was optimised by improving the I/O efficiency and computational efficiency. Second, with scheduling and controllers under cloud computation, high-performance massive RS data reprojection was achieved. The overall performance of the RS data reprojection module with the optimised algorithm in a single machine considerably outperformed that of the commercial software. The scheduling module can maximise the use of cloud resources to achieve high-performing reprojection calculations. The results indicate the stable performance of the proposed method, and we are working to expand the structure of OCRM to other computationally intensive remote sensing processes, such as data preprocessing and data mining with deep learning.

1. Introduction

Currently, RS data are widely used in the fields of earth science, disaster early warning, emergency dispatch, land management, etc. [1][2]. RS data are one of the most important data sources of the digital earth [3]. The era of remote sensing big data and digital earth big data has arrived, and thus, the demand for massive RS data computation has grown rapidly[4][5]. By this trend, specific regimes for RS big data preprocessing and cleansing [6] are expected to maintain the structural unity of RS data. Preprocessing includes geometric correction, radiometric correction, spatial and temporal reconstruction, quality evaluation, etc. [7]. Among them, the procedure that transfers multisource RS data of different projections into the same projection coordinate system is called reprojection. Reprojection is a typical computationally intensive and I/O intensive procedure that is widely used in various preprocessing steps. Its implementation efficiency deeply influences RS data preprocessing efficiency. However, different data providers use different projected coordinate systems. Different fields need RS data from different projection coordinate systems due to factors, such as technology stacks, demands, policies and equipment. The mismatch between providers and consumers needs services that can adapt highly dynamic data delivery demands and rapidly expand service capabilities. However, traditional commercial software (such as ArcGIS) can only support frame-level reprojection and can only run on a single machine. It is not able to cope with dynamic reprojection demands, process superlarge RS data, and rapidly expand to support large-scale data processing. To address these issues, this paper proposes a high-performance cloud-based RS data reprojection method called OCRM. To solve the problem caused by computational intensity, this research first optimised the calculation under basic efficiency and high dynamic demand, which improved the computational efficiency of reprojection computing. To minimize storage intensity, this research expanded the capacity through task scheduling and management, which optimised the migration ability and the efficiency of the transfer projection of superlarge data. This work is based on our previous work on ScienceEarth [8][9].

2. Methodology

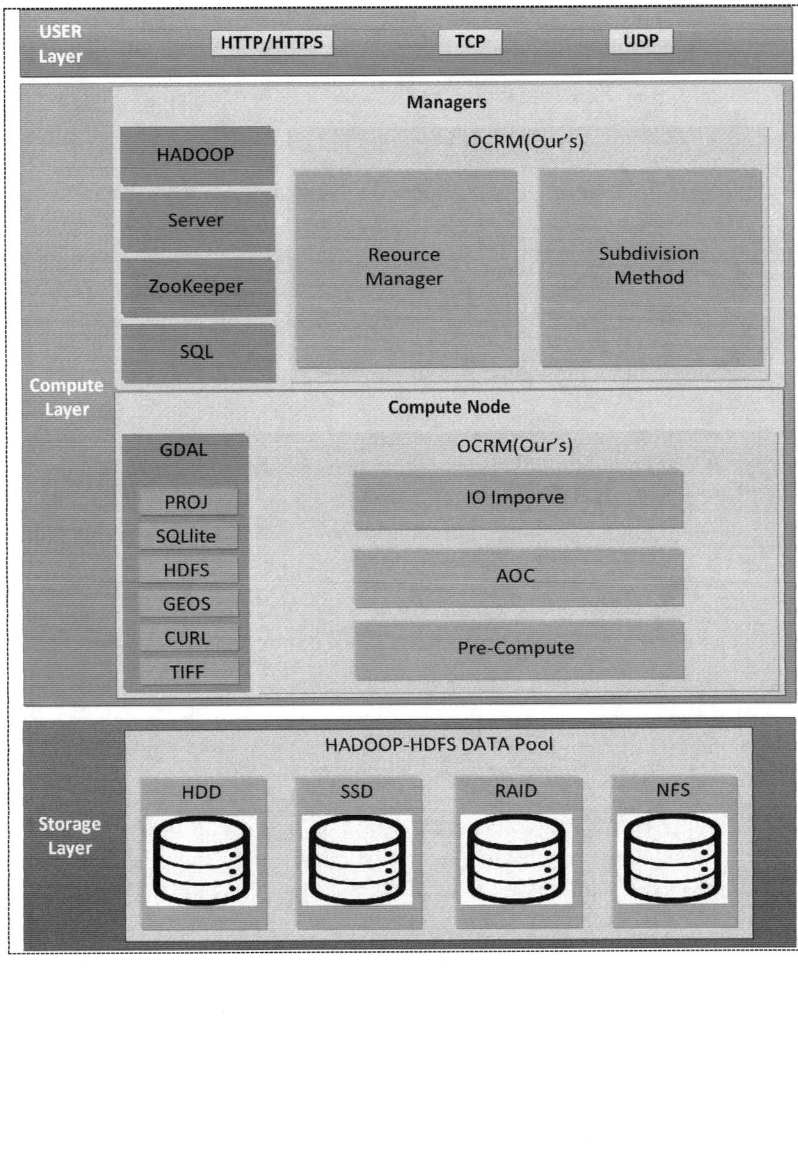

Figure 1. Technology Architecture of OCRM.

Our technology architecture included three layers: the user layer, the compute layer, and the storage layer. The user layer received the user's reprojection demands and then sent them to the computation layer. The compute layer used the subdivision method and divided the user's requests into task groups. Then, the resource manager scheduled those task groups and sent each task to the compute nodes. The compute node was responsible for the reprojection computing procedure. The whole compute layer was based on a storage layer structured by Hadoop [10]. Our technology architectural design achieved computational and storage distributions. OCRM was implemented in conjunction with the technical improvements mentioned below.

2.1. Improvements of reprojection computing algorithm

2.1.1. Data read/write optimisation

To combat the inefficiency of I/O when random reading is reprojected, we used the locality of the projection algorithm[1], which improved the aggregation of data reading. We performed a single point reprojection operation on the boundary required by the user to determine the original data range. The original data range was applied to read the range of image data into memory. Then, all calculations and copies of the data occurred in memory, and the reprojected data were written to disk after processing. This optimisation changed the random access to sequential reads and writes, which decreased the stress for the hard drive. This change greatly increased the speed of the I/O. To reduce the data size and better adapt to cloud computation, the COG format was also implemented in this study.

2.1.2. Memory addressing optimisation [2]

Since reprojection calculations involve a large number of image band operations, such as data filling and cell interpolation, converting data from AOB (array of band) structures to AOC (array of colour) structures can effectively reduce the memory addressing time.

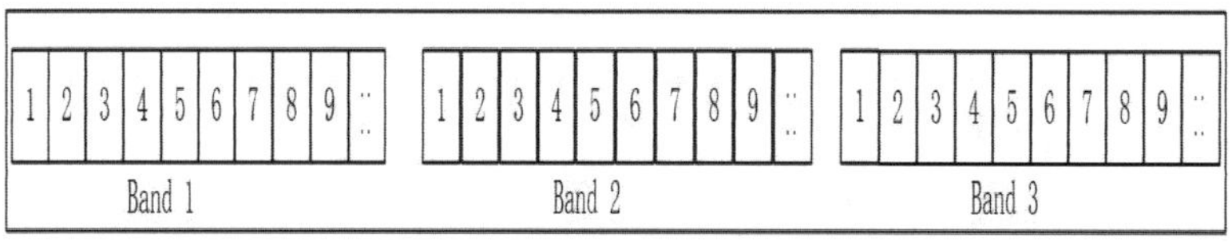

Figure 2. AOB data structure diagram.

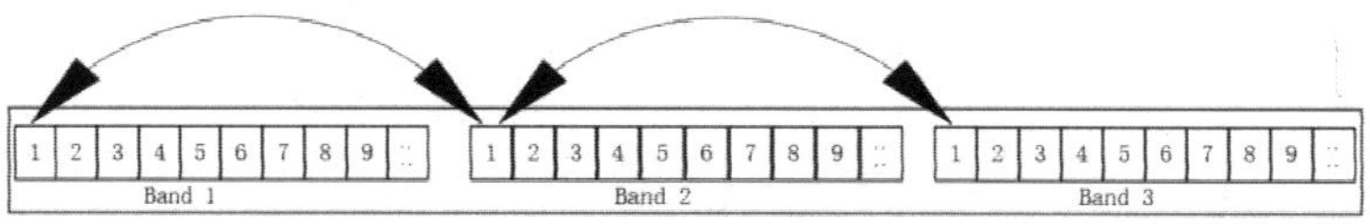

Figure 3. AOB data read schematic.

As shown in Figure 2, a typical array of AOB structures, which is very friendly for I/O reading and writing, can use the speed of hardware sequential read and write to improve I/O efficiency. However, for calculations such as reprojection, which require read and write data within the band frequently, using AOB structures can cause memory addresses to differ largely during calculations. As shown in Figure 3, this large difference caused addresses to exceed the range of a single memory page. The system frequently transformed page memory. Therefore, an additional expenditure of time would occur.

[1] Detailed discussion of locality at Appendix A.
[2] All examples in this section are 3 bands.

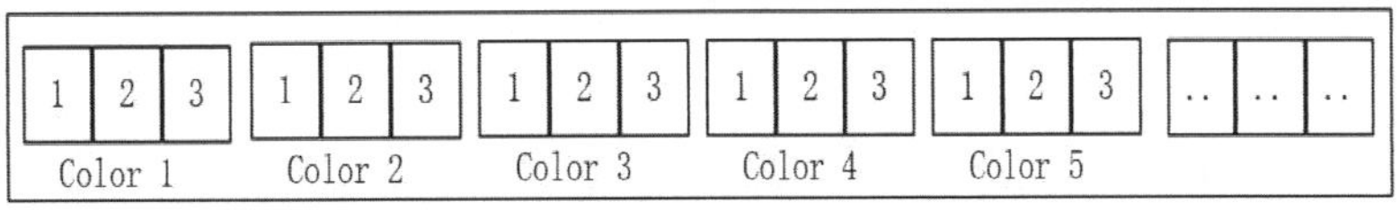

Figure 4. AOC data structure diagram.

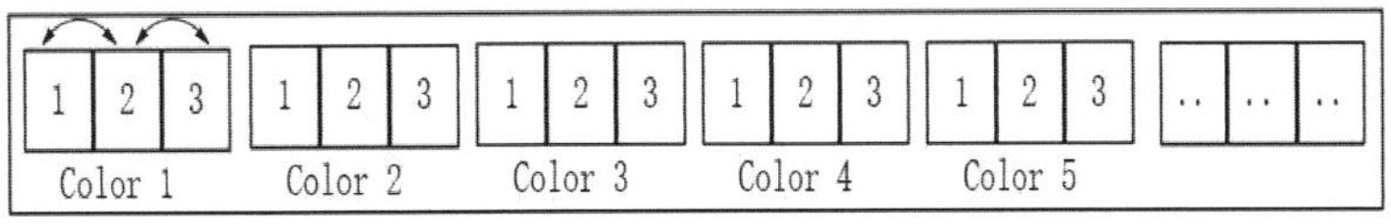

Figure 5. AOC data read schematic.

As shown in Figure 5, after rearranging the data according to AOC, each data address was near the neighbour. This process greatly reduced the possibility of page memory displacement and improved the efficiency of addressing.

2.1.3. Pre-building the environment

Building the environment while computing will consume time. However, the environment needed is stable for the same algorithm, and each computation only needs to change the input data. This feature leads us to try a prebuilt computation environment to compress the reprojection computation time. To ensure the correctness and generalisation use of the components, we chose GDAL [11] as the technical base.

2.2. Cloud-based tasks management design

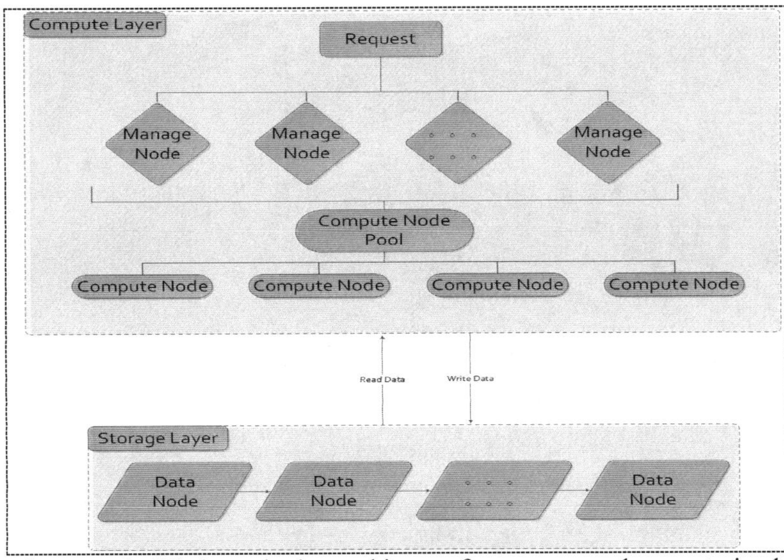

Figure 6. The logical structure of the system. The system is divided into two layers: the computing layer and the storage layer. The overall system is driven by the managing node.

For large-scale RS data, single machine performance may be constrained, which can be improved by calculating in parallel by scheduling. Since the range of user requirements does not often fall into individual RS data within the underlying data, converting the user's needs into several RS data reprojection subtasks is often needed. We proposed a division method for the reprojection of individual RS data. The file system HDFS and the task scheduling system, designed with the characteristics of reprojection tasks, achieved a dynamic expansion of resources and completed oversized reprojection. After the cluster was built according to this study, we proposed the following cloud-based task management methods to ensure the high availability, stability and overall controllability of the system. The overall structure is shown in Figure 6.

2.2.1. Resources use function

To meet the controller's resource management needs, the manager needed to estimate the memory resources used by the task before running. It also prevented running instability when the physical machine was running out of memory. We found that the memory use of the task execution had the following mathematical relationships:

Input data is:

- Starting coordinates under the target projection:$Proj_x$,$Proj_y$
- Amount of change per unit under the target projection:$Proj_{\Delta x}$,$Proj_{\Delta y}$
- Width and height of the target picture:$Proj_{width}$, $Proj_{height}$
- Reprojection function:$p(Proj_{x1}, Proj_{y1}) = (Ori_{x1}, Ori_{y1})$

The bounding box for the target projection can be calculated as follows:

$$Proj_{xEnd} = Proj_x + Proj_{\Delta x} * Proj_{width} \tag{1}$$
$$Proj_{yEnd} = Proj_y + Proj_{\Delta y} * Proj_{height} \tag{2}$$
$$Proj_{minx} = \min(Proj_x, Proj_{xEnd}) \tag{3}$$
$$Proj_{miny} = \min(Proj_y, Proj_{yEnd}) \tag{4}$$
$$Proj_{maxx} = \max(Proj_x, Proj_{xEnd}) \tag{5}$$
$$Proj_{maxy} = \max(Proj_y, Proj_{yEnd}) \tag{6}$$

As shown in Formula (7), the bounding box of the original data can be obtained by calculating the projection of the boundary points of the target projection:

$$Ori_{bounds}: \begin{bmatrix} p(Proj_x, Proj_y) & p(Proj_x + Proj_{\Delta x} * 1, Proj_y) & \cdots & p(Proj_{xEnd}, Proj_y) \\ p(Proj_x, Proj_y + Proj_{\Delta y} * 1) & & \ddots & \vdots \\ \vdots & & & \\ p(Proj_x, Proj_{yEnd}) & p(Proj_x + Proj_{\Delta x} * 1, Proj_{yEnd}) & \cdots & p(Proj_{xEnd}, Proj_{yEnd}) \end{bmatrix} \tag{7}$$

$$Ori_{minx} = \min(Ori_{bounds}[x]) \tag{8}$$
$$Ori_{miny} = \min(Ori_{bounds}[y]) \tag{9}$$
$$Ori_{maxx} = \max(Ori_{bounds}[x]) \tag{10}$$
$$Ori_{maxy} = \max(Ori_{bounds}[y]) \tag{11}$$

We use the original unit change read from the original data:$Ori_{\Delta x}$, $Ori_{\Delta y}$. We calculate the width and height of the original data required as follows:

$$Ori_{width} = \frac{(Ori_{maxx} - Ori_{minx})}{Ori_{\Delta x}} \tag{12}$$
$$Ori_{height} = \frac{(Ori_{maxy} - Ori_{miny})}{Ori_{\Delta y}} \tag{13}$$

The use of memory resources can be calculated with the following formula:

$$memory = (Ori_{width} * Ori_{height} + Proj_{width} * Proj_{height}) * B * U + C \tag{14}$$

B: number of bands

U: number of bytes occupied by data type

C: constant memory for the environment (in our environment C = 2MB)

2.2.2. Subdivision method

The task will be divided if memory consumption exceeds the threshold (we set it to 750 MB). The subtask will be divided later if the subtask still fails to meet the threshold. For the subdivision method, we divide the image according to the configuration, as shown in Figure 7.

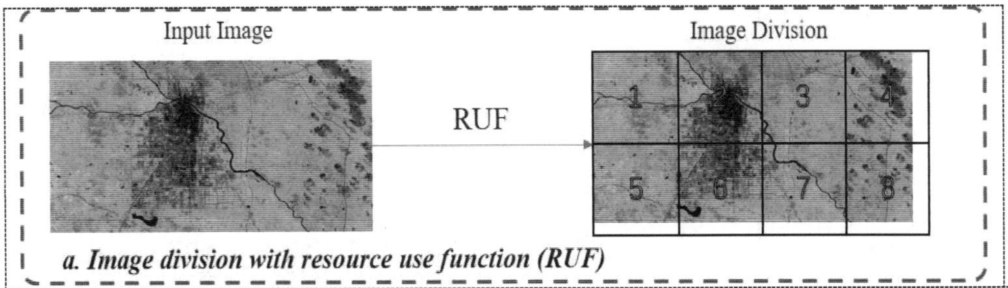

Figure 7. Schematic diagram of the task subdivision, according to configure (x-Axis, 4 evenly divided y-Axis, 2 equally divided).

By applying this method, the read/write conflict area in the image task can be controlled at the segmentation boundary. The parallel executability of the overall task can be improved by cooperating with the scheduling strategy.

2.2.3. Scheduling strategy

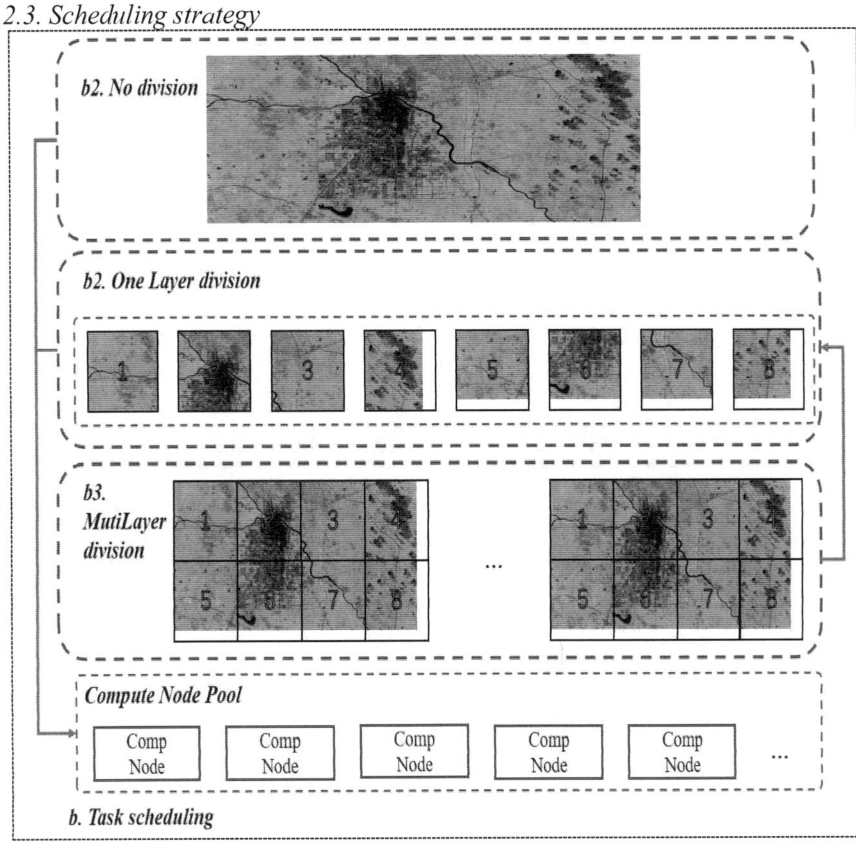

Figure 8. Task scheduling diagram

After the user's task goes through the division process, if there is no division, the task is directly put into the computing node pool to wait for the allocation of computing nodes. If the split task group has only one layer, all of the subtasks in this layer are sequentially placed into the computing node pool. If multilayer division occurs, the batches are divided into n layers, and then each batch is scheduled in a single-layer manner.

3. Result

3.1. Implementation environment introduction
This research has two implementation environments: Table 1 for a single machine and Table 2 for cloud computation.

Table 1. Implementation environment single machine.

System	Windows	Linux
CPU	11th Gen Intel(R) Core (TM) i5-1145G7	11th Gen Intel(R) Core (TM) i5-1145G7
RAM	32GB 3200	16GB 3200

Table 2. Implementation environment cloud computation.

Node type	Computing node	Management node	Storage Node
CPU	8 cores	8 cores	4 cores
RAM	16GB	8 GB	8GB

3.2. Single machine efficiency comparison
The single machine efficiency experiment using an RS data original projection is EPSG:4513, the size is: 4518 * 4732, and compares two common usage scenarios:

Table 3. Single machine efficiency comparison table.

	Mission1[a]		Mission2[b]	
	Time(ms)	Memory(MB)	Time(ms)	Memory(MB)
Arcgis	9226.99999809	NOT Available	NOT Available	NOT Available
GdalPY	4610.871826171	178.72	185.95874023	66.69
Gdal	4017.36	192.60	114.705	38.87
OCRM	6054.84	173.60	33.35	39.10

[a] Convert the overall original image to other projections

[b] Extract 256*256 target range from the original image to other projections

The efficiency comparison source code can be found at https://gitee.com/ddyy_1213/tile-extract. The project experiment can be reproduced under the Linux environment with Gdal installed.

In Mission 1, our implementation was better than that of ArcGIS, but it was slower than that of Gdal, which will be discussed in Section 4.2.

In Mission 2, our implementation outperformed Gdal and GdalPY by an order of magnitude in time consumption and was almost the same as the C version of that of Gdal in memory consumption.

Table 4. Reprojection time table.

	Read Time(ms)	ReProjection Time(ms)	Write Time(ms)
Mission1[a]	280.3697	6795.79	60.2263
Mission2[b]	3.1467	33.5523	1.139

[a] Convert the overall original image to other projections

[b] Extract 256*256 target range from the original image to other projections

It can be seen from Table 4 that after the above optimisation, the reprojection operation still occupies the main body of the overall execution.

3.3. Cloud computation experiment

During the study, we received a request to process the latest satellite remote sensing data from a certain place into usable RS products. We divided the satellite data into three batches, each of which contained one hundred images. The data volume per image was approximately 200 G (uint8, 5 band), and the data diversities were almost the same. We tested the traditional reprojection method with the three images in the first batch and processed them for an average of approximately 19 minutes. Traditional reprojection has difficulty achieving parallel processing. The processing efficiency of a batch is estimated to be far more than 60 hours if considering the single hard disk capacity limitation and data transfer time. This efficiency makes it difficult to complete the task on time. A large amount of manual participation is extremely error-prone. We successfully processed data after applying OCRM and provided the required RS products.

Table 5. Cloud computation experiment results

	Traditional Method (only 3)		OCRM	
	AvgTime(sec)	TotalTime(Hour)	AvgTime(sec)	TotalTime(Hour)
Batch 1	1187	>>60(estimate)	693.80892	19.27247
Batch 2	NOT Available	NOT Available	681.26184	18.92394
Batch 3	NOT Available	NOT Available	697.22208	19.36728

In this practice, we used 3 computing nodes and each node was configured with 7.5 GB of memory for the reprojection calculations. In the end, it took 57.56 hours to complete the reprojection calculations of the three batches of data.

4. Discussion

4.1. Optimisation effects

From Figure 9, after I/O improved, the execution efficiency on Mission2 steadily surpassed that of Gdal. However, the execution time was still generally more than 50 ms. Later, with the implementation of the prebuilt optimisation, the execution time was reduced to less than 50 ms, but the performance was still not stable enough. Finally, after the introduction of AOC optimisation, the execution time was stable in the 30-50 ms interval.

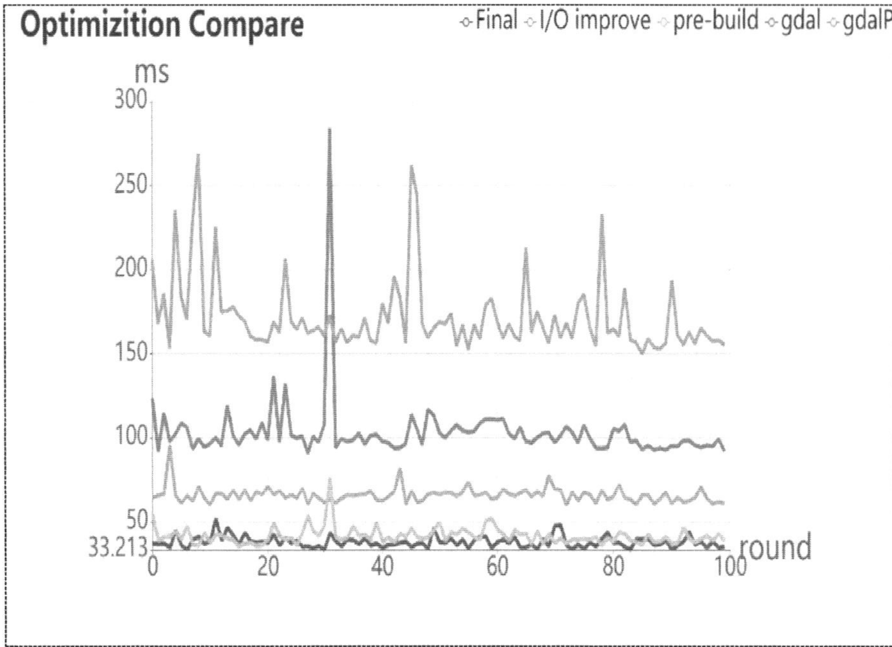

Figure 9. Detailed comparison of the optimisation items of the reprojection module.
The y-axis is the execution time and the
x-axis is the round of executions.

4.2. Why mission 1 OCRM had a worthy performance

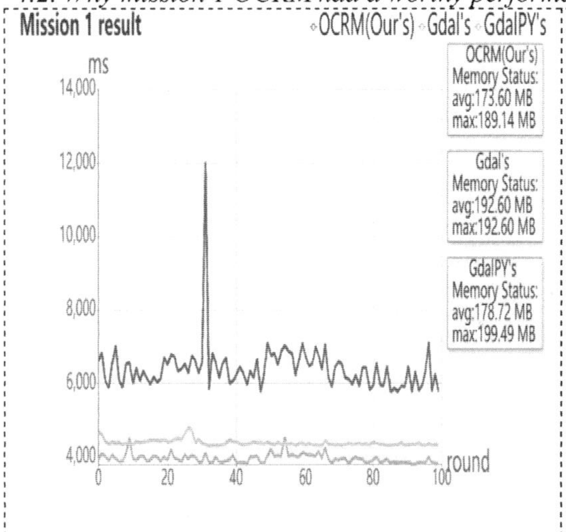

Figure 10. Mission1 comparison results, including the
y-axis as the execution time and the
x-axis as the round

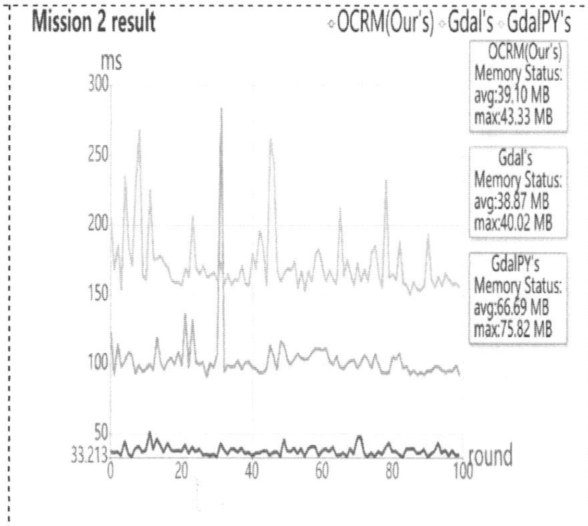

Figure 11. Mission2 comparison results, including the
y-axis as the execution time and the
x-axis as round

Mission 1 OCRM was slower than Gdal. The preliminary analysis shows that the implementation of the OCRM is point-by-point reprojection, while the implementation of Gdal is to take a certain interval for reprojection calculation, and the missing points in the middle are supplemented by interpolation. The reprojection calculation was most time-consuming, as shown in Table 4. Gdal achieved the overall efficiency improvement by reducing the number of reprojection calculations.

In Mission 2, the performance of OCRM was far ahead of Gdal. Therefore, we believe that purely from the perspective of the reprojection process, the optimisation scheme of this study already optimised the execution time of reprojection. If point-by-point reprojection is also used in Gdal, OCRM should perform better than Gdal on the entire image reprojection.

4.3. Cloud computation practice
In the cloud computation practice in Section 3.3, since the total data volume exceeds 120 TB, only two of our storage nodes could safely store project data. Due to the limitation of storage nodes, too many computing nodes would have been stalled due to I/O speed. After considering the calculation speed and the I/O speed, we only used 3 computing nodes to process the data. If there was no storage limitation, with the expansion of computing nodes and storage nodes, the task execution time could have been be shortened.

5. Conclusion

This paper proposes a high-performance cloud-based RS data reprojection method called OCRM. First, we optimised the process of the reprojection algorithm of RS data. After I/O, addressing, prebuilding and other optimisation methods, the time of reprojection calculation is effectively reduced. To ensure the accuracy and versatility of the results, as well as the ease of use in engineering, we used the reprojection algorithm provided by Gdal. Cloud-based computing is used to optimise the computing efficiency of a large amount of RS data, avoiding the difficulty of expanding computing resources and storage resources. The experiment shows that the overall performance of the optimised algorithm when running the RS data reprojection module on a single machine is considerably better than that of the current commercial software. Second, the dispatch system can maximise the use of cloud resources to achieve high-performance reprojection calculations. This research supports single-machine deployment or multi-node deployment, fully considering the availability issues in different computing environments and the scalability of the overall system. Possible follow-up optimisation directions include: 1. We may research the implementation of reprojection algorithms and try to propose more efficient algorithms. 2. We may try to implement the reprojection algorithm with more efficient software and hardware. 3. We may try to explore the method of taking points at intervals to reduce the number of reprojections.

6. Appendix

Appendix A: Spatial locality of reprojection
Spatial locality is defined as the feature that immediately accesses its neighbouring addresses when it is likely that a program accesses a memory address. The reprojection operation is executed sequentially in the original projection space, which has spatial locality. When performing interpolation operations in the postprojection space, it also visits 8 directional points of the postprojection point, which has spatial locality.

Appendix B: System migration ability
Different modules were separated, which enhanced the expansion capability of the overall system when interacting with each other by protocol. When facing other kinds of data sources, the mismatched management node or computing node can be replaced in a modular manner to adapt the diversity of the data. In addition, our reprojection method is only a module of this architecture to achieve cluster reprojection, and other reprojection methods or other calculation methods can also be incorporated into the system as cloud-based tasks after the control protocol has been extended.

Appendix C: Software version

Table C1. Software version table.

Software	Version	Software	Version	Software	Version
Arcgis	10.03	PROJ	8.0.0	Curl	7.75.0
gcc	9.3.0	SQLite	3350300	Python	3.8.10
Gdal	3.2.2	Tiff	4.2.0	PYGdal	3.2.2

Funding: This work is supported by the Innovation Drive Development Special Project of Guangxi (GuikeAA20302022), the National Natural Science Foundation of China (41974108) and the Strategic Priority Research Program of Chinese Academy of Sciences, Project title: CASEarth (XDA19080103).

References

[1] N. Pahlevan, S. K. Chittimalli, S. V. Balasubramanian, and V. Vellucci, "Sentinel-2/Landsat-8 product consistency and implications for monitoring aquatic systems," *Remote Sens. Environ.*, vol. 220, no. October 2018, pp. 19–29, Jan. 2019.
[2] H. Huang *et al.*, "Mapping major land cover dynamics in Beijing using all Landsat images in Google Earth Engine," *Remote Sens. Environ.*, vol. 202, pp. 166–176, Dec. 2017.
[3] M. F. Goodchild *et al.*, "Next-generation Digital Earth," *Proc. Natl. Acad. Sci.*, vol. 109, no. 28, pp. 11088–11094, Jul. 2012.
[4] H. Guo *et al.*, "Big Earth Data science: an information framework for a sustainable planet," *Int. J. Digit. Earth*, vol. 13, no. 7, pp. 743–767, Jul. 2020.
[5] B. Zhang *et al.*, "Remotely sensed big data: evolution in model development for information extraction [point of view]," *Proc. IEEE*, vol. 107, no. 12, pp. 2294–2301, Dec. 2019.
[6] C. Yang *et al.*, "Big Earth data analytics: a survey," *Big Earth Data*, vol. 3, no. 2, pp. 83–107, 2019.
[7] A. Lewis *et al.*, "The Australian Geoscience Data Cube — Foundations and lessons learned," *Remote Sens. Environ.*, vol. 202, pp. 276–292, Dec. 2017.
[8] C. Xu, X. Du, Z. Yan, and X. Fan, "ScienceEarth: A Big Data Platform for Remote Sensing Data Processing," *Remote Sens.*, vol. 12, no. 4, p. 607, Feb. 2020.
[9] C. Xu *et al.*, "A Modular Remote Sensing Big Data Framework," pp. 1–11, 2021.
[10] K. Shvachko, H. Kuang, S. Radia, and R. Chansler, "The Hadoop Distributed File System," in *2010 IEEE 26th Symposium on Mass Storage Systems and Technologies (MSST)*, 2010, vol. 454, no. 3, pp. 1–10.
[11] O. R. G. GDAL, "GDAL--Geospatial Data Abstraction Library." Open Source Geospatial Foundation, 2012.

IOP Conf. Series: Earth and Environmental Science 1004 (2022) 012006 doi:10.1088/1755-1315/1004/1/012006

An adaptive geographic meshing and coding method for remote sensing data

Xiaohui Huang[1,2], **Jiabao Li**[1,2], **Jining Yan**[1,2] **and Lizhe Wang**[1,2]

1. School of Computer Science, China University of Geosciences, Wuhan, Hubei, 430074, P. R. China
2. Hubei Key Laboratory of Intelligent Geo-information Processing, China University of Geosciences, Wuhan, Hubei, 430074, P. R. China

Corresponding author: Lizhe Wang (lizhe.wang@gmail.com)

Abstract. Spatial indexing techniques, inherently data structures, are generally used in portals opened by institutions or organizations to efficiently filter RS images according to their spatial extent, thus providing researchers with fast Remote Sensing (RS) image data discovery ability. Specifically, space-based spatial indexing approaches are widely adopted to index RS images in distributed environments by mapping RS images in two-dimensional space into several one-dimensional spatial codes. However, current spatial indexing approaches still suffer from the boundary objects problem, which leads to multiple spatial codes for a boundary-crossing RS image and thus alleviates the performance of spatial indexes built on top of these spatial codes. To solve this problem, we propose an adaptive geographic meshing and coding method (AGMD) by combining the famous subdivision model GeoSOT and XZ-ordering to generate only one spatial code for RS images with different spatial widths. Then, we implement our proposed method with a unified big data programming model, (i.e., Apache Beam), to enable its execution in various distributed computing engines (e.g., MapReduce, and Apache Spark, etc.) in distributed environments. Finally, we conduct a series of experiments on real datasets, the archived Landsat metadata collection in level 2. The results show that the proposed AGMD method performs well on metrics, including the following aspects: the effectiveness of the storage overhead and the time cost are up to 359.7% and 58.02 %, respectively.

1. Introduction

The advent of sensor technologies has resulted in the generation of Remote Sensing (RS) images at a staggering rate. Recently, institutions and organizations have open-sourced a huge number of RS images in their web portals to researchers to help them explore and gain insights from these images to better study the Earth. A spatial index is an essential means to filter data. It is widely adopted in the above portals to help researchers efficiently discover RS images by observing their desired space areas. Current approaches to building a spatial index for RS images can be categorized into data-based approaches and space-based approaches. Space-based approaches are widely adopted to build an efficient spatial index for successively generated RS images in a distributed environment [1].

Space-based approaches divide the entire space into a number of equal-sized grids and use space-filling curves to convert discrete grids in two-dimensional space into a number of spatial codes in one-dimensional space. In this way, each two-dimensional RS image is represented by a set of one-dimensional spatial codes that can be indexed with proper one-dimensional indices.

However, current space-based approaches face two main challenges: discrepancies in the widths of RS images and boundary RS images. The discrepancies in the widths of RS images make it hard to determine a unified subdivision level when dividing spaces for space-based approaches. Recently, a number of landmark works have been proposed to solve this problem. These works aim to generate the optimum subdivision level for each RS image and generate spatial codes. For example, both GeoSOT [2] and minGeoHash [3] are representative works that automatically determine the subdivision level according to the width of a spatial object, using polygon geometry. However, there is another problem caused by boundary RS images; multiple spatial codes are generated for boundary-crossing RS images when they are at their optimum subdivision level. Multiple spatial codes alleviate the performance of spatial indexing methods by introducing duplicated indexing records. However, current methods for solving boundary RS images are inefficient. For example, minGeoHash uses at most nine spatial codes to represent a spatial object. The code merging strategy proposed by Huang et al. [1] merges four spatial codes of an RS image generated by GeoSOT into one spatial code under a specific condition to limit the duplication of indexing records. However, the ratio of duplication in indexing records for these methods is still high. These experiments [1] indicate that the minimum ratio of duplication of indexing records in the Landsat RS image collection is 42.26%.

To solve the problem caused by boundary RS images, we first propose an adaptive geographic meshing and coding method. This method integrates the GeoSOT subdivision model and XZ-ordering [4]. The optimal subdivision level for an RS image is calculated based on its maximum longitudinal width and latitudinal width. Therefore, RS images with different widths have their own subdivision levels. However, each RS image is covered by up to four elements at its optimal subdivision level and has multiple spatial codes when the image crosses the grid. To solve this problem, we adopt XZ-ordering (i.e., a spatial encoding method that extends Z-ordering by doubling the size of the element) to get only one spatial code for a boundary RS image. The spatial code presents the element in which the lower-left corner of the RS image's minimum bounding rectangle (MBR) is located, at the optimum subdivision level, by adopting XZ-ordering. Considering that the volume of RS images is huge and constantly increasing, it is necessary to leverage the potential of current computing frameworks to efficiently generate spatial codes for the bulk of RS images. However, different computing frameworks exist (e.g., MapReduce and Apache Spark). To ensure that the proposed adaptive geographic meshing and coding method (AGMD) method can run on the above frameworks, we adopt Apache Beam, a project that aims to provide common programming interfaces to enable applications to run on various distributed environments [5, 6], to encapsulate the proposed method as a task to generate spatial codes for massive RS images in different environments.

In summary, our technique contributions are as follows.

1) We propose an adaptive geographic meshing and coding method to generate only one spatial code for an RS image. This method is a combination of the GeoSOT subdivision model and XZ-ordering. Our aim is to solve problems caused by discrepancies in the widths of RS images and boundary RS images. Therefore, it can improve the efficiency of spatial indices by reducing the ratio of duplicated indexing records. In addition, this method preserves better spatial proximity.
2) We encapsulate our proposed method as a task that can run on various computing environments (e.g., local, MapReduce, Spark) to generate spatial codes in bulk for massive RS images with the help of Apache Beam.

2. Methodology

In this section, we first introduce the proposed adaptive geographic meshing and coding method in Section 2.1 and present how to encapsulate the method with concepts in Apache Beam in Section 2.2.

2.1. Adaptive geographic meshing and coding method

The proposed AGMD method aims to solve the challenge caused by boundary RS images by integrating the GeoSOT subdivision model and the XZ-ordering technique. Boundary-crossing RS images have multiple spatial codes because they cross multiple axes of grids. XZ-ordering [4], as a type of space-filling curve, maps spatial objects in multi-dimensional space into order-preservation one-dimensional integer values by extending the z-curve. XZ-ordering is designed to solve problems when applying the one-value approximation method or the optimized redundancy method to generate one-dimensional keys for spatial objects (i.e., lines and polygons) that intersect with grid lines. XZ-ordering is applied in numerous applications, such as GeoMesa [7] and JUST [8], an urban spatiotemporal data engine developed by JD. Considering this, we adopt XZ-ordering, which provides a one-value representation for an RS image. Considering that XZ-ordering requires the entire space to be subdivided into four elements, we adopt the global subdivision model, namely GeoSOT, which logically expands the geographical space three times. This model integer then subdivides the Earth's entire surface into many discrete, seamless, and hierarchical grids using the quad-tree method. Currently, GeoSOT is widely used in spatial data management [9], especially in RS data management [10]. However, GeoSOT generates multiple spatial codes for boundary RS images.

For an RS image I, the proposed AGMD method generates the integer number (i.e., the spatial code), also known as the "XZ-value," with the following steps:

Step 1. Get the MBR of I. Hence, the longitudinal and latitudinal widths of the RS image, denoted by $Width_{lon}$ and $Width_{lat}$ respectively, can be obtained.

Step 2. Get the maximum longitudinal and latitudinal widths, denoted by $maxWidth = max\{Width_{lon}, Width_{lat}\}$.

Step 3. Calculate the minimum subdivision level, denoted by l_{min}, based on Equation x:

$$l_{min} = \lfloor \log_{0.5} maxWidth \rfloor \quad (1)$$

Step 4. Determine the optimum resolution level, denoted by l. $l = l_{min}$ if the condition is satisfied in Equation (2), wherein Lon_{min} and Lat_{min} denote the lower boundary of the RS image in the longitude direction and the latitude direction, respectively. Otherwise, $l = l_{min} + 1$.

$$\left\lfloor \frac{Lon_{min}}{l_{min}} + 2 \right\rfloor \cdot l_{min} \leq Lon_{min} + Width_{lon} \; and \; \left\lfloor \frac{Lat_{min}}{l_{min}} + 2 \right\rfloor \cdot l_{min} \leq Lat_{min} + Width_{lat} \quad (2)$$

Step 5. Calculate the XZ-value of the RS image based on Equation xx, wherein g is the maximum resolution level. In this paper, the value of g is set to 12.

$$XZ(I) = \sum_{0 \leq i < l} \left(q_i \cdot \frac{4^{g-i} - 1}{3} + 1 \right), \; q_i \in \{0, 1, 2, 3\} \quad (3)$$

For example, assume that the lower left corner and the upper right corner of the MBR of I is (-170, 10) and (-40, 30) and that the maximum resolution level $g = 2$. There are 21 elements in total when the whole space is recursively partitioned until g. The optimal resolution l for this RS image is 2, based on Step 3. The XZ-value for this RS image is then calculated according to Step 4.

When $i = 0, q_0 = 2, XZ(I)_{i=0} = 2 \cdot 5 + 1 = 11$,

When $i = 1, q_1 = 0$, thus $XZ(I)_{i=1} = 0 \cdot 1 + 1 = 1$,

Therefore, the XZ-value for this RS image, denoted by $XZ(I)$, is calculated by $XZ(I) = XZ(I)_{i=0} + XZ(I)_{i=1} = 12$.

2.2. Abstraction of AGMD with Apache Beam

Remote sensing data is regarded as a type of "Big data" [11] and is generated continuously. To generate spatial code for massive RS images, it is preferable to leverage current computing architecture that is evolving distributed and high parallel systems (e.g., MapReduce [12] and Apache Spark [13]). However, the diversity of distributed engines results in repeated work to implement the spatial encoding method adopting Application Programming Interfaces (APIs) provided by these engines. To avoid duplication of effort, we adopt Apache Beam and provide a high-level abstraction of our proposed spatial encoding task.

Apache Beam is open sourced by Google Inc. and provides a common programming interface to encapsulate batch or streaming processing tasks. These tasks can be executed with various runners (i.e., distributed engines). In Apache Beam, a data processing task is abstracted as a Pipeline, which includes a number of PCollection objects representing a distributed dataset on which Pipeline operates and a series of PTransform functions representing data processing operations. At the start of the task, Beam reads data from external sources or in-memory sources with the I/O transforms to generate primarily PCollection, which is immutable. Then, a PTransform function with user-defined code is designated to operate on the primary PCollection to generate a new PCollection object, which can serve as the input of the next PTransform function. The data processing task, which is composed of PCollection objects and PTransform functions, logically forms a Directed Acyclic Graph (DAG). The DAG is submitted to specific runners (i.e., distributed computing engines) to accomplish the task.

We create a class called *XZEncoderDoFn* that extends the DoFn abstraction class provided by Apache Beam. *XZEncoderDoFn* contains the processing logic of generating the XZ-value for RS images and is applied to each element in the input PCollection. *XZEncoderDoFn* receives the four values (i.e., the minimum and maximum values of the RS image along the longitude direction and latitude direction), generates the XZ-value with the input based on the five steps in Section 2.1, and finally outputs the XZ-value for the RS image.

Our proposed AGMD is simply abstracted as a three-step DAG with Apache Beam. It receives the metadata of a bulk of RS images, processes them with the defined *XZEncoderDoFn* function, and finally generates the XZ-value for each RS image as the output. The primary metadata is read from external or in-memory data sources with rich I/O connectors provided by Apache Beam.

3. Experiments

In this section, we conduct a series of experiments to evaluate the efficiency of the proposed method. We first give an overview of the experimental setting in Section 3.1. Next, we evaluate the proposed AGMD method in Section 3.2.

3.1. Experimental setup

We first present the experimental environment (i.e., hardware and software) in Section 3.1.1 and introduce details about the real datasets used in our experiments in Section 3.1.2.

3.1.1. Hardware and software

We conduct the experimental evaluation on a seven-node cluster. Each node is a physical machine with two Intel(R) Xeon(R) eight-core Central Processing Unit (CPUs) with version E5–2609 and 16 GB of memory. Each node has three disks that are 2 TB and thus provide 6 TB of local storage in total. The operating system running in each node is CentOS 7 in version 1511 x86 64.

To evaluate the performance of our designed AGMD method abstracted with Apache Beam, we deploy a Hadoop cluster with version 2.8.5 and a Spark cluster with 2.4.0 in the above cluster. One physical machine is chosen as the master node, and the others are designated as slaves. All codes are written in Java 8.

3.1.2. Datasets

We use a real dataset, freely downloaded via the USGS's bulk metadata service [14]. This dataset is composed of a massive scale of metadata of Landsat RS images generated through observation of the Landsat program. We choose this dataset to evaluate the performance of the proposed method for two reasons. First, the Landsat program has provided long-term monitoring of the Earth's surface by launching several satellites since 1972 [15]. Therefore, it generates RS images and abundant metadata at a staggering rate. Second, all metadata of RS images obtained by satellites in the Landsat program are well recorded in the CSV format.

We download four files in the CSV format [14] and process all files to generate only one CSV file as our experimental data. The final CSV file contains six attributes (i.e., Landsat product identifier, acquisition date [in yyyy-mm-dd format], satellite, sensor, WKT string of the center point of the image, and WKT string of the polygon geometry of the image), each of which is separated by a comma. The statistics of the experimental data are listed in Table 1.

Table 1. Statistics of metadata of Landsat remote sensing images

Satellite	Sensor	Time range	Volume
Landsat1	MSS	1972–1978	162,111
Landsat2	MSS	1975–1982	309,914
		1987	31
Landsat3	MSS	1978–1988	144,492
Landsat4	MSS	1982–1992	178,230
	TM	1982–1984	4,035
		1987–1993	42,946
Landsat5	MSS	1981–1999	612,944
		2012–2013	19,496
	TM	1984–2012	2,869,698
Landsat7	ETM	1999–2020	2,913,426
Landsat8	OLI	2013–2015	1,258
		2019–2020	6,939
	TIRS	2013–2020	3,948
	OLI_TIRS	2013–2020	1,923,136
Volume of metadata in total			9,192,604

3.2. Evaluation of time cost and space overhead

In this section, we generate spatial codes for each RS image in the experimental dataset listed in Table 2 using the proposed AGMD method, GeoSOT, and minGeoHash, respectively. We use two metrics (i.e., time spent on generating spatial codes and space overhead) to evaluate the efficiency of the proposed method. In addition, we calculate $T_{improvement}$ and $S_{improvement}$ based on Equation 4 and 5 to investigate the efficiency of the proposed AGMD method by comparing it to GeoSOT and minGeoHash. Note that T_i refers to the time cost spent on generating spatial codes by the GeoSOT method or the minGeoHash method. S_j refers to the number of spatial codes that are generated by the GeoSOT method or the minGeoHash method.

$$T_{improvement} = \frac{T_i - T_{AGMD}}{T_{AGMD}} \quad (4)$$

$$S_{improvement} = \frac{S_j - S_{AGMD}}{S_{AGMD}} \quad (5)$$

The experimental results are shown in Table 1. It can be observed that the number of generated spatial codes equals the number of RS images of our proposed AGMD method. However, the other two methods generate more spatial codes for RS images. This is because our proposed method generates only one spatial code (i.e., the XZ-value) for each RS image. The other two methods generate multiple spatial codes for boundary-crossing RS images. GeoSOT generates two or four spatial codes for boundary-crossing RS images. Another method, minGeoHash, generates at most nine spatial codes for boundary-crossing RS images. The effectiveness of the storage overhead for the proposed AGMD method is up to 3.597. It can also be observed that the time cost of our proposed method is smaller than those of the other two methods. The effectiveness of the time cost for the AGMD method is up to 58.02%.

Table 2. Time cost and space overhead

Method	# of RS images	# of spatial codes	$S_{improvement}$ (%)	Time cost (ms)	$T_{improvement}$ (%)
AGMD	9,192,604	9,192,604	-	389,670	-
GeoSOT		22,136,770	140.8	615,762	58.02
minGeoHash		42,256,797	359.7	536,654	37.77

4. Conclusion

This paper sets out to resolve the discrepancies that arise in the widths of RS images and boundary RS images when adopting space-based approaches; our aim is to build a spatial index to support efficient spatial range query-based RS image discovery. To do this, we first propose an AGMD method that integrates the GeoSOT subdivision model and XZ-ordering to determine the optimal subdivision level and to generate only one spatial code for each RS image. In addition, we encapsulate the AGMD method with Apache Beam to enable its running on widely used, distributed frameworks to generate spatial codes for ever-increasing RS images effectively. The insights gained from this study may be of assistance in generating one-dimensional spatial code for RS images (one code per RS image) effectively in a distributed environment. One of the more significant findings to emerge from this paper is that both the time cost and the space overhead are significantly improved when the number of spatial codes for an RS image is limited.

5. Acknowledgements

This work is funded by the National Natural Science Foundation of China (U1711266 and No. 41925007) and the Strategic Priority Research Program of the Chinese Academy of Sciences, grant number XDA19090128.

References

[1] Huang X, Deng Z, Yan J, Li J, Chen Y and Wang L 2020 A high-performance spatial range query-based data discovery method on massive remote sensing data via adaptive geographic meshing and coding *IEEE Journal on Miniaturization for Air and Space Systems* p 1 doi:10.1109/JMASS.2020.3035649

[2] Cheng C, Tong X, Chen B and Zhai W 2016 A subdivision method to unify the existing latitude and longitude grids *ISPRS International Journal of Geo-Information* 5 (9) (Multidisciplinary Digital Publishing Institute) p 161 doi:10.3390/ijgi5090161

[3] Le, HV and Takasu A 2018 G-HBase: A high performance geographical database based on HBase *IEICE Transactions on Information and Systems* E101.D (4) pp 1053–1065 doi:10.1587/transinf.2017DAP0017

[4] Böhm C, Klump G, and Kriegel HP 1999 Xz-ordering: A space-filling curve for objects with spatial extension *Proceedings of the 6th International Symposium on Advances in Spatial Databases* 1999 Jul 20 (Berlin, Heidelberg: Springer-Verlag) pp 75–90 doi:10.1007/3-540-48482-57

[5] Karau, H 2017 Unifying the open big data world: The possibilities of Apache BEAM *IEEE International Conference on Big Data (Big Data)* 2017 Dec 1 p 3981 doi:10.1109/BigData.2017.8258410

[6] Li S, Gerver P, MacMillan J, Debrunner D, Marshall W and Wu KL 2018 Challenges and experiences in building an efficient Apache Beam runner for IBM streams *Proceedings of the VLDB Endowment* 11 (12) pp 1742–1754 doi:10.14778/3229863.3229864

[7] Hughes JN, Annex A, Eichelberger CN, Fox A, Hulbert A and Ronquest M 2015 GeoMesa: A distributed architecture for spatio-temporal fusion *Geospatial Informatics, Fusion, and Motion Video Analytics* vol. 9473 (International Society for Optics and Photonics) p 94730F doi:10.1117/12.2177233

[8] Li R, He H, Wang R, Huang Y, Liu J, Ruan S, He T, Bao J and Zheng Y 2020 JUST: Jd urban spatio-temporal data engine *IEEE 36th International Conference on Data Engineering (ICDE)* (Dallas, TX, USA) pp 1558–1569 doi:10.1109/ICDE48307.2020.00138

[9] Qian C, Yi C, Cheng C, Pu G, Wei X and Zhang H 2019 Geosot-based spatiotemporal index of massive trajectory data *ISPRS International Journal of Geo-Information* 8 (6) (Multidisciplinary Digital Publishing Institute) p 284 doi:10.3390/ijgi8060284

[10] Wang L, Cheng C, Wu S, Wu F and Teng W 2015 Massive remote sensing image data management based on HBase and GeoSOT. *IEEE International Geoscience and Remote Sensing Symposium (IGARSS)* pp 4558–4561 doi:10.1109/IGARSS.2015.7326842

[11] Casu F, Manunta M, Agram PS and Crippen RE 2017 Big remotely sensed data: Tools, applications and experiences. *Remote Sensing of Environment* 202 (1) pp 1–2 doi:10.1016/j.rse.2017.09.013

[12] Dean J and Ghemawat S 2008 MapReduce: Simplified data processing on large clusters *Communications of the ACM* 51 (1) pp 107–113 doi:10.1145/1327452.1327492

[13] Tang S, He B, Yu C, Li Y and Li K 2020 A survey on spark ecosystem: Big data processing infrastructure, machine learning, and applications *IEEE Transactions on Knowledge and Data Engineering* p 1 doi:10.1109/TKDE.2020.2975652

[14] USGS 2021. Landsat Missions: Bulk Metadata Service. https://www.usgs. gov/land-resources/nli/landsat/bulk-metadata-service.

[15] Wulder MA, Loveland TR, Roy DP, Crawford CJ, Masek JG, Woodcock CE, Allen RG, Anderson MC, Belward AS, Cohen WB and Dwyer J 2019 Current status of Landsat program, science, and applications *Remote Sensing of Environment* 225 (May) pp 127–147 doi:10.1016/j.rse.2019.02.015

HBase-based spatial-temporal index model for trajectory data

Hong Liu[1,2], Jining Yan[1,2] and Xiaohui Huang[1,2]

1. School of Computer Science, China University of Geosciences, Wuhan, Hubei, 430074, P. R. China
2. Hubei Key Laboratory of Intelligent Geo-information Processing, China University of Geosciences, Wuhan, Hubei, 430074, P. R. China
Corresponding author: Jining Yan (yanjn@cug.edu.cn)

Abstract. The development of global positioning technology and the popularization of smart mobile terminals has led to a trend of rapid growth in the data volume and coverage of trajectory data. This type of data has the characteristics of fast update speed, high dimensional characteristics, and a large amount of information that can be mined. Many technology companies will use trajectory data to provide location-based services, such as vehicle scheduling and road condition estimation. However, the storage and query efficiency of massive trajectory data have increasingly become bottlenecks in these applications, especially for large-scale spatiotemporal query scenarios. This paper solves this problem by designing a trajectory data index model based on GeoSOT-ST spatiotemporal coding. Based on this model, the HBase-based trajectory data storage scheme and spatiotemporal range query technology are studied, and MapReduce is used as the calculation engine to complete the query attribute condition filtering in parallel. Comparative experiments prove that the index model proposed in this paper can achieve efficient trajectory data query management.

Keywords. Trajectory data; Index; GeoSOT-ST; Spatiotemporal range query

1. Introduction

In recent years, the ubiquity of location-aware mobile devices and widespread deployment of wireless networks has led to a rapidly increasing popularity of location-based services (LBSs) [1]. The volume of location data has accumulated exponentially to TB, even PB or EB [2]. These positions are increasingly recorded as track data. The term trajectory data refers to the geographic location, motion state, and related attributes of one or more moving objects collected by a positioning device at different times during the movement. The mathematical formula can usually be expressed as a series of trajectory points $T = \{P_1, P_2, P_3, \cdots, P_n\}$ arranged in chronological order, where $P_i = \{x, y, a, t\}$, (x, y) is the coordinates of the track point, a is the attribute value of the track point, and t is the information collection time [3]. These trajectory data provide a large amount of useful data that can be mined. Therefore, how to efficiently organize and index massive trajectory data is becoming a research hotspot.

Intuitively, a trajectory is the path (or track) that a moving object follows through space as a function of time [4]. The trajectory data records the whole process change of the position and state of the moving object. Once stored, it will be retained for a long time. A large number of query operations are provided to the outside world, and generally there will be no modification or deletion operations. This has led to a continuous increase in the amount of stored data, especially with the blowout growth of trajectory data, and traditional relational databases have been unable to cope with the current large-scale data access requirements. As a representative of non-relational databases, HBase has the characteristics of a simple data model, good horizontal scalability, and fast reading and writing speed, and it has been widely used in all walks of life. Therefore, HBase has become a trend for managing huge amounts of real-time trajectory data [5].

As with the storage of trajectory data, the query of trajectory data is also diverse and includes, for example, precise query of the trajectory, query of the spatiotemporal range, and query of K nearest neighbors. For these classic trajectory query problems, effective data indexing is the key. The essence of the construction of the trajectory data indexes is to map the multi-dimensional query attribute conditions to a one-dimensional row key (RowKey) of the HBase database. An index is usually used to optimize data retrieval and is an important means of improving data access efficiency [6]. Building efficient spatiotemporal index on trajectories can significantly facilitate query processing in trajectory databases [7]. In addition, multiple indexes can be built on a single data set.

The existing trajectory index construction methods are mainly divided into two categories: object-oriented approaches and spatial-partitioning-based approaches. Object-oriented approaches usually have structures similar to R-tree, such as 3D R-tree [8], SEB-tree [9], and TB-tree structures [10]. They use the Minimum Bounding Box (MBB) of an individual trajectory as an indicator of spatial variation. There is lots of overlap and coverage in MBB when the amount of data grows rapidly, which leads to excessive index maintenance overhead and results in significant performance degradation [11]. Spatial-partitioning-based approaches mainly include quadtree [12] and grid indexes. These indexes divide the space in advance so that the indexes do not need updating when moving objects keep moving within the same spatial unit. Therefore, we designed an efficient spatiotemporal index and retrieval method based on a spatial-partitioning-based method to achieve rapid query for large-scale trajectory data.

This paper mainly focuses on the query operation of the time and space range of trajectory data; that is, given a time region and space region, we return all trajectories in these two regions [13], design a space-time index model based on GeoSOT-ST, and propose a plan for trajectory data partition optimization storage. In addition, a parallel computing framework, MapReduce, is used to support an efficient spatiotemporal range query for massive trajectory data.

2. Spatiotemporal index model and storage of trajectory data

2.1 GeoSOT-based spatiotemporal index

The trajectory data are typical spatiotemporal data, which have two-dimensional characteristics of space and time. Therefore, we consider adding a time dimension to extend the two-dimensional spatial division to the GeoSOT-ST encoding of the three-dimensional division.

GeoSOT is a new global grid division method based on 2^n one-dimensional integer array geographic coordinates [14]. Its core idea is to expand the latitude and longitude space of the earth three degrees, minutes, and seconds to form a quadtree division grid of full degrees, full minutes, and full seconds. After 3 expansions and 32 subdivisions, the entire earth will be divided into a global quadtree recursively divided grid structure as large as the entire earth and as small as centimeter-level patches. At the same time, according to the default Z-space filling curve, the whole-degree, whole-minute, and whole-second level patch coding is performed, and each patch code corresponds to a unique grid unit. GeoSOT reduces the two-dimensional latitude and longitude space to a one-dimensional entity.

GeoSOT-ST recursively divides time into two halves on the basis of space. First, the time is expanded four times by year, month, day, and hour, and the expanded one-dimensional time is then strictly recursively divided into two classes. Specifically, the time division zone is first established. We count from 1970 (the beginning time of the timestamp) and then select 128 years, to 2097, for recursive division. This 128-year period is chosen because it guarantees the division of the whole year. The 128-year time zone is now defined as the 0-level period, and it can be divided into a 1-year period by 7 recursive subdivisions. The first extension expands the original 1-year period to 16 months, and the extended 16-month period can then be divided into 1-month period by 4 recursive subdivisions. The second extension expands the original 1-month period to 32 days, and the extended 32-day period can then be divided into 1-day periods by 5 recursive subdivisions. The third extension expands the original 1-day period to 32 hours, and the extended 32-hour period can then be divided into 1-hour periods by 5 recursive subdivisions. Finally, in the same way, by extending the original 1-hour period to 64 minutes and performing 6 recursive subdivisions, a 1-minute period resolution can be obtained. In this way, after four extensions and 27 subdivisions, the 128 years will be divided into a minimum of 1-minute time segments, thereby basically meeting the accuracy requirements of the time dimension subdivision of the trajectory data. The space-time division scales corresponding to different subdivision levels are shown in Table 1.1.

Table 1.1. Time and space division scale corresponding to different levels.

Level	Scale	Level	Scale	Level	Scale	Level	Scale
0	128 year 512°	7	1 year 4°	14	4 day 2′	21	1 h 1″
1	64 year 256°	8	8 month 2°	15	2 day 1′	22	32 min 1/2″
2	32 year 128°	9	4 month 1°	16	1 day 32″	23	16 min 1/4″
3	16 year 64°	10	2 month 32′	17	16 h 16″	24	8 min 1/8″
4	8 year 32°	11	1 month 16′	18	8 h 8″	25	4 min 1/16″
5	4 year 16°	12	16 day 8′	19	4 h 4″	26	2 min 1/32″
6	2 year 8°	13	8 day 4′	20	2 h 2″	27	1 min 1/64″

The time dimension coding calculation is shown in Figure 2.1. The root node represents the division area, the maximum range of the time dimension (1970–2097), and it then recursively divides the nodes in the tree into two equally sized children. The binary bit is then allocated: if the position is on the left, it is allocated as 0; otherwise, it is allocated as 1. For example, for 2021-08-20 08:05:00 in the 9-level subdivision period, the binary code is 011001110.

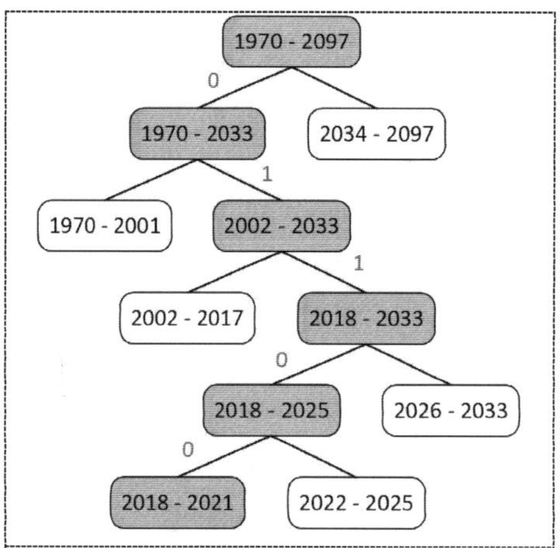

Figure 2.1. Time tree.

The calculation of the space-time code (STCode) of the track points based on GeoSOT-ST is mainly divided into the following steps.

1) Calculate the GeoSOT spatial coding

GeoSOT has quaternary one-dimensional, binary one-dimensional, and binary two-dimensional encoding forms. For example, the latitude and longitude coordinates of a certain track point are 114°23'E, 30°40'N. When the grid is divided into the 9th level, its quaternary one-digit code is G001132230, and the corresponding binary one-dimensional code is 00 00 01 01 11 10 10 11 00; the latitude code of the corresponding binary two-dimensional code is 000011110, and the longitude code is 001110010.

2) Calculate the time dimension code

The time dimension code of the trajectory data is generated according to the time division rule; for example, the trajectory point at the time of 2021-08-20 08:05:00, in the 9th level of the division period, has an obtained binary code of 011001110.

3) Spatiotemporal cross coding

According to the order of the latitude code, longitude code, and time code, the Morton cross code is obtained by bit-wise interleaving. Since the time and space maintain the same hierarchical level, the encoding length of the two is the same, with no extra bit for bit-by-bit crossover. For example, crossing the latitude code 000011110, the longitude code 001110010, and the time code 011001110 gives 000 001 011 010 110 101 101 111 000.

4) Three-dimensional integration to form the STCode

The binary encoding result of step 3) is converted to an octal to generate the final result 013265570. The reason for converting to an octal is to avoid encoding too long. A second reason is because conversion of every three digits yields a value that represents a three-dimensional space-time cube, which contains complete space-time information and is of practical significance. At the same time, for any hierarchical level L of the time-space section, the octal STCode can be obtained without bit supplementation, and the code length is L.

According to these four steps, the trajectory point is $P_i = \{114°23'\text{E}, 30°40'\text{N}, \text{a}, 2021-08-20\ 08{:}05{:}00\}$ (where a represents the attribute value of the trajectory point) in the 9th time-space section. The STCode under the hierarchical level is 013265570. Referring to Table 1.1 for the time-space division scale, we can see that the spatial resolution of this STCode is 1°, and the time resolution is 4 months.

For any given hierarchical level of the time-space profile, a trajectory point has a unique STCode, which represents a certain time-space region in which it is located, and all trajectory points in this region have the same STCode. This means that frequent updating of the index is unnecessary when a trajectory object moves within the tolerance range (the space-time cube under a certain level), and this then reduces the cost of index maintenance. In addition, the three-dimensional division of each space-time cube will result in 8 sub-level space-time cubes, as shown in Figure 2.2. The third-order Z curve ensures that the trajectory data is organized and indexed in a space-time adjacent manner, making it convenient for subsequent data storage and query. Note that the upper and lower space-time cubes have a common coding prefix, which will be an important basis for the spatiotemporal range query.

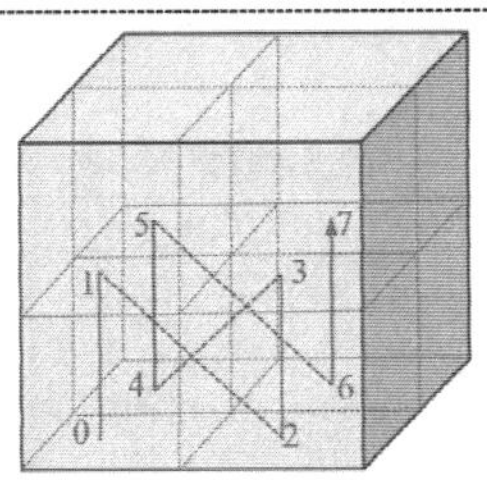

Figure 2.2. 3D subdivision of the space-time cube.

2.2 *Pre-partitioned storage*

In HBase, a good row key design will not only overcome the shortcomings of low storage performance caused by row key redundancy, but it will also solve the problem of writing hotspots to a certain extent [15]. For the query operation of the spatiotemporal range of trajectory data, the row key can be designed as (STCode + TrajectoryID). The STCode fixes a certain subdivision level, and TrajectoryID is added to distinguish different trajectory objects in the space-time cube of this level.

Because the data in HBase are stored in the order of the ASCII value of the row key, and most of the time and space adjacent trajectory objects have similar STCodes, they are also adjacent in physical storage, which allows full use of the sequential I/O of the disk and ensures high trajectory data query efficiency.

When data are written into the HBase table, only one Region exists at the beginning. As the track data are continuously inserted, the Region will continue to increase until it reaches the set threshold, and the Region will be divided into two regions of basically the same size. The Region is the most basic unit used to implement HBase distributed storage and load balancing [16]. Different Regions will be assigned to different Region servers. If data are concentrated on a Region server during writing, then Region splitting and merging will occur continuously. Frequent partition splitting will consume substantial system I/O resources and affect the efficiency of data writing and reading.

To solve these problems, we performed a pre-partitioning operation when building the table. First, we created several empty Region partitions and set the data range to be stored in each partition. The data would be written to the corresponding Region when it was stored, thereby greatly reducing the number of divisions of the Region. We consider that a good pre-partitioning strategy not only distributes data write requests to the Region, but it also ensures that the trajectory data in a single Region have temporal and spatial relevance. Therefore, we set the pre-partition separator based on GeoSOT-ST space-time coding. For example, when the time-space profile hierarchical level is 2, the global geographic space-time will be equally divided into 64 space-time cubes corresponding to 64 STCodes. Then, using these 64 STCodes as the pre-partition separator, 65 empty Regions will be generated.

However, this pre-partitioning scheme does not take into account the uneven temporal and spatial distribution of the trajectory data. Trajectory objects usually move frequently in a certain spatio-temporal range, while the amount of data in other ranges is relatively small. This creates difficulties for some regions to be hit, so some regions will have write hotspots and will finally present a phenomenon of uneven data distribution and data skew. In this way, data access is always concentrated on servers with a few Regions. However, this causes excessive load pressure on the host where it is located, thereby affecting the efficiency of the data query.

Our solution to this problem is to obtain statistics on historical trajectory data and then divide the research area into space-time cubes of different sizes according to the statistics. The space-time cube is smaller in a hotspot area with dense data volume, whereas the space-time cube in the edge area with sparse data volume is larger. The specific method is to divide the space-time cube with densely distributed trajectory data again to obtain eight new child space-time cubes. Then, we add the STCode corresponding to these eight cubes as the separator before the partition to ensure the relative balance of the data volume of each partition to achieve load balancing.

3. Spatio-Temporal Range Query Based on GeoSOT-ST

The spatio-temporal range query of trajectory data takes the spatial range and time interval as inputs, and the goal is to find all trajectory objects within the range. The spatio-temporal range query based on GeoSOT-ST code can convert the three-dimensional (latitude, longitude, and time) query to a one-dimensional query for STCode, thereby greatly improving the efficiency of data retrieval.

3.1 Spatio-temporal range query framework

The query framework of the spatio-temporal range of trajectory data is shown in Figure 3.1. First, the user enters the query space range and time range, and then, based on the GeoSOT-ST space-time coding query algorithm, the user maps the space-time query range to STCode, generates the HBase filter list, and extracts the data to obtain the candidate result set. This period is the filtering stage. To filter out the trajectories that do not meet the query conditions in the candidate set, we used the MapReduce parallel computing framework to filter the results in the candidate set once more using spatio-temporal attribute values. This period is the refinement stage. After these two stages, the final

IOP Conf. Series: Earth and Environmental Science 1004 (2022) 012007 doi:10.1088/1755-1315/1004/1/012007

and accurate results of the spatio-temporal range query are obtained. Next, we describe in more detail the methods used in these two stages.

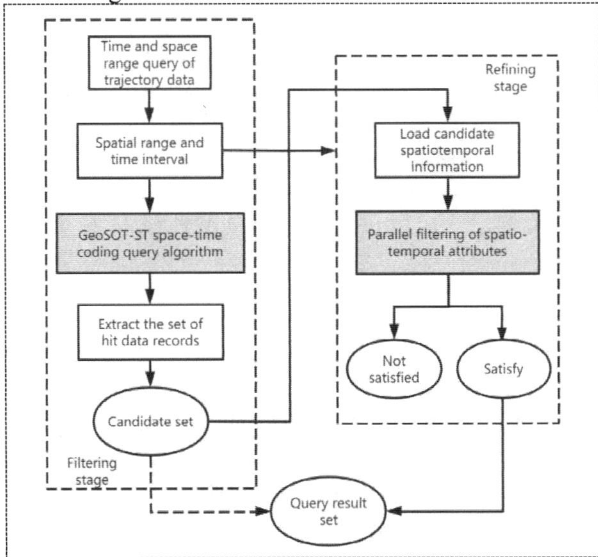

Figure 3.1. Spatio-temporal range query framework.

3.2 Filtering stage

The purpose of the spatio-temporal range query is to convert the query conditions into STCode, and the key is to determine the level of the time-space division. As shown in Figure 3.2, if the level is too high, the query range will be mapped to multiple small grids, making the filter conditions too trivial and reducing the query efficiency. By contrast, when the level is too low, the query process will return a large number of absents. Querying "fake data" within the scope will bring a larger number of tasks to the subsequent refining stage.

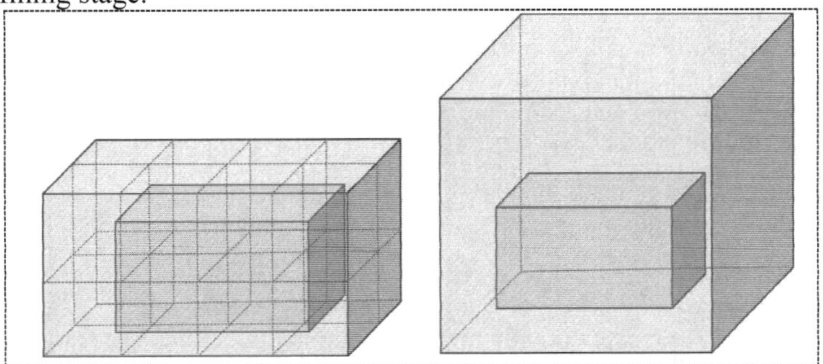

Figure 3.2. Grids of different sizes cover the space-time cube.

To solve these problems, we designed a scheme for adaptive adjustment of the level of spatiotemporal division according to the scope of the query. We first choose an optimal level of spatiotemporal division $optimalLevel$. Under this level, the spatiotemporal query area can be mapped to 1, 2, 4, or 8 spatiotemporal cubes, so that several spatiotemporal cubes can completely cover the spatiotemporal query range.

The determination of the level is mainly to find the minimum division level from the scale table, so that the corresponding space-time cube is greater than or equal to the size of the query space-time area. Referring to Table 1.1, we assume that when $Level = N$, the time resolution is tScale_N and the

IOP Conf. Series: Earth and Environmental Science 1004 (2022) 012007 doi:10.1088/1755-1315/1004/1/012007

spatial resolution is $sScale_N$. Using the inequality group (1), the optimal spatiotemporal division level $optimalLevel$ can be calculated. Among these, Dis_S is the larger of the length and width of the Minimum Bounding Rectangle (MBR) corresponding to the query space range, and Dis_T is the query time interval value.

$$\begin{cases} sScale_{sLevel+1} < Dis\_S \leq sScale_{sLevel} \\ tScale_{tLevel+1} < Dis\_T \leq tScale_{tLevel} \\ optimalLevel = Min(sLevel, tLevel) \end{cases} \quad (1)$$

The following is the specific implementation process of the GeoSOT-ST space-time coding query, as shown in Algorithm 1. Under the $optimalLevel$ level, we calculate the STCode for the 8 vertices of the query space-time region, and each STCode corresponds to a query space-time cube. We use these STCodes as the prefix to generate the HBase prefix filter list, which can be indexed to the $optimalLevel$ level, and all the trajectory data in the space-time cube that intersect, contain, or are included in the query area.

Algorithm 1 GeoSOT-ST space-time coding query algorithm.

Input: query time interval $[Time_1, Time_2]$, query space range I.

Output: All trajectory that meet the query conditions.

1 Calculate the Minimum Bounding Rectangle MBR^I of the space range I.

// $wLon(MBR^I), wLat(MBR^I)$ are the length in the longitude direction and the latitude direction respectively.

2 $Dis\_S \leftarrow Max(wLon(MBR^I), wLat(MBR^I))$

3 $Dis_T \leftarrow (Time_2 - Time_1)$

4 Obtain the optimal division level $optimalLevel$ through the inequality group (1).

5 $STCodeSet \leftarrow \emptyset$

// $Query^{st}$ is the space-time cube corresponding to the query region.

6 $\boldsymbol{for}\ coordinate \in eight\ Coordinates\ Of\ Query^{st}\ \boldsymbol{do}$

  // Calculate the STCode of the current space-time coordinate point under $optimalLevel$.

  $STCodeSet \leftarrow STCode$

// $STCodeSet's\ Size\ is\ 1、2、4 or 8.$

7 $\boldsymbol{for}\ STCode\ in\ STCodeSet\ \boldsymbol{do}$

  // The initial value of $FilterList$ is $\emptyset$.

  $FilterList \leftarrow PrefixFilter(STCode)$

8 $CandidateResult \leftarrow table.scan(new\ Scan().setFilter(FilterList))$

3.3 *Refinement stage*

Parallel filtering of the spatiotemporal attributes takes candidate results as input and performs an accurate calculation of the spatiotemporal relationship between each trajectory object in the candidate result set and the query range. Space-time relational calculation is a very time-consuming operation, so the parallel computing framework MapReduce is used to complete this work in a distributed manner. The core idea of MapReduce is to split large-scale data processing jobs into multiple independently runnable Map tasks, distribute them to multiple processors to run, produce a certain amount of intermediate results, and then shuffle and merge the reduced tasks to produce the final output file [17].

The parallel filtering process of spatiotemporal attributes based on MapReduce is as follows: for the candidate result set obtained by the GeoSOT-ST spatiotemporal coding query algorithm, the filtering task is distributed to multiple distributed nodes for parallel execution through a custom map function, with the distributed nodes based on the spatiotemporal containment relationship. We determine whether the location of the trajectory data is inside the query area and then filter the trajectory data that meet the query conditions, yielding the final result of the spatiotemporal range query.

4. Experiment and Analysis

4.1 Experimental data and operating environment

This article uses the GPS trajectory data of 10357 taxis in Beijing from February 2 to February 8, 2008. The total distance of the trajectory reaches 9 million kilometers, and the number of trajectory data is about 17 million.

The experimental computer cluster is composed of 3 servers: one is the central management node and the other 2 are the distributed data processing nodes. The main software environment versions of the cluster are Apache Hadoop 2.7.3, Apache Zookeeper 3.4.10, and Apache HBase 1.2.6. The main server hardware configuration is an AMD Ryzen 7 4800H CPU with Radeon Graphics, 4 cores, and 6 GB memory.

4.2 Spatiotemporal index comparison experiment

In intelligent transportation applications, an efficient trajectory data query is not only reflected in the spatial range query, but it is also reflected in the time dimension. Therefore, this paper designs two row key structures and compares and analyzes the query performance of their temporal and spatial ranges. 1) The Geo-Time row key, with the structure "GeoSOT + time + TrajectoryID", first searches for the trajectory that meets the requirements of the spatial location and then filters the time. 2) The GeoSOT-ST row key, with the structure "STCode + TrajectoryID", is the scheme introduced in this article. We determine the size of several groups of space-time query areas, and each group uses two schemes to query the random spatiotemporal range. The experimental results are shown in Figure 4.1.

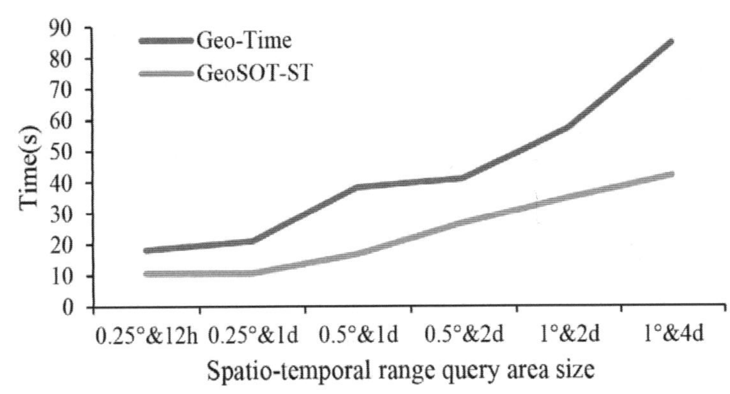

Figure 4.1. Comparison of index efficiency in different time and space ranges.

An important observation is that as the query range increases, GeoSOT-ST indexes require less time than Geo-Time indexes. Based on the left-hand matching principle of the HBase row key index, the Geo-Time row key search for trajectories that meet the spatial query conditions, and we then perform time filtering. The index effect on the time dimension can be demonstrated only when the spatial query range is not larger than the spatial span corresponding to the coding level of the spatiotemporal index. Normally, filtering two query attributes at one time is not possible. By contrast, the GeoSOT-ST row key designed in this article can play an index role in both time and space, and as the query range increases, the index effect becomes more obvious.

5. Conclusions

To support faster response services, including data mining, analysis, etc., for a huge volume of trajectories stored in distributed environments, a trajectory data index model based on GeoSOT-ST spatial-temporal coding is designed in this paper. To do this, we first generate a one-dimensional representation of each trajectory point based on the GeoSOT-ST global geospatial-temporal subdivision model, and we then store the trajectory in different partitions in groups. We then design an adaptive method to adjust the optimal subdivision level, which makes full use of the index function of the linear key and realizes rapid extraction of the trajectory data set for the forthcoming reference. Finally, the distributed computing framework MapReduce is used to refine the query and realize an efficient query of the spatiotemporal range for trajectory data. The insights gained from this study may help to realize efficient storage and query management of real-time and massive trajectory data.

6. Acknowledgements

This work was supported by the National Natural Science Foundation of China under Grant 41801363 and Grant U1711266.

References

[1] Bai J W, Wang J Z and Huang J L 2013 Spatial query processing on distributed databases *Advances in Intelligent Systems and Applications* vol 1 (Berlin, Heidelberg: Springer) pp 251–260

[2] He S W, Chu L X and Li X Y 2017 Spatial query processing for location based application on Hbase *Proc. 2017 IEEE ICBDA*. Liverpool: IEEE Beijing Section, p 5

[3] Li J, Liu Q J, Zhao X S, Huang Q, Shun W B, Xu Z H and Wang H 2021 Trajectory data management and analysis framework supported by geographic grid model: methods and applications. *J. Wuhan University* **46** 640-649

[4] Lin X, Jiang J, Ma S, Zuo Y and Hu C 2019 One-pass trajectory simplification using the synchronous Euclidean distance *The VLDB Journal* **28** 897–921

[5] Zhang C, Yu Z Y, Song X X, Hu P and Zhang W F 2016 Design method of distributed real-time trajectory access system *Command Information System and Technology* **7** 70–75

[6] Lu N, Cheng C, Jin A and Ma H 2013 An index and retrieval method of spatial data based on GeoSOT global discrete grid system *Proc. IEEE Int. Geosci. Remote Sensing Symp - IGARSS* 4519–4522

[7] Zheng K, Zhao Y, Lian D, Zheng B, Liu G and Zhou X 2020 Reference-based framework for spatio-temporal trajectory compression and query processing *IEEE Transactions on Knowledge and Data Engineering* **32** 2227–2240

[8] Zhu Q, Gong J and Zhang Y 2007 An efficient 3D R-tree spatial index method for virtual geographic environments *ISPRS Journal of Photogrammetry and Remote Sensing* **62** 217–224

[9] Song Z and Roussopoulos N 2003 SEB-tree: An approach to index continuously moving objects *Int. Conf. on Mobile Data Management*. (Berlin, Heidelberg: Springer) pp 340–344

[10] Pfoser D, Jensen C S and Theodoridis Y 2000 Novel approaches to the indexing of moving object trajectories *Proc. 26th VLDB Conf.* (Burlington MA: Morgan Kaufman) 395–406

[11] Li and Tang J 2011 A new R-tree spatial index based on space grid coordinate division *Proc. 2011 Int. Conf. Informatics, Cybernetics, and Computer Engineering (ICCE2011)* (Berlin, Heidelberg: Springer) pp 133-140

[12] Ding R and Meng X 2001 A quadtree based dynamic attribute index structure and query process *Proc 2001 Int Conf Computer Networks and Mobile Computing* (Piscataway NJ: IEEE) pp 446–451
[13] Shan J W 2019 *Research on large-scale trajectory data processing system* (Dalian: Dalian University of Technology)
[14] Cheng C Q, Ren F H and Pu G L 2012 *An Introduction to Spatial Information Subdivision Organization* (Beijing: SciencePress) pp 111–116
[15] Lu T, Fang J and Qiao Y K 2015 Real-time storage system of traffic flow data based on HBase. *Computer Application* **35** 103-107
[16] Xie Z Y 2014 *Research and application of storage and mining methods for massive on-board information.* (Xi'an: Xidian University)
[17] Wu X D and Ji S W 2018 Comparison of MapReduce and Spark for big data analysis. *J. Software* **29** 1770–1791

IOP Conf. Series: Earth and Environmental Science 1004 (2022) 012008 doi:10.1088/1755-1315/1004/1/012008

Study on the Expression of Spatiotemporal Patterns in Marine Ecosystems

Kang Linchong[1,2], Huang Haiyan[1,*] Jiang Xiaoyi[1,2], Zuo Guocheng[1], Wang Yi[1,2], Liu Chang[1]

1. National Marine Data and Information Services, Tianjin, 300171, China
2. Marine Information Technology Innovation Center, Ministry of Natural Resources, Tianjin, 300171, China;

Abstract: The ocean is a three-dimensional ecosystem with a wide range of temporal and spatial scales, thus making the visual expression of marine ecological characteristics extremely challenging. What types of marine ecosystems exist in a region, where are they distributed, what is their current status, and what are their changing trends? These are important questions faced by marine managers and decision makers. Based on the analysis of the relationship between biology and environment, this paper conducts ecological classification and data production for various marine elements. This paper expresses spatiotemporal marine ecological characteristics using a four-level general framework, including sea zone, bathymetry space, water space, seabed space. The bathymetry reflects the relevant marine topographic and landscape characteristics, and the water space and seabed space are composed of water environment, sediment environment and living organisms respectively. The framework's main functions are the display, description, and statistical analysis of ecological characteristics, and display the spatiotemporal distribution pattern. These methods were combined with the construction of a database and R&D of system functions to fill the gap in this field in China and provide decision support for marine ecosystem management.

Keywords: Ecological characteristics; Spatiotemporal patterns; Characteristics expression; System construction

1. Introduction

Marine ecosystems feature a high degree of complexity; they are a natural whole with interrelation, interaction, and automatic regulation mechanisms formed by the biological and abiotic environment through energy flow and material cycles in a certain time and space. Marine ecosystems have three-dimensional characteristics with a wide range of temporal and spatial scales, thus making the visual expression of marine ecological characteristics extremely challenging.

At present, the mainstream method of visually expressing marine ecological characteristics is habitat mapping on the basis of a general ecological classification system, displaying the mapping results after system integration. In 2012, the United States issued the Coastal and Marine Ecological Classification Standard [1] , applying classification mapping in the waters of the Northwest Atlantic [2], the northern Gulf of Mexico [3], and the survey of abyssal benthos in the Northeast Pacific [4], formulating a plan for comprehensive marine and coastal mapping [5]. Australia developed habitat mapping based on CMECS and released Sea Map [6, 7], and the EU released seabed habitats [8] based on the UNIX classification system, which mainly uses survey data such as sediment, benthos, and macroalgae to detail and display habitat types.

In China, this research is in its infancy and there are no similar results, though the country has spent manpower and material resources to carry out marine monitoring and investigation and the construction of a marine information system [9,10] in recent years. The organization and design of the obtained data and products are mainly set according to the nature of the discipline, without taking the ecosystem as a whole into consideration, and with the lack of description of the relationship between biology and environment and its temporal and spatial distribution pattern, it is impossible to have a comprehensive understanding of the marine ecosystem.

The China Sea spans three dimensional zones: temperate zone, subtropical zone and tropical zone, and the water depth gradually increases from the mainland to the sea. In this study, the bathymetry connects the water space and seabed space of the ocean, so as to establish a three-dimensional framework and integrate the regional water body, geological form, sediment and biological data. The study establishes a unified base map reflecting the marine ecological status as a multi-disciplinary, three-dimensional expression. It provides technical support for promoting the unified management of natural resources based on ecosystems.

2. System architecture

The ocean is a three-dimensional, dynamic system composed of marine organisms and their marine environment. The expression of marine ecological characteristics shows the matching characteristics of the physical marine environment and biological distribution. The system architecture for expressing spatiotemporal characteristics in marine ecology includes basic data layers, an ecological characteristics layer, a function layer, and an application layer.

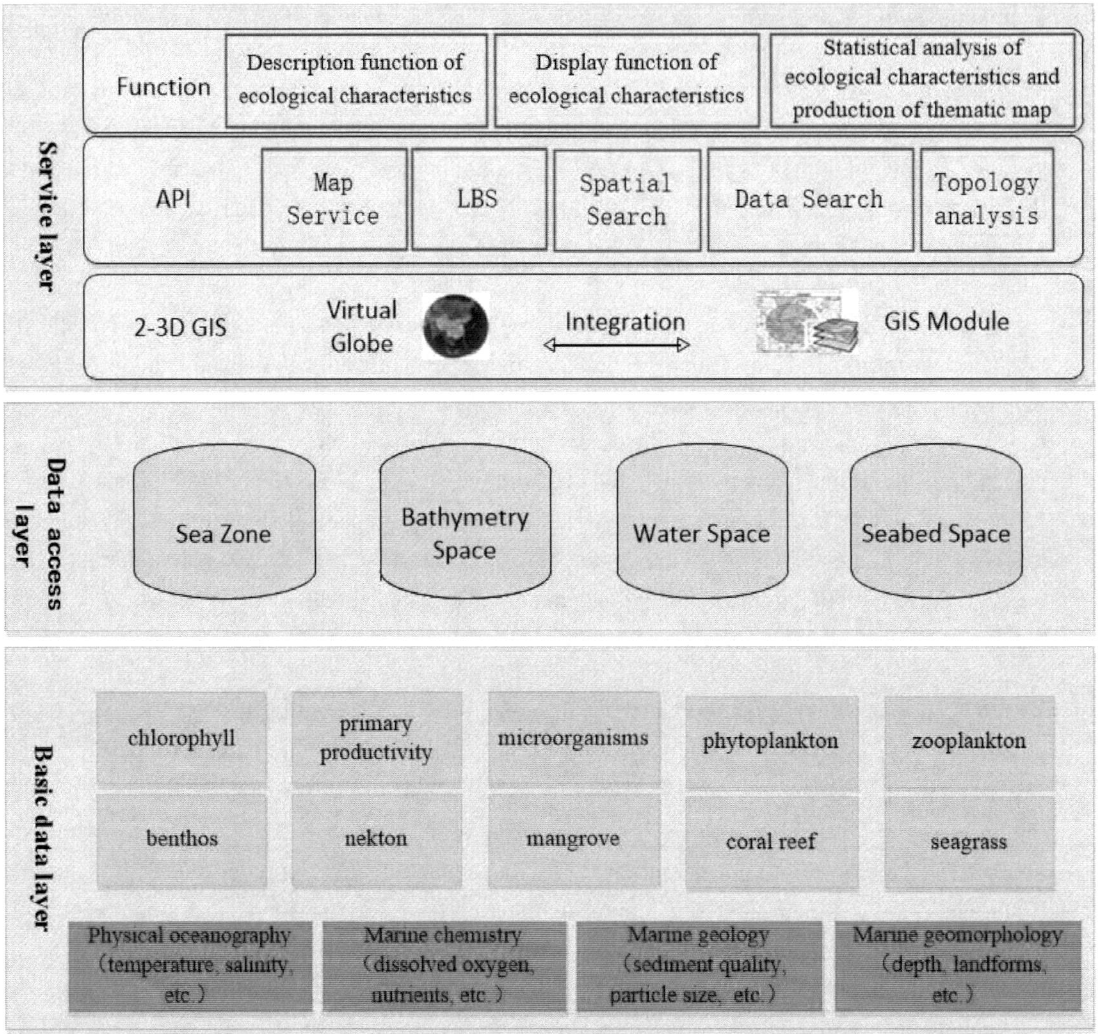

Figure 1. Architecture of a spatiotemporal patterns expression system for marine ecosystem

2.1 Basic data layer

Basic marine ecology data refers to biological and environmental data obtained from marine monitoring, observation, and surveys.

(1) Biological data includes structured tabular data on chlorophyll, primary productivity, microorganisms, phytoplankton, zooplankton, benthos, nekton, and other marine life, as well as vector spatial data on biological distribution such as the locations of mangroves, coral reefs, seagrass beds, and oyster reefs.

(2) Environmental data involves many disciplines such as physical oceanography, marine geology, marine geomorphology, and marine chemistry. It includes such measures as water depth, temperature, salinity, flow velocity, flow direction, sediment particle size, sediment quality, dissolved oxygen, and nutrients. The data formats mainly include structured tabular data, vector spatial data, and raster spatial data.

2.2 Data access layer

The existing data resources were sorted, classified, and graded. Considering the spatiotemporal expression needs of marine ecology, a multi-component, multi-dimensional general framework was adopted to build databases for the sea zone, bathymetry, water space (including water environment and aquatic organisms), seabed space (including sediment environment and benthos) based on the unified marine data domain model. Database access and a call interface was developed according to a unified standard.

2.3 Service layer

Since marine ecosystems feature a high degree of complexity, their management should not only be planned step-by-step from the macro level, but also dynamically adjusted by paying close attention to local monitoring. The spatiotemporal expression of marine ecological characteristics provides a basic unified framework to display marine ecological characteristics at different levels in a three-dimensional way. The framework acts as a decision support system that allows comprehensive data utilization, collaboration, and sharing by superimposing various types of data resources. For example, it incorporates data on development, utilization, restoration, the ecological red line, carrying capacity, and protected areas of marine ecosystems.

2.3.1 Describing ecological characteristics

For the defined regional or spatial sites, the description of ecological characteristics was given including the information of each level in the sea zone, bathymetry, water space (including water environment and aquatic organisms), seabed space (including sediment environment and benthos). The information was appropriately increased for surrounding marine protected areas, the ecological red line, typical nearby ecosystems, and other characteristics.

2.3.2 Displaying ecological characteristics

Basic data and ecological data were checked according to the bathymetry, water space, seabed space. Then the ecological types of each group were displayed on a map according to the data extent or predetermined regions (waters under the jurisdiction of the whole country, national nearshore and sea areas, or waters under the jurisdiction of provinces and cities).

2.3.3 Statistical analysis of ecological characteristics and production of a thematic map

The area, quantity, and proportion of each type were counted according to the bathymetry space, water space, seabed space, and the quantities of marine ecological products in the region were counted with each group as the label. The map elements were designed in a standard-sized map sheet and thematic map, including a map name, legend, picture frame, longitude and latitude annotation, and scale to form a base map template. Users can customize the above basic data or ecological characteristics layer, range, proportion, and template type to make thematic maps online.

IOP Conf. Series: Earth and Environmental Science 1004 (2022) 012008 doi:10.1088/1755-1315/1004/1/012008

3. Database construction

The ocean system is a dynamic, continuous, three-dimensional system with a wide range of spatial and temporal scales. The ecological characteristics layer adopts a multi-component, multi-dimensional general framework to integrate and display data related to various ecological processes on large and small scales. The framework includes sea zone, bathymetry, water space and seabed space. Each group shows the spatial distribution of each ecological type, in which spatial data plays the leading role, including vector and raster spatial data.

3.1 Sea Zone

The sea zone provides location and background information for further understanding bathymetry, water space and seabed space characteristics. Assuming a site is located in a certain sea zone, its general ecological characteristics can be inferred. For example, if it is located in the Yangtze River zone in a temperate zone on a continental shelf, the site will be mainly affected by the runoff of the Yangtze River, with relatively low water salinity, high seawater turbidity and medium tidal intensity. Therefore, it can be inferred that the geomorphic type is an estuary delta, which may be affected by upwelling and rich in aquatic resources.

China has four traditional sea areas: the Bohai Sea, the Huanghai Sea, the Donghai Sea and the Nanhai Sea, reflecting the heterogeneity of latitude zones and geographical contours. From the continental coast in the direction of the open sea includes estuarine, nearshore, shallow sea, bathyal regions, and abyssal regions, reflecting the zonal distribution characteristics of of water depth. In addition, typical marine ecosystem types such as estuaries, bays, islands, sandy coasts, muddy coasts, mangroves, coral reefs and seagrass beds are also distributed in different sea area

3.2 Bathymetry Space

The bathymetry connects the water space and seabed space to establish a three-dimensional framework of marine ecosystem. It refers to the terrain and its related marine landscape characteristics, presenting the physical structure of the environment related to the distribution of biological communities or leading to their differences. The bathymetry space provide structure, controls the transportation of energy and materials, and provides places for organisms to find food and shelter, reproduce, and rest. Different bathymetry space show heterogeneity in material structure, energy structure, and spatial structure, which plays an important role in the formation of biological communities.

The bathymetry space could be divided into four layers showing the spatial distribution data of all ecological types at this level. The first layer shows the tectonic characteristics includes the divergent active continental margin, ocean basin, fault zone, etc.; the second layer shows the physiographic characteristics includes continental seas, coastal geoforms, large open bays, intertidal zones, continental shelves, etc.; and the third layer and fourth layer are divided according to spatial scale, with intersected types, including islands, coves, bedrock coasts, beaches, and shelf plains.

3.3 Water Space

The vast ocean water body provides a wide range of habitats for marine organisms. The "barriers" formed by temperature, salinity, and depth will affect the distribution of organisms. Ocean currents not only affect the ecological environment of marine organisms, but also play an important role in the community structure, distribution, and diffusion of species. The impact of tidal environments on the coastal zone, where only marine organisms with very strong adaptability can survive, is particularly important. The three-dimensional structure of water bodies, high temporal variability, and the extensive spatial scale of water environmental characteristics pose great challenges to the expression of marine ecological characteristics, especially when coupled with the inherent difficulties in the measurement of various parameters.

This paper presents the ecological characteristics of the ocean in terms of vertical level, temperature, salinity, hydrology, and biogeochemistry. The abiotic environmental characteristics in water space is divided into five layers including water layer (surface layer, upper layer, thermocline, lower layer, deep layer, etc.), salinity layer (oligohaline, mesohaline, polyhaline with low salinity, polyhaline with high salinity, euhaline, and high-concentration saline), temperature layer (cold, cool, warm, hot, etc.), Ocean phenomenon layer (ocean current,, water mass ocean front, ocean fluctuation, and biogeochemical characteristics layer (chlorophyll, dissolved oxygen, nutrients, etc.).

The biotic environmental characteristics in water space includes plankton and benthic/attached biota. The former includes phytoplankton, zooplankton, and floating/suspended plants, macroalgae, and microbes. The biotic characteristics shows the spatial distribution of dominant life forms currently observed or monitored. The display object is not all marine organisms, but groups that have been observed, have specific ecological significance, and can be used as research objects and management objectives. For example, algal blooms which is the ecological anomaly formed by the explosive proliferation and aggregation of phytoplankton.

3.4 Seabed Space

The seabed space includes sediments and living benthos. The sediments is the substrate for organisms to survive. Providing living space for benthos and periphyta, it is the most important physical factor affecting the community structure of benthos. Hard bottom sediments include bedrock and reef. Strongly stable, hard, and porous, they are very suitable for the attachment of mollusks and can provide shelter and breeding places for some crustaceans. For example, reef corals must grow on hard substrates such as rock, reef segments, gravel, or dead coral bone. Loose sediments mainly include gravel, sand, and mud, which have relatively poor stability. Gravel provides attachment areas for periphyta. Sandy sediment is suitable for burrowing animals such as crabs, and the nutrient-rich sediment is suitable for the growth of aquatic plants and other epiphytes.

The sediments can be divided into four levels in a nested structure. The layer data of the next level can be processed to produce the data of the previous layer. In addition, there are also biogenic substrate and anthropogenic substrate in the seabed space, and the distribution data of various types is objectively displayed. The organisms in seabed space mainly includes reef biota, faunal beds, aquatic vegetation beds, emergent wetland, scrub-shrub wetland, forest wetland, microbial communities, moss, and lichen communities. The biotic group establishes 9 layers according to the above biological categories, and the distribution data of corresponding ecological types are displayed in each layer.

4. System development and demonstration

4.1 Visual expression of regional characteristics

The visual expression of marine characteristics in different regions aims to show the sea zone it is located in and the layers of various characteristics types in the bathymetry space, water space, seabed space. Taking the Bohai Sea as an example, we adopted the system architecture mentioned above, collected basic data on organisms and the environment in the Bohai Sea, made maps of various ecological types, and loaded them into the developed system to display their ecological characteristics. The figure below visually expresses the sea zone, the ocean current layer and the layer of phytoplankton algal bloom in the water space, the terrain layer in the bathymetry space and the benthic community in the seabed space, Different types of ecological data can be gradually added to this framework, enriching the content expressed by the system.

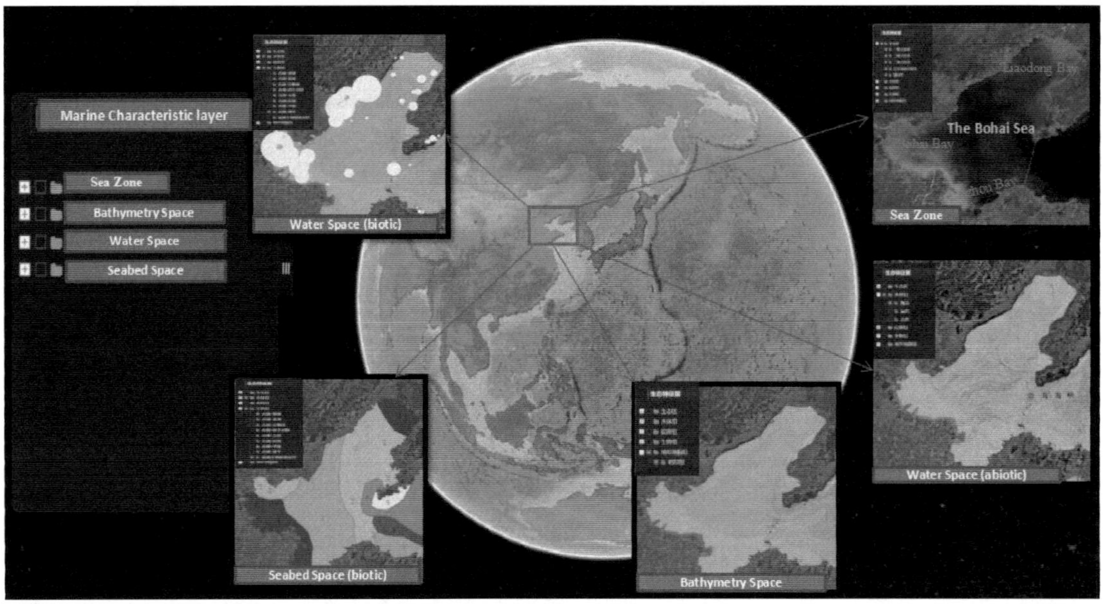

Figure 2. Visual expression of marine characteristics in the Bohai Sea

4.2 Description of regional characteristics

Taking the Bohai Sea as an example, according to the structure of the table below to describe the spatiotemporal patterns characteristics. The content can be extracted from the corresponding layer in the system, and the missing data can be added and improved according to the existing basic data and research results.

IOP Conf. Series: Earth and Environmental Science 1004 (2022) 012008 doi:10.1088/1755-1315/1004/1/012008

Table 1 Example description of the spatio-temporal patterns characteristics of the Bohai Sea

Described region		The Bohai Sea
Sea zone	Natural sea zone	The Bohai Sea
	Seaward zone	Nearshore
	Character zone	The Bohai Bay, Laizhou Bay, Liaodong Bay, Bohai central basin
Bathymetry Space	Bathymetry characteristics	18m in average, 85m in maximum
	Types	
	Level 2	Continental sea
	Level 3	Cove; estuarine delta; tidal sand ridge group; shelf plain, etc.
Water Space	Abiotic component	
	Temperature	There are great seasonal differences, from frozen water (< 0° C) to very warm water [25°, 30°)
	Salinity	Upper polyhaline
	hydroform	Coastal water of Liaodong Bay, coastal water of Bohai Bay and Laizhou Bay, water mass of the Bohai Sea
	Biotic component	
	Plankton	Phytoplankton
		Diatom phytoplankton
		Diatom bloom
		Skeletonema bloom
Seabed Space	Abiotic component	
	Level 1	Unconsolidated mineral substrate
	Level 2	Fine unconsolidated substrate
	Level 3	Sandy mud: mud
	Level 4	Sandy silt: silt
	Biotic component	
	Benthic/attached biota	Faunal bed
		Soft sediment fauna
		Mobile mollusks on soft sediments
		Basket whelk community

The structure of the table describes the characteristics of the region from four aspects: sea zone, bathymetry space, water space, seabed space. Among them, each aspect adopts different architectures and uses the the ecological types for feature expression, so that the data from different sources can be unified into the same framework.

5. Summary

The ocean has a wide range of spatial and temporal scales, thus making visual expression of its ecological characteristics extremely challenging. A single ecological characteristic presents the situation of specific ecosystem units at a specific time point. The existing measurement tools can only capture a part of the space-time continuum. Only by integrating large-scale and small-scale components related to geological, biological, water, and sediment processes into a general hierarchical framework can the status and extent of changes in the entire system be presented by comparing a single characteristic at different times. In this paper, the spatiotemporal characteristics of marine ecology are expressed by using the general framework of four levels—the sea zone, bathymetry space, water space, seabed space—through the functions of display, description, and statistical analysis of ecological characteristics. By superimposing management data on marine development, utilization, ecological red line, and protected areas, the framework provides decision support for marine ecosystem management.

References

[1] Federal Geographic Data Committee, Coastal and Marine Ecological Classification Standard (FGDC-STD-018-2012), 2012.07

[2] Weaver, K.J., E.J. Shumchenia, K.H. Ford, M.A. Rousseau, J.K. Greene, M.G. Anderson a nd J.W. King. 2013. Application of the Coastal and Marine Ecological Classification Sta ndard (CMECS) to the Northwest Atlantic. The Nature Conservancy, Eastern Division C onservation Science, Eastern Regional Office. Boston, MA. http://nature.ly/EDcmecs.

[3] Kelly Kingon. Applying the Coastal and Marine Ecological Classification Standard (CMECS) to Nearshore Habitats in the Northeastern Gulf of Mexico. Geosciences 2018, 8, 22; doi:10.3390/geosciences8010022

[4] Bassett, R.D., M. Finkbeiner, and P.J. Etnoyer. 2017. Application of the Coastal and Marine Ecological Classification Standard (CMECS) to Deep-Sea Benthic Surveys in the Northeast Pacific: Lessons from Field Tests in 2015. NOAA Technical Memorandum NOS NCCOS 228, NOAA National Ocean Service, Charleston, SC 29412. 49 pp. doi:10.7289/V5/TM-NOS-NCCOS-228

[5] Integrated Ocean and Coastal Mapping. https://iocm.noaa.gov/cmecs/applying-cmecs.html

[6] Seamap Australia – National Marine Benthic Habitat Map, https://seamapaustralia.org/

[7] Seamap Australia [Version 1.0] the development of a national benthic marine classification scheme for the Australian continental shelf, 6th November 2017.

[8] European Marine Observation and Data Network (EMODnet). https://emodnet.ec.europa.eu/en/seabed-habitats

[9] Guo Xue; Jiang Xiaoyi and Kang Linchong, 2020 *Marine Information* vol 2 Research and Realization of the Marine Map Service Platform pp 48-56

[10] Xiao Peng; Wang Yue-dong; Li An-long, 2020 *Marine Information* vol 1 'Digital Seabed' :the Management and Visualization of Seabed Data pp 40-45

Acknowledgments
This work was funded by the National Key R&D Program of China (No. 2017YFC1405305) and received data support from the National Marine Data Center, National Science & Technology Resource Sharing Service Platform of China.

IOP Conf. Series: Earth and Environmental Science 1004 (2022) 012009 doi:10.1088/1755-1315/1004/1/012009

A Decision Support System for Marine Regulation

Kang Linchong[1,2], Jiang Xiaoyi[1,2], Zhao Longfei[1,2], Wang Yi[1,2], He Long[1,2],Guo Xue[1,2] and Huang Haiyan[1,*]

1. National Marine Data and Information Service, Tianjin, 300171, China;
2. Marine Information Technology Innovation Center, Ministry of Natural Resources, Tianjin, 300171, China;

Abstract: The implementation of marine spatial data management and decision support based on GIS provides a powerful tool for marine regulation. This study constructed a three-dimensional spatiotemporal data model of marine resources that is hierarchical, partitioned, and classified according to the management principle of using an underground resource layer, surface matrix layer, surface overburden layer, and management layer, taking the marine resource entity as the unit. Key research topics include marine multi-source information integration, visual expression, and spatiotemporal topology analysis for typical application scenarios such as sea area use, island development and utilization, and marine environmental monitoring and protection. The study developed a marine information decision support system with comprehensive integration and visual analysis of information on maritime islands, early warning and monitoring, the marine economy, and the marine environment. The system was applied to dynamic regulation of sea area use, spatiotemporal characteristic analysis of marine resources, the regulation of human activities in marine protected areas, and the monitoring of sea level rise. Practice has proved that a three-dimensional spatiotemporal data model of marine resources can meet the application requirements of "One Map", the basic land and spatial information platform.

Key words: Marine regulation; Decision support system; Three-dimensional spatiotemporal database;

1. Introduction

As an important space for human survival and development, the ocean is rich in spatial resources, biotic resources, mineral resources, and marine energy. More and more marine development and utilization by humankind has greatly promoted local marine economies in coastal areas, but also caused problems such as the destruction of marine ecosystems, marine environmental pollution, and the decline of the retention rate of natural coastlines. Therefore, it is urgent to strengthen the ability to comprehensively manage and control marine development and utilization. A research hotspot has emerged in coordinating the development of marine economies and resources and environmental carrying capacity from the perspective of comprehensive marine regulation.

Integrating marine information based on geographic location and building a decision support system can provide display, query, and spatial analysis capabilities of the ocean and its development and utilization in their current state. Furthermore, spatial land use planning can improve the scientificity, accuracy, and timeliness of marine development, utilization, and environmental assessment while improving the level of comprehensive management. Many scholars in China have carried out a series of studies and applications focusing on offshore oil spill emergencies, sea level rise, coastal erosion, and regional risk assessment.

This paper proposes a three-dimensional spatiotemporal model based on research on more general such models for natural resources but tailored to the characteristics of marine resources. Oriented to the typical application, this paper develops a marine information decision support system for marine regulation, integrating marine information, spatial topology analysis, and visual analysis applied to scenarios such as sea area use, island development and utilization, environmental monitoring, and protection of marine ecosystems.

2. Three-dimensional spatiotemporal model of marine resources

The Ministry of Natural Resources has issued the "Overall Plan for the Construction of Natural Resources Investigation and Monitoring System", which proposes a hierarchical classification of various natural resource information. Three-dimensional spatial location is the basic link for organizing and connecting all natural resources, basic surveying and mapping results are the framework, digital elevation models are the base, and high-resolution remote sensing imagery is the background. The system scientifically organizes each natural resource entity in an orderly distribution on, above, and below Earth's surface, forming a complete three-dimensional spatiotemporal model of natural resources supporting production, life, and ecology.

As an important part of natural resources, marine resources are hierarchically organized and managed according to category and content, such as an underground resource layer, surface matrix layer, surface overburden layer, and management layer. The system sorts out, cleans, refines, and transforms data with natural attributes and management attributes among data at each layer, then combines and merges them according to region, discipline, specialty, element, application field, and management theme.

- Underground resource layer: includes the location, distribution, and area data of marine mineral resources such as marine sand, oil, and gas.
- Surface matrix layer: includes seabed water depth, landform, and seabed sediment data.
- Surface overburden layer: includes marine ecosystem, marine water environment, marine biotic resources, and other data. The marine ecosystem includes the type, distribution, and area data of coastal wetlands and coastal tidal flats; the type, distribution, and area data of the sea, the number, location and area data of islands; the distribution, area, and composition data of coral reefs, mangroves, seagrass beds, and other ecosystems; and the marine water environment, including attribute data of the physical ocean, marine biology, marine chemistry, marine acoustics, marine optics, and elements of other disciplines at different water depths.

- Management layer: includes data such as the type, length, spatial distribution, and change in the marine ecological red line and coastline; current protection and utilization of coastal wetlands and coastal tidal flats; the protection and utilization of coastal zones and sea reclamation in China; present situation and change in the development, utilization, and protection of islands and island resources; and the development and utilization of marine mineral resources.

3. Construction of a decision support system (DSS)

3.1 Technology framework

This study carried out the overall architecture design of the system, which is divided into a data layer, function layer, and application layer. The three layers proceed in an orderly way and are interrelated with one another. Through data and function encapsulation, the coupling between each layer and between modules is reduced, so that the system function is easy to expand and maintain.

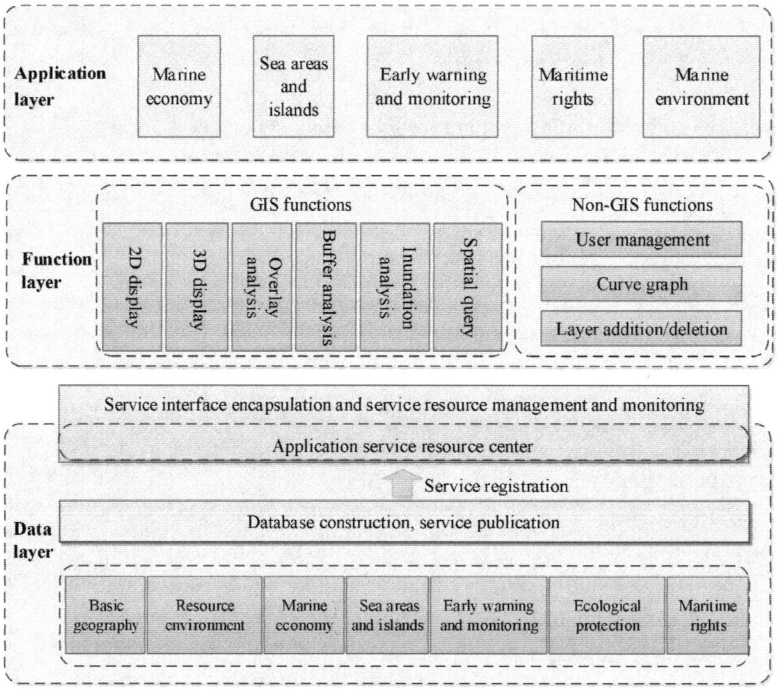

Figure 1. Architecture of DSS

- Data layer: processes and manages various kinds of basic data and thematic data such as basic geography, marine environment, marine economy, sea areas and islands, marine early warning and monitoring, ecological protection, and marine equity, both spatial and non-spatial data, structured and unstructured. Service encapsulation and publication is carried out in accordance with the Webservices protocol, and applications are registered in the unified application service resource pool.

IOP Conf. Series: Earth and Environmental Science 1004 (2022) 012009 doi:10.1088/1755-1315/1004/1/012009

- Function layer: function migration and transformation, including GIS functions and non-GIS functions, are carried out for different thematic applications based on the data layer and guided by application requirements. The GIS functions include two-dimensional display, three-dimensional display, overlay analysis, buffer analysis, inundation analysis, arbitrary spatial queries, and so on. The non-GIS functions mainly include user management, statistical charts (histograms, pie charts, curve graphs), tree lists, background service management (addition, deletion), and other chart visualization, statistical analysis, and service configuration functions.
- Application layer: data from all fields of study related to the ocean are integrated for the comprehensive management requirements of sea area and island regulation, monitoring and evaluation of marine economics, marine early warning and monitoring, maintenance of marine equity, and environmental analysis. Various interfaces are called based on the function layer, allowing the decision support system to provide collaborative services for comprehensive regulation and decision support functions based on One Map.

3.2 Database construction

This study adopted PostgreSQL database management software to construct and manage the database. The business database logically includes two parts: a spatial database and non-spatial database. The spatial database records the business information and spatial information of six aspects, including marine rights and interests, early warning and monitoring, marine environment, spatial planning, sea areas and islands, and marine economy, while the non-spatial database records the business information on disaster prediction and reduction and the marine economy. The background configuration database includes related database table contents such as layer, function, menu, service, and template configuration data tables. These comprise the functions of the background configuration management tool for the function menu, thematic tree, function content, web reference service and HTML template, and user configurations. In the background configuration database, the association between related table fields of the configuration at all levels is realized by means of foreign keys.

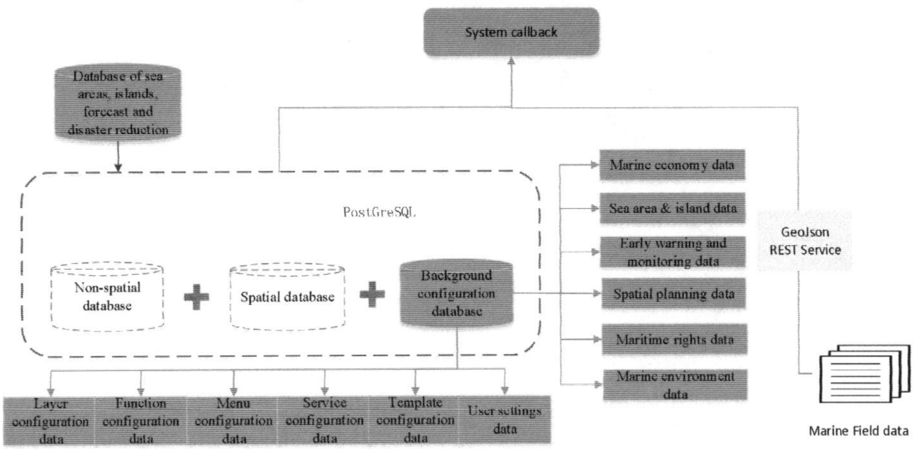

Figure 2. Database for DSS

3.3 System function R&D

An integrated marine management decision-making platform was established with unified interface standards, unified resource management, a unified basic platform, and a unified service portal, upgrading and organically integrating the existing application systems to provide a "one-stop" marine information query, retrieval, and visualization service.

3.3.1. Analysis of spatiotemporal characteristics of marine resources

(1) Analysis of interannual variation of coastline resources

Based on the continental coastlines extracted from remote sensing images in 1990, 2000, 2008, 2010, and 2014, the system provides overlay analysis of continental coastlines and remote sensing images over the years, which can visualize changes and provide support for further investigation and identification of the causes of changes.

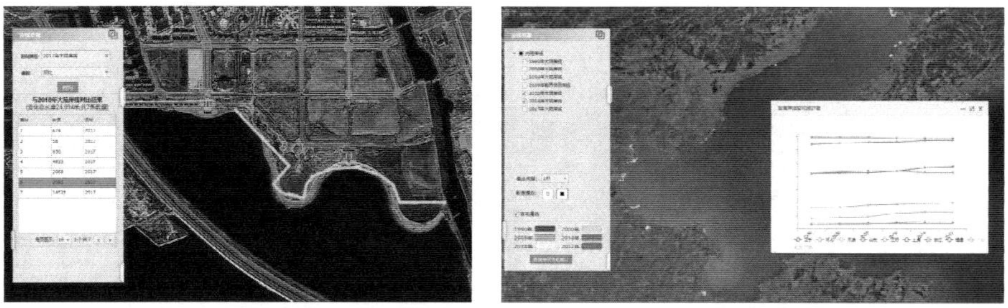

Figure 3. Shoreline change analysis and statistical analysis of length in coastal provinces and cities in different years

(2) Spatial distribution analysis of marine biotic resources

Based on the statistical yearbook data of offshore marine biological species in China, the system has functions for the spatial distribution of phytoplankton, zooplankton, and benthos; species station data monitored in a spatial grid; a thermal map of species; and a morphological view of species.

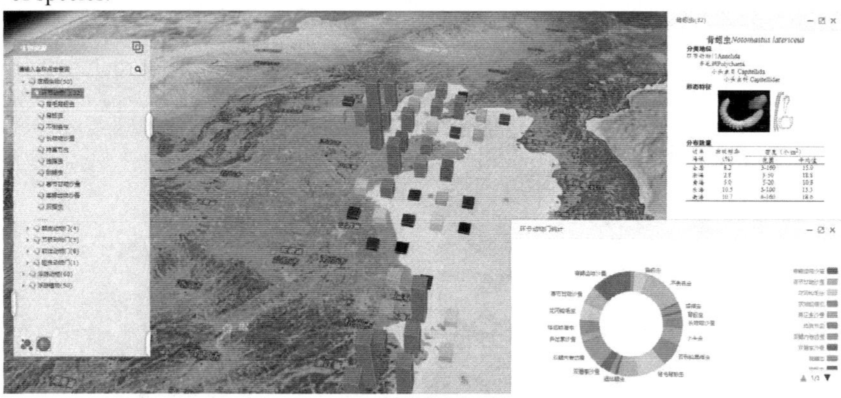

Figure 4. Spatial distribution, species proportion, and morphology of Annelida

3.3.2. Statistical analysis of marine economy

The system integrates the overall operation of the marine economy, the development of major marine industries, and the development of regional marine economies in China since 2000. It analyzes the proportion of different industries, development situations, the marine economic aggregate at the city level, and the sea area level, providing important support for the monitoring and evaluation of marine economic operation.

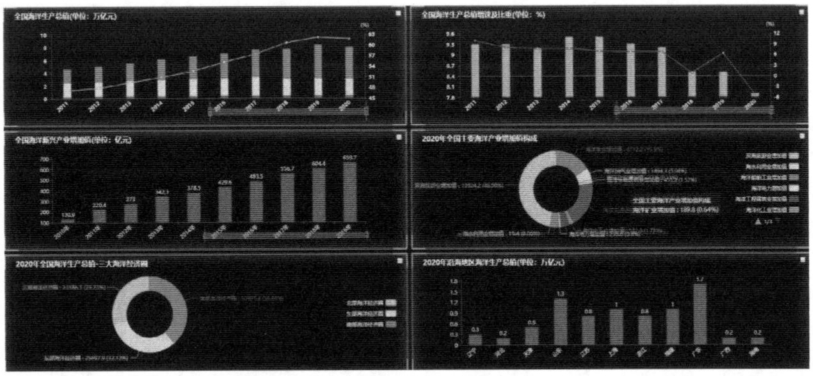

Figure 5. Statistical analysis of the proportion, growth rate and added value of emerging industries of the marine economy

3.3.3. Analysis of sea area use and island development

- Analysis of current sea area use and island utilization

The system integrates data on the current use of sea areas and data on the development and utilization of islands; produces the classification statistics, comprehensive query, and spatial topology analysis of different types of sea area use; and conducts the statistical analysis of island coastline types, vegetation coverage, and interannual comparative analysis of development and utilization.

- Analysis of the approval of new sea area use projects

The system provides overlap analysis of newly added sea use projects and current sea use data, compliance analysis of marine functional regionalization, and comparative analysis with applied length or area. Meanwhile, it is possible to query the status of the marine environment and marine protection planning around the sea use projects, providing support for analyzing the approval of sea use projects.

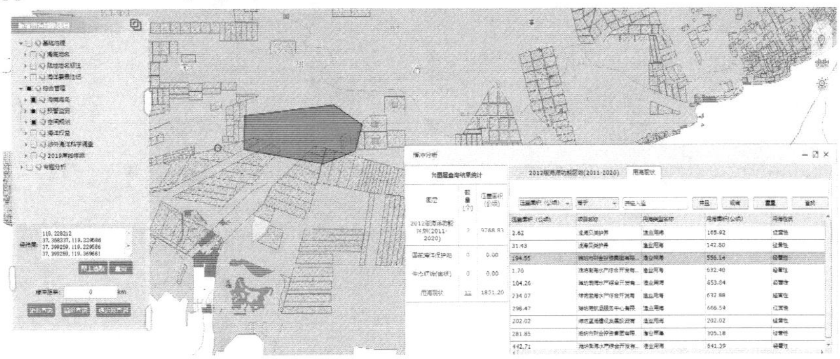

Figure 6. Spatial topology analysis of new sea use projects in any area

IOP Conf. Series: Earth and Environmental Science 1004 (2022) 012009 doi:10.1088/1755-1315/1004/1/012009

3.3.4. Early warning and monitoring analysis

- Vulnerability analysis of sea level rise

Based on the vulnerability index calculation model of the impact of sea level rise, this paper integrates vulnerability regionalization data in coastal cities and counties nationwide, and analyzes the natural environment vulnerability, socio-economic vulnerability, coastal zone vulnerability index, and vulnerability level in coastal cities and counties.

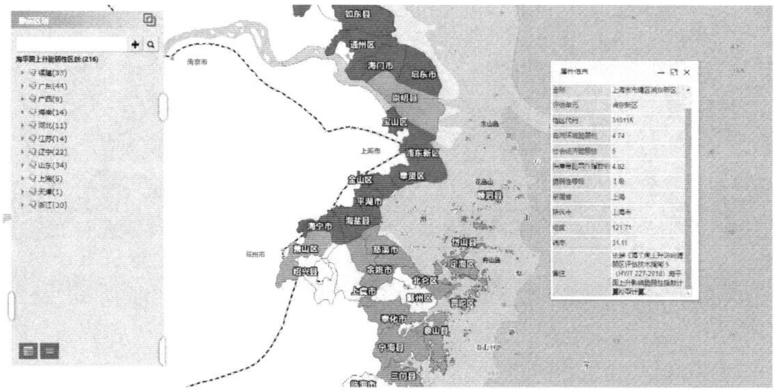

Figure 7. Vulnerability analysis of sea level rise in sea-related cities and counties

- Analysis of sea use in the marine ecological red line area

A GIS spatial topology analysis was conducted for current national sea use, the marine ecological red line, and marine protected areas, and the analysis of suspected illegal sea use is provided according to the types and spatial scope of sea use.

- Analysis of vessel behavior in marine protected areas

Information mining of vessels passing in and out of ports and channels can help judge whether vessels are possibly involved with illegal operation in protected areas. The information is based on target radar data with 5, 10, and 15 nautical mile buffers and the establishment of electronic fences, path pickup and target behavior analysis, and overlay analysis algorithms.

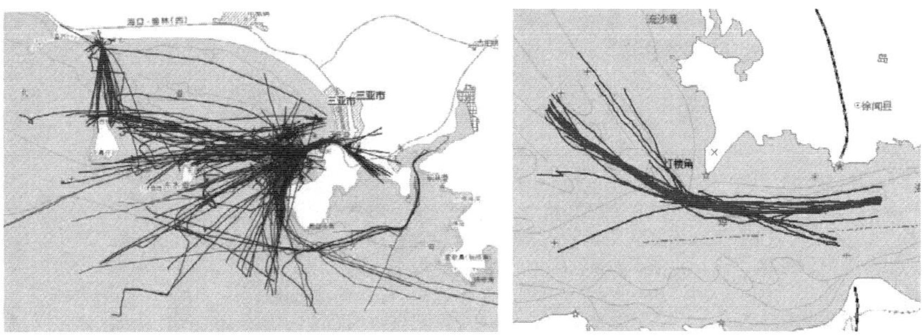

Figure 8. Track diagram of a typical port and channel based on target radar data

3.3.5. Analysis of marine environment

A "One Map" visual analysis of marine environment information is provided to display and analyze the spatiotemporal change and regional characteristics, focusing on change in the marine environment. It includes a large-scale map, single point profile curve, spatiotemporal dynamic sequence map, profile map, and time remapping, providing information support services for further study of change mechanisms. Through the analysis of the topographic environment in coastal zones and nearshore and offshore areas, functions are provided for regional isoline analysis, maximum/minimum analysis, and tensile transformation analysis.

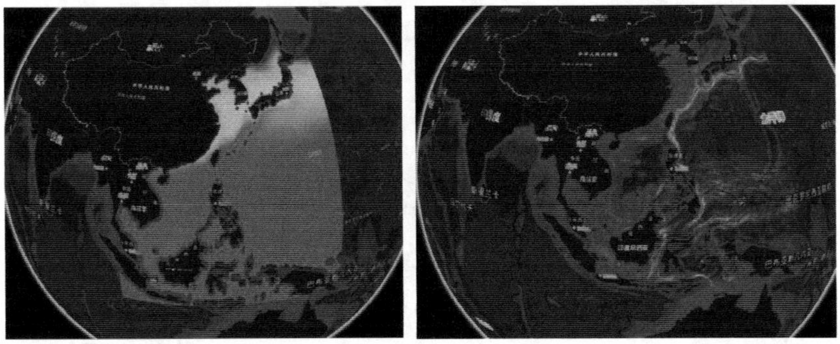

Figure 9. Visualization of sea surface temperature and flow field of the Northwest Pacific

4. Key technologies

4.1 Data synchronization and service technology

The decision support system for marine regulation, which maintains the freshness, continuity, and integrity of data, is an important link in information integration and decision support. It is the fundamental guarantee for the implementation of a business collaborative application. Data synchronization technology dynamically updates the database established in this study according to data type, data content, update frequency, and other factors, focusing on various data generated in the business application systems such as sea areas and islands, marine economies, disaster prevention and reduction, and marine equity. Based on the concept of SOA, the system conducts service encapsulation and processing of various database contents according to the WebService protocol, allowing for real-time registration in the service resource pool and unified management and publication of service resources.

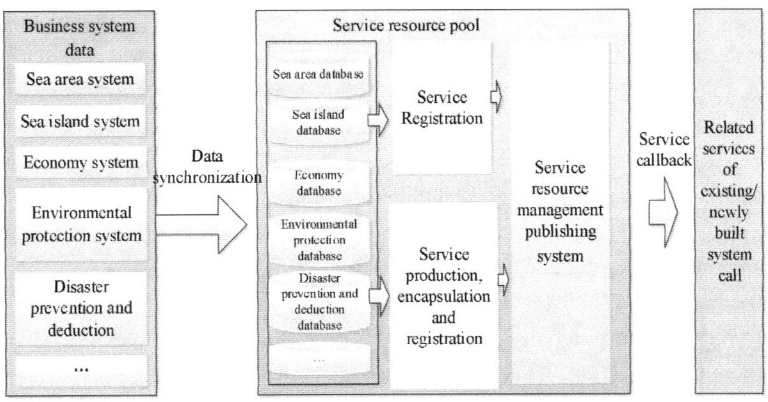

Figure 10. Technology framework for data synchronization of the business system

This study published services for raster data, vector data, and map data according to OGC standards: WCS, WFS, WMS, and TMS\TFS tile map. Service encapsulation and publication, based on SOA, were done for multimedia data such as marine environment data and web photos, providing users with data service registration, publication, and call interfaces.

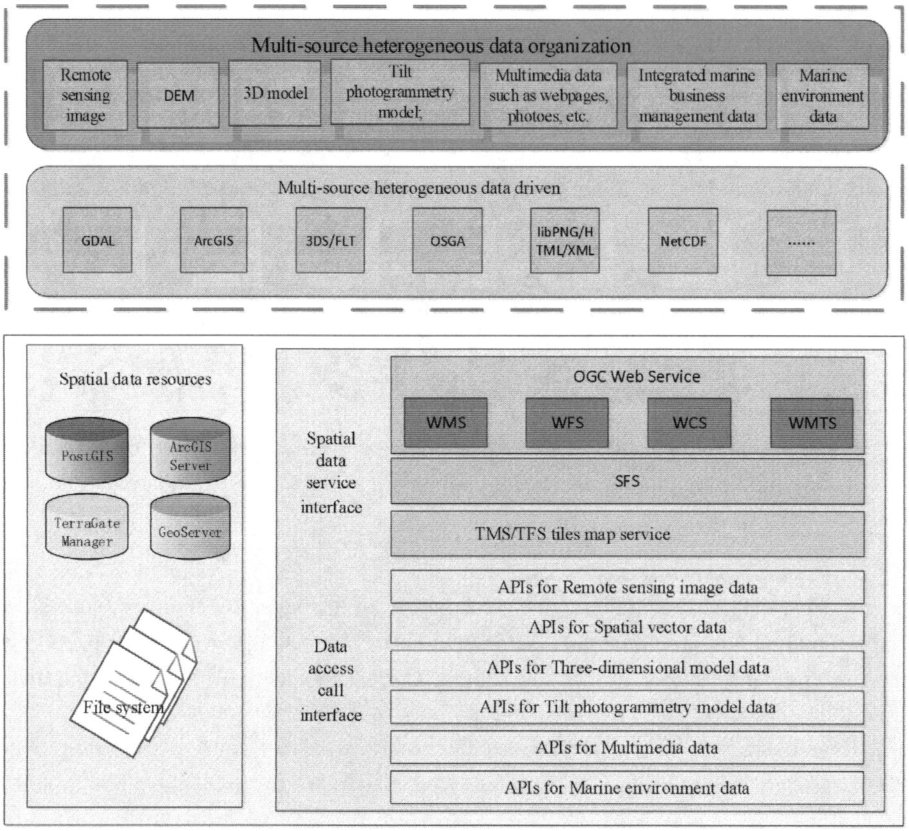

Figure 11. Publication and reuse of multi-source heterogeneous data services

4.2 Data visualization technology

This paper visualizes marine information in three aspects: basic geographic marine data, marine management business data, and environmental data.

Based on the virtual sphere model, the basic geographic marine data constructs image and terrain pyramids, carries out LOD layering and tile map service calls according to the data spatial resolution, and publishes services as a data flow. Multi-scale images and terrain pyramids are composed of original multi-scale image terrain files, index files, data logic processing information, unified pyramid parameters, coordinate system parameters, environmental parameters, and other data files. This model stores all the original data, encodes them with a spatial index, and performs logical processing on the original data, such as clipping, removing outliers, local value editing, edge smoothing, image color equalization, and LOD setting, thus generating a unified pyramid model.

Data visualization is integrated into the sphere system according to its spatial geographic information. Data can include, for example, marine management data in the form of numerical values and text, such as the area and perimeter of sea use for fisheries, transportation, and so on; data on the management of sea areas; and the gross economic product in marine economies. Meanwhile, according to the actual business management

IOP Conf. Series: Earth and Environmental Science 1004 (2022) 012009 doi:10.1088/1755-1315/1004/1/012009

requirements, classification visualization is conducted towards various business needs in the form of statistical charts (such as histograms, pie charts, curve graphs, and scatter diagrams).

Marine environmental data are mainly scalar field and vector field. The visualization of scalar data such as ocean temperature and salinity provides methods such as large-area maps, single point process curves, and profile maps. For the drawing of flow field, the core content is the linear flow effect, which mainly calculates the movement amount and direction within T starting from point A, calculates all positions within T, draws this line segment, and gives the corresponding colors according to the length of the line segment. The specific implementation is as follows.

Through the calculation of ($x_0 = \lceil i/dx \rceil * dx,\ y_0 = \lceil j/dy \rceil * dy$),($x_1 = \lfloor i/dx \rfloor * dx\ \ y_1 = \lfloor j/dy \rfloor * dy$), the quadrangular coordinates are obtained for the data grid where B is located, the corresponding data according to the coordinates are obtained, giving the vector starting from point B through the bilinear difference algorithm $f(i,j) = f(x_0,y_0)\left(1-\frac{i-x_0}{dx}\right)\left(1-\frac{j-y_0}{dy}\right) + f(x_1,y_0)\left(\frac{i-x_0}{dx}\right)\left(1-\frac{j-y_0}{dy}\right) + f(x_0,y_1)\left(1-\frac{i-x_0}{dx}\right)\left(\frac{j-y_0}{dy}\right) + f(x_1,y_1)\left(\frac{i-x_0}{dx}\right)\left(\frac{j-y_0}{dy}\right)$. The trajectory of point B is drawn within T, and the motion trajectory of point A within T is obtained by continuous drawing.

To meet the requirement for allowing flexible configurations of system functions, menus, and data, a configurable system was developed with interaction between the logical design of the configurable database and the template customization and module packaging. The configurable database, which includes tables for the system menu, system function, layer-tree, and marine data, plays a logical supporting role in the system development and implementation process. Based on the customizable configuration, the system gives a choice of templates for front-end function-pages, function-modules, and layer-tree according to marine data type, attributes, and contents.

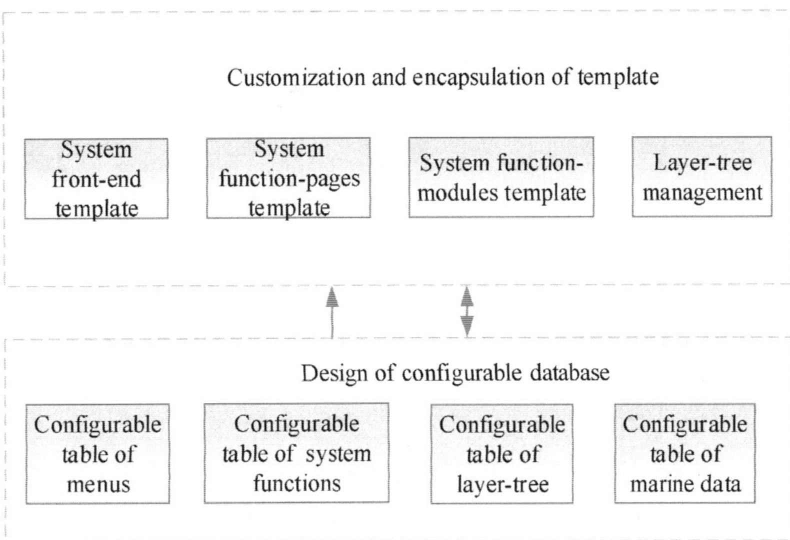

Figure 12. Logical flow of configuration and encapsulation for marine data visualization

5. Conclusion

The ocean is a complex, giant system, including the marine environment, marine development and utilization, marine targets, and human activity, bringing great challenges to marine regulation. Through the construction of a three-dimensional spatiotemporal model of marine resources, data for an underground resource layer, surface matrix layer, surface overburden layer, and management layer are uniformly organized and visually displayed. The system meets requirements for rapid access, statistical analysis, and auxiliary application of marine information. It is a platform for collaborative service, comprehensive regulation, and intelligent decision-making based on One Map. The platform offers comprehensive cross-system and cross-business information services, providing systematic support for refined, comprehensive management of marine resources.

References

[1]Xie J. 2016 Research on the Decision Support System of Offshore Space Resource Utilization. Dalian University of Technology,

[2]Bao L M. 2019 ZheJiang University Development of Emergency Command and Auxiliary Decision-Making System for Marine Oil Spill.

[3]Zhaoyang Yang, Zhi Chen. 2021 *Marine Pollution Bulletin* vol 167 Decision support tools for oil spill response (OSR-DSTs): Approaches, challenges, and future research perspectives.

[4]Lien-Kwei Chien, Wen-Chien Tseng, et al. 2012 *Ocean & Coastal Management* vol 69 A study of ocean zoning and sustainable management by GIS in Taiwan pp 35-49

[5]Naser Ahmed, Newton Howlader, et al. 2021 *Ocean and Coastal Management* Coastal erosion vulnerability assessment along the eastern coast of Bangladesh using geospatial techniques

[6]Silvia Torresan, Andrea Critto, et al. 2016 *Ocean & Coastal Management* vol 120 DESYCO: A decision support system for the regional risk assessment of climate change impacts in coastal zones, pp 49-63

[7]Mehebub Sahana, Haroon Sajjad. 2019 *Remote Sensing Applications: Society and Environment* vol 13 Vulnerability to storm surge flood using remote sensing and GIS techniques: A study on Sundarban Biosphere Reserve, India pp 106-120

[8]Barbara Zanuttigh, Dario Simcic, et al. 2014 *Coastal Engineering* 87THESEUS decision support system for coastal risk management pp 218-239

[9] ZHANG Miaomiao;ZHAO Rui;WANG Liyan and XING Zhe.2020, *Marine Information* vol 2 Study on the Composite Mapping of Marine Spatial Planning based on 'Multiple Plans Integration pp 11-16

[10] ZHANG Feng, HOU Chen, YANG Xiaotong, WANG Jing,JIAO Hongbo,WANG Liyan and LI Yanwen; 2020 *Marine Information* 4The Operational Mechanism of RS Image Rapid Processing & Analyzing Oriented for Marine Supervise pp 9-14

Acknowledgments

This work was funded by the National Key R&D Program of China (No. 2017YFC1405305). Data support was provided by the National Marine Data Center of the National Science & Technology Resource Sharing Service Platform of China.

Global distribution of the cold-water coral *Lophelia pertusa*

R Tong[1*], A J Davies[2], A Purser[3], X Liu[4] and F Liu[4]

[1] Fujian University of Technology, Fuzhou, China
[2] University of Rhode Island, Kingston, USA
[3] Alfred-Wegener-Institut, Bremerhaven, Germany
[4] Fuzhou University, Fuzhou, China

Abstract. *Lophelia pertusa* plays an important role as a major contributor to many cold-water coral reefs, supporting a high diversity of associated benthic and benthopelagic species. Due to the high sensitivity of *L. pertusa* to human activity, it has been classified as indicator species for Vulnerable Marine Ecosystems. However, the global spatial distribution of *L. pertusa* is far from well known. In this study, a database of *L. pertusa* presence data was compiled derived from the large number of *L. pertusa* occurrence records appearing in recent years. In conjunction with data layers covering a range of environmental drivers, habitat suitability for *L. pertusa* was predicted using the Random Forest approach. Suitable habitat for *L. pertusa* was predicted to occur primarily on continental margins, with the most suitable habitat likely to occur in the North East Atlantic and South Eastern United States of America. Aragonite saturation state, temperature and salinity were identified as the most important contributors to the habitat suitability model. Given the high vulnerability of reef-forming cold-water corals to anthropogenic impacts, habitat suitability models are critical in developing worldwide conservation and management strategies for biodiverse and biomass rich cold-water coral ecosystems.

Key words- habitat suitability modelling, framework-forming cold-water coral, random forest

1. Introduction

Framework-forming cold-water corals play important roles in the ecological functioning of marine ecosystems (Corbera *et al.*, 2019; Matos *et al.*, 2021). They construct complicated 3-dimensional skeletons of calcium carbonate that may form mounds or reefs, which can fulfil the role of providing an appropriately hard substrate for the settlement of larvae of sessile animals. Cold-water coral reefs also serve as feeding sites, breeding and nursery habitats, and refuges from predators for benthic species, significantly enhancing local biomass and biodiversity (Capezzuto *et al.*, 2018; Freiwald and Roberts, 2005; Linley *et al.*, 2017; Buhl-Mortensen, 2017; Buhl-Mortensen *et al.*, 2017). *Lophelia pertusa* is a scleractinian framework-forming cold-water coral species with a cosmopolitan distribution. It is mainly found at depths from 39 to 3380 m, and is particularly abundant in North Atlantic waters (Roberts et al., 2009; Mortensen, 2000). *Lophelia pertusa* reefs support a high diversity of benthic species, with more than 1300 species found at such reefs in the North East Atlantic (Buhl-Mortensen et al., 2010; Kutti et al., 2014).

Due to physical fragility and slow recovery rates, reef-forming cold-water corals are very sensitive to anthropogenic disturbance, particularly bottom trawling, and climate change (Pecl *et al.*, 2017; Georgian *et al.*, 2016; Movilla *et al.*, 2014). Spatial distribution information for cold-water corals is necessary for the development of conservation strategies, but at present there remain considerable gaps in knowledge on the distribution of these species. Habitat suitability modelling is a particularly useful approach for charactering the potential distribution of cold-water corals in deep-sea areas (Georgian *et al.*, 2019; Reiss *et al.*, 2015; Bargain *et al.*, 2018; Burgos *et al.*, 2020; Matos *et al.*, 2021; Tong *et al.*, 2013; Wiltshire *et al.*, 2018; Lecours, 2017), providing important input for the creation of conservation and management plans for deep-sea ecosystems (Matos *et al.*, 2021; Reiss *et al.*, 2015). The global habitat suitability of *L. pertusa* has been predicted in several previous studies; using 2060 records, coarse-resolution 1° environmental data and the Ecological-niche factor analysis (ENFA) method (Davies *et al.*, 2008), and with 863 logged occurrences at 30-arc second resolution environmental data and the Maximum Entropy modelling method (Davies and Guinotte, 2011).

This study aims to create an updated database of occurrence records of the framework-forming cold-water coral *L. pertusa*, and develop new global habitat suitability maps for *L. pertusa*, with the model outputs used to better understand its global potential distribution more accurately than was previously possible with the previously utilized smaller databases.

2. Methods

2.1. Species data

Georeferenced presence-only records were obtained from peer-reviewed scientific outputs and public databases, including the NOAA Deep Sea Coral Data Portal, the ICES Vulnerable Marine Ecosystems data portal and the Ocean Biogeographic Information System portal (OBIS). In order to reduce potential errors in spatial position of the occurrence records included for subsequent modelling, presence records were processed in four steps. Firstly, for records with neither position accuracy data nor depth information provided were excluded. Secondly, records with position accuracy > 1000 m of linear distance were also excluded. Thirdly, for those records with no position accuracy information, but had depth or depth range data, the provided depth or depth range was compared to that extracted from the gridded bathymetry data, and any record where these depths, or depth range and depth, differed by more than 50 m in absolute depth were excluded. Lastly, the remained records with absolute depth values extracted from bathymetry grids < 40 m were excluded. In order to decrease the influence of sampling bias, only one occurrence record was retained in a single cell (after Davies and Guinotte 2011). In addition, similar numbers of cells were randomly chosen to act as pseudo-absence background data.

2.2. Environmental variables

Thirty three environmental variables were applied as candidate predictors in this study, with a cell size of 30 arc-sec in a WGS 84 coordinate system (after Davies and Guinotte 2011). These environmental variables can be summarized into four ecologically relevant groups, including 9 terrain variables and gridded bathymetry, 14 water chemistry variables, 7 biological variables, and 2 hydrographic variables (Table 2).

Five environmental variables, namely depth, bottom temperature, aragonite saturation state, bottom horizontal current velocity and vertical current velocity were a priori selected for inclusion in the model, due to their proven high ecological relevance in previous habitat suitability studies for framework-forming cold-water corals published in recent years (Matos *et al.*, 2021; Davies and Guinotte, 2011; Sundahl *et al.*, 2020; Bargain *et al.*, 2018; Dolan *et al.*, 2021; Georgian *et al.*, 2019; Rengstorf *et al.*, 2012).

Since inclusion of highly correlated environmental variables may inhibit model performance and interpretation (Huang et al., 2011), the correlation of environmental variables was investigated by each variable group using analysis of the variance inflation factor (VIF) approach, with depth, temperature, aragonite saturation state, bottom horizontal current velocity and vertical current velocity excluded from the VIF calculations. In addition to the 5 environmental variables, these variables with VIF < 5 indicating a low level of co-linearity, were included in final model (Yesson *et al.*, 2017).

2.3. Predictive modeling approach

In this study, the Random Forest for classification and regression approach (RF) was used to predict the global habitat suitability for *L. pertusa*. RF is a regression algorithm, which builds a number of regression trees, and further averages these trees (Breiman, 2001). These trees grow based on different training data subsets created by the bagging procedure to avoid the correlation of trees (Rodriguez-Galiano *et al.*, 2015). RF has been applied to predict benthic species distributions in a series of studies, demonstrating good predictive performances (Matos *et al.*, 2021; Morato *et al.*, 2020; Georgian *et al.*, 2019; Rowden *et al.*, 2017; Gullage *et al.*, 2017).

The species presence and pseudo-absence dataset was randomly split into two sections, 80% as calibration data, with the remaining 20% as evaluation data. Three widely used statistics methods, TSS (true skill statistics), AUC (area under the receiver operating characteristic curve) and kappa statistics, were adopted to evaluate model performance (Allouche *et al.*, 2006). To better understand the predicted habitat suitability, three thresholds were applied to reclassify the predicted map, including sensitivity and specificity balanced threshold, a 90% sensitivity threshold and a 95% sensitivity threshold (Thuiller *et al.*, 2003; Tong *et al.*, 2013).

3. Results

In total, 31635 records were obtained from papers and public databases (Table 1, Figure 1). Through processing via the four steps to remove inaccurate records, 3660 records were retained for analysis, with duplicate occurrences within a particular cell deriving from two or more input sources removed. In addition, 3660 cells were randomly chosen as pseudo-absence data.

Table 1. The number of occurrence records of *L. pertusa* from different resources.

Resources	OBIS	ICES	NOAA	Papers
Occurrence records	16726	433	12654	1822

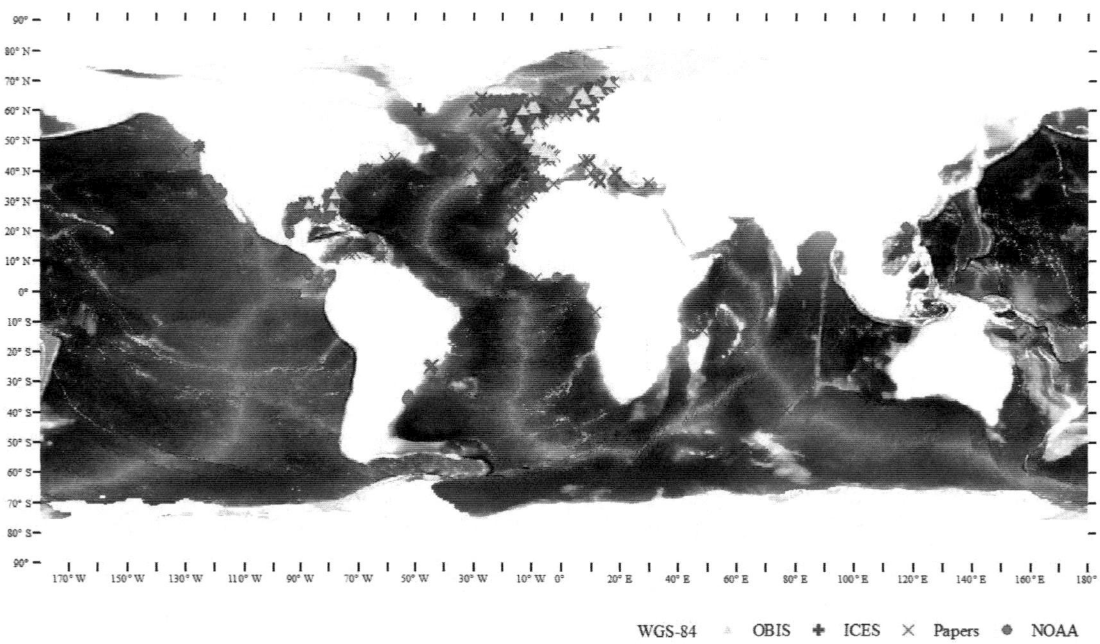

Figure 1. Spatial distribution of the 31635 *L. pertusa* occurrence records used in this study, with data sources illustrated.

IOP Conf. Series: Earth and Environmental Science 1004 (2022) 012010 doi:10.1088/1755-1315/1004/1/012010

Table 2. The 33 environmental variables of four groups used in this study after Davies and Guinotte (2011). The VIF values of environmental variables of three groups tested with VIF<5 in bold. And the variable importance values of the 13 environmental variables used in habitat suitability modelling illustrated.

Groups	Environmental variables	Reference	VIF	Variable importance
Depth & Terrain variables	**Depth**	Becker *et al.* (2009)		0.046
	Slope	Jenness (2013)	20.0	
	Aspect	Jenness (2013)	**1.0**	0.000
	Plan curvature	Jenness (2013)	38.5	
	Profile curvature	Jenness (2013)	**2.4**	0.000
	Tangential curvature	Jenness (2013)	36.6	
	Roughness	Wilson *et al.* (2007)	40.3	
	Rugosity	Jenness (2013)	**2.2**	0.009
	Terrain ruggedness index	Wilson *et al.* (2007)	42.9	
	Topographic position index	Wilson *et al.* (2007)	**2.7**	0.000
Chemistry variables	**Temperature**	Boyer *et al.* (2005)		**0.088**
	Aragonite saturation states	Steinacher *et al.* (2009)		**0.067**
	Alkalinity	Steinacher *et al.* (2009)	37.2	
	Apparent oxygen utilization	Garcia *et al.* (2006a)	176.9	
	Dissolved inorganic carbon	Steinacher *et al.* (2009)	54.1	
	Dissolved oxygen	Garcia *et al.* (2006a)	31.3	
	Nitrate	Garcia *et al.* (2006b)	43.2	
	Calcite situation states			
	calcite_stein	Steinacher *et al.* (2009)	18.3	
	calcite_orr	Orr *et al.* (2005)	11.4	
	Salinity	Boyer *et al.* (2005)	**1.5**	0.050
	Percent oxygen saturation	Garcia *et al.* (2006a)	185.3	
	Phosphate	Garcia *et al.* (2006b)	35.5	
	Silicate	Garcia *et al.* (2006b)	12.4	
	Seasonal variation index	Lutzs *et al.* (2007)	**2.4**	0.013
Biological variables	**Particulate organic carbon**	Lutzs *et al.* (2007)	**2.7**	0.053
	Vertically generalized productivity model	Behrenfeld and Falkowski (1997)		
	VPGM_min		**2.5**	0.003
	VPGM_mean		19.6	
	VPGM_max		6.8	
	Primary productivity	NASA Ocean Color		
	PR_min		Inf	
	PR_mean		15.0	
	PR_max		Inf	
Hydrographic variables	**Regional current velocity**	Carton *et al.* (2005)		0.007
	Vertical current velocity	Carton *et al.* (2005)		0.001

Eight out of 28 environmental variables from the three groups were selected with VIF <5, indicating low co-linearity, including aspect, profile curvature, rugosity, topographic position index, salinity, seasonal variation index, particulate organic carbon and vertically generalized productivity model VPGM_min (Table 2).

The model evaluation showed that the model performed well with TSS 0.974, AUC 0.997 and kappa statistics 0.974. Bottom temperature and aragonite saturation state were the most important variables contributing to the RF model (Table 2). RF predicted suitable habitat for *L. pertusa* throughout the world's oceans (Figure 2a). The suitable habitat was predicted to occur primarily on continental shelves and slopes (Figure 2). The most suitable habitat for *L. pertusa* was predicted to occur in the North East Atlantic and South Eastern America (Figure 2b).

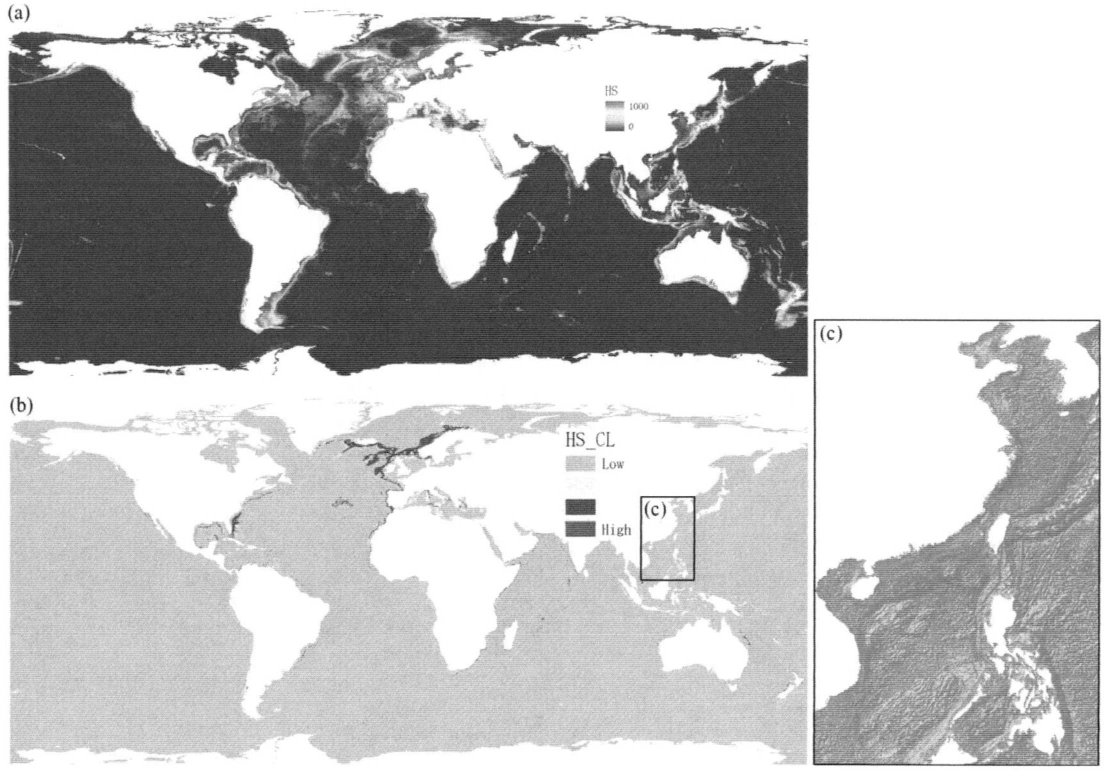

Figure 2. (a) The predicted global habitat suitability for *L. pertusa*. (b) The classified habitat suitability map using 3 thresholds, including the sensitivity and specificity balanced threshold, the 90% sensitivity threshold and the 95% sensitivity threshold. (c) the detailed map of the rectangle section of (b) with hillshade.

4. Discussion and Conclusions

The habitat suitability model developed for *L. pertusa* in this study was influenced primarily by temperature and aragonite saturation state, followed by particulate organic carbon and salinity, which is consistent with findings from previous studies into global stony cold-water coral distributions (Davies *et. al.*, 2008; Davies and Guinotte, 2011; Tittensor *et al.*, 2009). The most important variable, temperature, is often reported as one of the most important factors for contributing towards a modelled distribution output for cold-water corals (Matos *et al.*, 2021; Yesson *et al.*, 2017; Sundahl *et al.*, 2020; Bargain *et al.*, 2018). Calcium carbonate in form of aragonite is essential for construction of the skeletal framework of reef-forming cold-water corals and strongly limits the spatial distribution of *L. pertusa* (Georgian *et al.*, 2019; Morato *et al.*, 2020). High-resolution current velocity as a variable can significantly enhance the performance of a predictive model, often correlated with cold-water coral distributions (Rengstorf *et al.*, 2012; Dolan *et al.*, 2021; Bargain *et al.*, 2018). Habitat suitability models at global scales for cold-water corals have shown a weak response to bottom currents, particularly when current conditions are averaged over large spatial extents, indicating resolution of the current is critical if the most accurate results are to be achieved (Davies *et. al.*, 2008; Davies and Guinotte, 2011; Tittensor *et al.*, 2009; Yesson *et al.*, 2017; Yesson *et al.*, 2012).

In this study, the database of 31635 *L. pertusa* records was constrained by limitations of position accuracy and/or depth data. The increase in published *L. pertusa* occurrence records since 2011 resulted in significantly more records being included in the model runs than has been possible in previous studies (from 863 in Davies and Guinotte (2011) to 31645, almost a 40-fold increase). Global habitat suitability for *L. pertusa* predicted in this study appears greater than found in the previous study by Davies and Guinotte, 2011 (Figure 2a). A number of areas of the global ocean, such as large seamounts, ridges and canyons, were predicted to have higher habitat suitability than in Davies and Guinotte (2011) study (Figure 2a). However, the high suitability of continental shelves and slopes, particularly in the North East Atlantic and South Eastern United States of America, was similar to previous predictions (Davies and Guinotte, 2011) (Figure 2b). Additionally, the high habitat suitability at North East Atlantic and along continental margin was similar to that demonstrated for other cold-water corals, such as Antipatharia, octocorals and other framework-forming corals (Davies *et. al.*, 2008; Davies and Guinotte, 2011; Yesson *et al.*, 2017; Yesson *et al.*, 2012). There was a narrow region along the continental shelf of China predicted to have relatively low habitat suitability for *L. pertusa* (Figure 2c).

The spatial resolution of environmental variables is very important for developing useful and accurate habitat suitability models. The global bathymetric dataset (SRTM30; Becker et al. 2009) used in this study, had a resolution of 30-arc seconds, approximately 1 km^2 cell resolution. It successfully captures large terrain features such as large canyons and seamounts (>1 km in diameter), whilst missed smaller terrain features such as mounds and ridges < 1 km in diameter (Davies and Guinotte, 2011). However, smaller terrain relief structures are also known to be important areas for deep-sea scleractinian framework-forming corals. The inclusion of the highest resolution bathymetry dataset SRTM15+ available at present (Tozer *et al.*, 2019) should be considered in any future studies, which may largely improve accuracy of habitat suitability modelling and allow for the incorporation of more species records as resolution increases. The predicted habitat suitability in this study may still be influenced by the sampling bias of *L. pertusa* records. Selecting background data reflecting the same bias as the occurrence data may be useful to reduce the effects of sampling bias (Phillips *et al.*, 2009). Using a kernel density estimate of sampling effort to reduce the influence of sampling bias could significantly improve the model performance (Burgos *et al.*, 2020; Georgian *et al.*, 2019). Such methods to reduce the influence of sampling bias should be considered in future to improve the performance of global habitat suitability model predictions.

Acknowledgments

This research was supported by the National Key R&D Program of China (NO. 2019YFE0127100), National Natural Science Foundation of China (NO. 42006140), Natural Science Foundation of Fujian Province of China（No.2019I0006).

References

[1] Allouche O, Tsoar A and Kadmon R 2006 *Journal of Applied Ecology* **43** 1223-32
[2] Bargain A, Foglini F, Pairaud I, Bonaldo D, Carniel S, Angeletti L, Taviani M, Rochette S and Fabri M C 2018 *Progress in Oceanography* **169** 151-68
[3] Becker J, Sandwell D, Smith W, Braud J and Binder B 2009 *Marine Geodesy* **32** 355-71
[4] Behrenfield M and Falkowski P 1997 *Limnology & Oceanography* **42** 1-20
[5] Boyer T, Levitus S, Garcia H, Locarnini R A, Stephens C and Antonov J 2005 *International Journal of Climatology* **25** 931–45
[6] Breiman L 2001 *Machine Learning* **45** 5–32
[7] Buhl-Mortensen L, Serigstad B, Buhl-Mortensen P, Olsen M N, Ostrowski M, Blazewicz-Paszkowycz M and Appoh E 2017 *Deep-Sea Research Part II-Topical Studies in Oceanography* **137** 148-56
[8] Buhl-Mortensen L, Vanreusel A, Gooday A, Levin L, Priede I, Buhl-Mortensen P, Gheerardyn H, King N and Raes M 2010 *Mar. Ecol.* **31** 21-50

[9] Buhl-Mortensen P 2017 *Marine Biology Research* **13** 1027-40
[10] Burgos J M, Buhl-Mortensen L, Buhl-Mortensen P, Olafsdottir S H, Steingrund P, Ragnarsson S A and Skagseth O 2020 *Frontiers in Marine Science* **7** 1-25
[11] Capezzuto F, Ancona F, Carlucci R, Carluccio A, Cornacchia L, Maiorano P, Ricci P, Sion L, Tursi A and D'Onghia G 2018 *Rendiconti Lincei. Scienze Fisiche e Naturali* **29**
[12] Carton J A, Giese B S and Grodsky S A 2005 *Journal of Geophysical Research Oceans* **110**
[13] Corbera G, Iacono C L, Gracia E, Grinyo J, Pierdomenico M, Huvenne V, Aguilar R and Gili J M 2019 *Progress In Oceanography* **175**
[14] Davies A, Wisshak, M, Orr, JC, Roberts, JM 2008 *Deep Sea Research Part I: Oceanographic Research Papers* **55** 1048-62
[15] Davies A J and Guinotte J M 2011 *PLoS ONE* **6** e18483
[16] Dolan M, Ross R, Albretsen J, Skardhamar J, Gonzalez-Mirelis G, Bellec V, Buhl-Mortensen P and Bjarnadóttir L 2021 *Geosciences* **11** 48
[17] Freiwald A and Roberts J M 2005 *Cold-Water Corals and Ecosystems*: (Springer)
[18] Garcia H E, Locarnini R A, Boyer T P and Antonov J I 2006b *World Ocean Atlas 2005* 4
[19] Garcia H E, R.A. L, T.P. B and J.I. A 2006a *World ocean atlas 2005* 3
[20] Georgian S E, Anderson O F and Rowden A A 2019 *Fish Res.* **211** 256-74
[21] Georgian S E, Dupont S, Kurman M, Butler A, Stromberg S M, Larsson A I and Cordes E E 2016 *Marine Ecology* **37** 1345-59
[22] Gullage L, Devillers R and Edinger E 2017 *Marine Ecology Progress Series* **582** 57-77
[23] Jenness J 2012 DEM Surface Tools. http://www.jennessent.com/arcgis/surface_area.htm
[24] Kutti T, Bergstad O A, Fosså J H and Helle K 2014 *Deep Sea Research Part II: Topical Studies in Oceanography* **99** 122-33
[25] Lecours V 2017 *Frontiers in Environmental Science* **4** 1-18
[26] Linley T D, Lavaleye M, Maiorano P, Bergman M, Capezzuto F, Cousins N J, D' Onghia G, Duineveld G, Shields M A, Sion L, Tursi A and Priede I G 2017 *Deep Sea Research Part II: Topical Studies in Oceanography* **145** 8-21
[27] Lutz M, Caldeira K, Dunbar R and Behrenfeld M 2007 *Journal of Geophysical Research* **112** C10011
[28] Matos F L, Company J B and Cunha M R 2021 *Deep-Sea Res. Part I-Oceanogr. Res. Pap.* **170** 12
[29] Morato T, Gonzalez-Irusta J-M, Dominguez-Carrio C, Wei C-L, Davies A, Sweetman A K, Taranto G H, Beazley L, Garcia-Alegre A, Grehan A, Laffargue P, Murillo F J, Sacau M, Vaz S, Kenchington E, Arnaud-Haond S, Callery O, Chimienti G, Cordes E, Egilsdottir H, Freiwald A, Gasbarro R, Gutierrez-Zarate C, Gianni M, Gilkinson K, Wareham Hayes V E, Hebbeln D, Hedges K, Henry L-A, Johnson D, Koen-Alonso M, Lirette C, Mastrototaro F, Menot L, Molodtsova T, Duran Munoz P, Orejas C, Pennino M G, Puerta P, Ragnarsson S a, Ramiro-Sanchez B, Rice J, Rivera J, Roberts J M, Ross S W, Rueda J L, Sampaio I, Snelgrove P, Stirling D, Treble M A, Urra J, Vad J, van Oevelen D, Watling L, Walkusz W, Wienberg C, Woillez M, Levin L A and Carreiro-Silva M 2020 *Global Change Biology* **26** 2181-2202
[30] Mortensen P B 2000 *Department of Fisheries and Marine Biology,* (Norway: University of Bergen) pp 1-42
[31] Movilla J, Orejas C, Calvo E, Gori A, López-Sanz À, Grinyó J, Domínguez-Carrió C and Pelejero C 2014 *Coral Reefs* **33** 675-86
[32] Orr J C, Fabry V J, Aumont O, Bopp L, Doney S C, Feely R A, Gnanadesikan A, Gruber N, Ishida A, Joos F, Key R M, Lindsay K, Maier-Reimer E, Matear R, Monfray P, Mouchet A, Najjar R G, Plattner G-K, Rodgers K B, Sabine C L, Sarmiento J L, Schlitzer R, Slater R D, Totterdell I J, Weirig M-F, Yamanaka Y and Yool A 2005 *Nature* **437** 681
[33] Pecl G T, Araújo M B, Bell J D, Blanchard J, Bonebrake T C, Chen I C, Clark T D, Colwell R K, Danielsen F, Evengård B, Falconi L, Ferrier S, Frusher S, Garcia R A, Griffis R B, Hobday A J, Janion-Scheepers C, Jarzyna M A, Jennings S, Lenoir J, Linnetved H I, Martin V Y, McCormack P C, McDonald J, Mitchell N J, Mustonen T, Pandolfi J M, Pettorelli N, Popova E,

Robinson S A, Scheffers B R, Shaw J D, Sorte C J, Strugnell J M, Sunday J M, Tuanmu M N, Vergés A, Villanueva C, Wernberg T, Wapstra E and Williams S E 2017 *Science* **355**
[34] Phillips S J, Dudík M, Elith J, Graham C H, Lehmann A, Leathwick J and Ferrier S 2009 *Ecological Applications* **19** 181-97
[35] Reiss H, Birchenough S, Borja A, Buhl-Mortensen L, Craeymeersch J, Dannheim J, Darr A, Galparsoro I, Gogina M, Neumann H, Populus J, Rengstorf A M, Valle M, Van Hoey G, Zettler M L and Degraer S 2015 *ICES Journal of Marine Science* **72** 297-315
[36] Rengstorf A M, Mohn C, Brown C, Wisz M S and Grehan A J 2012 *Deep Sea Research Part I: Oceanographic Research Papers* **93** 72-82
[37] Roberts J M, Wheeler A, Freiwald A and Cairns S 2009 *Cold-Water Corals: The Biology and Geology of Deep-Sea Coral Habitats.* (Cambridge University Press)
[38] Rodriguez-Galiano V, Sanchez-Castillo M, Chica-Olmo M and Chica-Rivas M 2015 *Ore Geologial Reviews* **71** 804-18
[39] Rowden A A, Anderson O F, Georgian S E, Bowden D A, Clark M R, Pallentin A and Miller A 2017 *Frontiers in Marine Science* **4** 1-19
[40] Steinacher M, Joos F, Frölicher T L, Plattner G K and Doney S C 2009 *Biogeosciences* **6** 515-33
[41] Sundahl H, Buhl-Mortensen P and Buhl-Mortensen L 2020 *Frontiers in Marine Science* **7**
[42] Thuiller W, Vayreda J, Pino J, Sabate S, Lavorel S and Gracia C 2003 *Global Ecology and Biogeography* **12** 313-25
[43] Tittensor D P, Baco A R, Brewin P E, Clark M R, Consalvey M, Hall-Spencer J, Rowden A A, Schlacher T, Stocks K I and Rogers A D 2009 *Journal of Biogeography* **36** 1111-28
[44] Tong R, Purser A, Guinan J and Unnithan V 2013 *Ecological Informatics* **13** 123-32
[45] Tozer B, Sandwell D, Smith W, Olson C, Beale J and Wessel P 2019 *Earth & Space Science* **6** 1847-64
[46] Wilson M F J, O'Connell B, Brown C, Guinan J C and Grehan A J 2007 *Marine Geodesy* **30** 33
[47] Wiltshire K H, Tanner J E, Althaus F, Sorokin S J and Williams A 2018 *Deep-Sea Research Part II-Topical Studies in Oceanography* **157** 121-33
[48] Yesson C, Bedford F, Rogers A D and Taylor M L 2017 *Deep Sea Research Part II: Topical Studies in Oceanography* **145** 79-86
[49] Yesson C, Taylor M L, Tittensor D P, Davies A J, Guinotte J, Baco A, Black J, Hall-Spencer J M and Rogers A D 2012 *Journal of Biogeography* **39** 1278-92

IOP Conf. Series: Earth and Environmental Science 1004 (2022) 012011 doi:10.1088/1755-1315/1004/1/012011

A Two-dimensional Attitude Steering Method to Compensate for the Doppler Centroid in Moon-based SAR

G Q Chen, Y X Ding, M Y Lv, H D Guo, and J Wu

[1] Key Laboratory of Digital Earth Science, Aerospace Information Research Institute, Chinese Academy of Sciences, Beijing 100094

Correspondence: chengq@radi.ac.cn

Abstract. The rising attention to Moon-based Earth observation provides a new way to monitor Earth. Among Moon-based detectors, the SAR method has the advantage of long-term, stable, unique observations compared to space-borne SAR. Unlike space-borne SAR, however, the complicated relative motion of a Moon-based sensor makes its Doppler centroid estimation more difficult. This paper proposes a method for two-dimensional attitude steering to compensate for the Doppler centroid of Moon-based SAR. In the periodic movement between Earth and the Moon, the track speed of sites distributed around the near side of the Moon are almost perpendicular to vectors from the sites to the Earth centroid. This situation indicates the feasibility of attitude steering to guide antennas into a zero Doppler plane intersecting Earth's surface. Therefore, this paper proposes a method for two-dimensional attitude steering to compensate the Doppler centroid to zero. In this method, the Moon is no longer deemed as a point but a natural celestial body with an average radius. Sites distributed around the hemisphere of the near side are also considered separately because of differences in spatial position and speed. Through the proposed method, the pointing of SAR antennas will be steered to a place on Earth where the Doppler centroid is compensated to zero. The JPL DE430 and EOP data are closer to reality and are used for accurate simulation. In this paper, the two-dimensional attitude steering method is compared with no attitude steering and a one-dimensional attitude steering method, and the results show its effectiveness.

1. Introduction

During the past decades, humankind's appetite for lunar exploration and public interest in lunar research has risen again [1]. Several countries and astronomical institutions, including CNSA in China, NASA in the USA, and ESA in Europe, have announced programs (e.g., Chang'e, the Artemis landing mission, and the Moon Village program) for lunar exploration. Among these lunar exploration programs, better performance in Earth observation from the Moon has been one of the main tasks. Earth observation research, including that from Moon-based SAR (MBS), is of great scientific potential in deep space exploration [2, 3], with potential to have a huge role in the sustainable development of humanity [4]. As the only natural satellite of Earth, the Moon has an average distance of 3.8×10^8 km, giving it a field of view (FOV) of the entire Earth disc at a small angle of just 1.8°. An MBS distributed around the Moon's near side would have advantages over space-borne SAR in long-term, stable monitoring of global phenomena, e.g., solid tides, radiation balance, and plasma/magnetosphere, [5, 6].

The literature mainly focuses on the system structure and potential applications of MBS; few studies have analyzed the Doppler centroid steering of MBS. Similar to space-borne SAR, especially geostationary satellites, the rotation of Earth also contributes to the Doppler centroid. Unlike space-borne SAR, which has free rotation in space [7], the Moon is a celestial body with an average radius of 1,738 km [8, 9], and SAR antennas may have occlusion constraints that mainly come from the spherical curvature and irregular topography [10]. The Moon's rotation and its revolution also contribute to the Doppler centroid. The long synthetic aperture time caused by the long distance between Earth and the Moon, together with the nonzero Doppler centroid, will result in high coupling of the azimuth and range signal, increased complexity in signal processing, and lower quality of the imaging formulation [11,12,13], all of which are unfavorable for MBS observation. Therefore, calculating and minimizing the Doppler centroid are the inevitable first steps in further research on signal processing of MBS. In order to simulate the real motion of MBS with an Earth ground target (EGT), the MBS is no longer considered to coincide with the Moon centroid. A high-precision ephemeris named JPL DEM430 and EOP data will be used [14,15].

This paper proposes an analytical approach to estimate the Doppler properties of MBS according to the geometrical model. On the surface of the Moon itself, craters, hills, and other uplifts are unevenly distributed, thus making the topography very complicated [16, 17]. The FOV of MBS might be partly sheltered, the slope of the lunar site might be too steep to construct MBS, and a site's buffer area might not satisfy the need for avoiding falling rocks. In order to reduce computational costs, the influence of topography on MBS is ignored in this study. Generally speaking, the Moon curvature will have an adverse influence on attitude steering because of its occlusion

The rest of this letter is organized as follows. Section 2 gives a description of the Doppler centroid and attitude steering method of MBS; section 3 studies the performance of the Doppler centroid on the near side of the Moon and the differences between two-dimensional attitude steering, no attitude steering, and one-dimensional attitude steering; and section 4 gives a brief conclusion.

2. Method

To study an MBS, it is necessary to conduct attitude steering to reduce its Doppler centroid, because a nonzero Doppler centroid will increase the complexity of signal processing and lower the quality of image formulation, all of which are unfavorable for MBS. In this section, three parts are presented to indicate the performance of MBS attitude steering. Section 2.1 indicates the calculation and decomposition method of Doppler. Section 2.2 indicates the MBS attitude steering method, in which the Doppler centroid can be reduced to zero to satisfy the purpose of this paper. Section 2.3 indicates the probability distribution of attitude steering for the near side of the Moon.

In the paper, vectors are expressed in bold; scalars are expressed in italics; position and speed vectors are expressed in the Earth Central Rotational Coordinate System (ECR), which is related to the Earth Central Inertial Coordinate System (ECI), the Moon Central Moon Fixed Coordinate System (MCMF), and others [18,19]; the symbol '$\| \|$' indicates the norm of a vector; the symbol '$\langle , \rangle$' indicates the inner product of two vectors; and the symbol '$\times$' indicates the cross product of two vectors.

2.1 Calculation and decomposition method of Doppler

As shown in Figure 1, the dark Earth surface coverage area, determined by angle in range, angle in azimuth, and look angle of MBS, changes along relative motions between Earth and the Moon. The synthetic aperture length (e.g., the 'Ls' in Figure 1) can reach hundreds of kilometers because of its longer synthetic aperture time than space-borne SAR. Considering the revolution of the Moon and rotation of the Moon and Earth, an MBS space trajectory relative to Earth's centroid is not exactly a straight line but an arc of smaller curvature. The real zero Doppler line, shown as a green dotted line in Figure 1, moves on the curved Earth surface; that is to say, the zero Doppler pointing and squint angle is changing along synthetic aperture time, and the zero Doppler plane rotates along the MBS space trajectory. Thus, for an accurate estimation of MBS Doppler performance, the motion between Earth and the Moon cannot be determined from air-borne flight or space-borne orbit. Once the pointing of MBS intersects with the zero Doppler line, the frequency of the Doppler centroid can be reduced to the desired value: zero.

In this paper, the position and speed of researched objects are calculated based on JPL DE430 and EOP data.

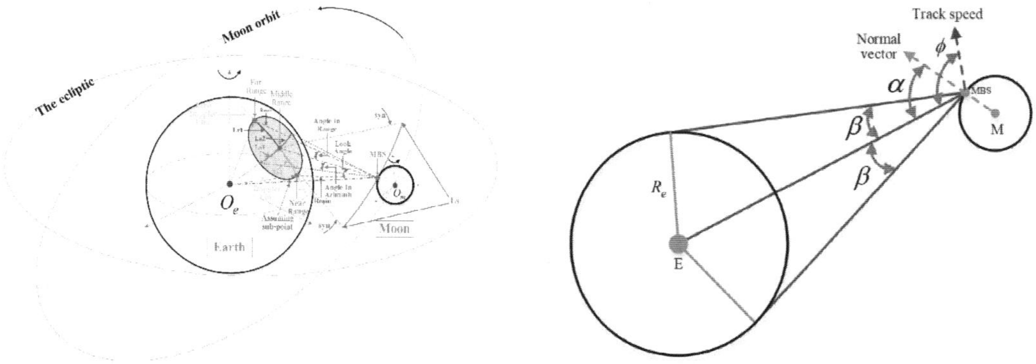

(a) (b)

Figure 1. Space trajectory of MBS (a), and the angle of MBS parameters (b).

For MBS and its achievable EGT, the frequency of the Doppler centroid caused by the relative motion between them are expressed as:

$$f_{dc} = -\frac{2}{\lambda}\cdot\frac{dR}{dt} = -\frac{2\cdot(\mathbf{R}_{mp}-\mathbf{R}_{ep})\cdot(\mathbf{V}_{mp}-\mathbf{V}_{ep})}{\lambda\cdot\|\mathbf{R}_{mp}-\mathbf{R}_{ep}\|}$$
$$= -\frac{2\cdot\left(\mathbf{R}_{mp}\cdot\mathbf{V}_{mp}-\mathbf{R}_{mp}\cdot\mathbf{V}_{ep}-\mathbf{R}_{ep}\cdot\mathbf{V}_{mp}+\mathbf{R}_{ep}\cdot\mathbf{V}_{ep}\right)}{\lambda\cdot\|\mathbf{R}_{mp}-\mathbf{R}_{ep}\|} \quad (1)$$

where $\mathbf{R}_{mp}$ and $\mathbf{R}_{ep}$ are the positions of MBS and GET; $\mathbf{V}_{mp}$ and $\mathbf{V}_{ep}$ are the corresponding speed in the same coordinate system, such as ECR; and λ is the wavelength of MBS.

The speed of EGT is perpendicular to its position, so $\mathbf{R}_{ep}\cdot\mathbf{V}_{ep}=0$. Through properties of vector operations, Equation (1) can be simplified as:

$$f_{dc} = -\frac{2\cdot(\mathbf{R}_{mp}-\mathbf{R}_{ep})\cdot(\mathbf{V}_{mp}+(\mathbf{R}_{mp}\times\mathbf{w}_e))}{\lambda\cdot\|\mathbf{R}_{mp}-\mathbf{R}_{ep}\|} = -\frac{2}{\lambda}\mathbf{A}\cdot\mathbf{B} \quad (2)$$

$$\mathbf{A} = \frac{\mathbf{R}_{mp}-\mathbf{R}_{ep}}{\|\mathbf{R}_{mp}-\mathbf{R}_{ep}\|} \quad (3)$$

$$\mathbf{B} = \mathbf{V}_{mp}+\mathbf{R}_{mp}\times\mathbf{w}_e \quad (4)$$

where $\mathbf{w}_e$ is the angular velocity vector of Earth's rotation, $\mathbf{A}$ is the pointing vector of the MBS center, and $\mathbf{B}$ is the track speed caused by relative motion.

Through trigonometric formulas, Equations (2) ~ (5) can be rewritten as Equation (5):

$$f_{dc} = -\frac{2}{\lambda}\|\mathbf{A}\|\cdot\|\mathbf{B}\|\cdot\cos\left(\frac{\pi}{2}-\theta\right) \quad (5)$$

where θ is called the equivalent squint angle between $\mathbf{A}$ and $\mathbf{B}$. If parameter θ is reduced to zero, the frequency of the Doppler centroid can be minimized to zero.

The above equations explain the frequency of the Doppler centroid from MBS pointing and track speed, but does not reflect the direct influence of MBS motion and Earth's rotation. The Doppler centroid in Equation (2) can also be rewritten as:

$$f_{dc} = -\frac{2\cdot(\mathbf{R}_{mp}-\mathbf{R}_{ep})\cdot\mathbf{V}_{mp}}{\lambda\cdot\|\mathbf{R}_{mp}-\mathbf{R}_{ep}\|} - \frac{2\cdot(\mathbf{R}_{mp}-\mathbf{R}_{ep})\cdot(\mathbf{R}_{mp}\times\mathbf{w}_e)}{\lambda\cdot\|\mathbf{R}_{mp}-\mathbf{R}_{ep}\|} = -\frac{2}{\lambda}\cdot\left(\frac{\langle\mathbf{R}_{me},\mathbf{V}_{mp}\rangle}{\|\mathbf{R}_{me}\|}+\frac{\langle\mathbf{w}_e,\mathbf{R}_{mp}\times\mathbf{R}_{ep}\rangle}{\|\mathbf{R}_{me}\|}\right) \tag{6}$$

where $\mathbf{R}_{me}$ is the vector from the EGT position to MBS position. The first term on the right side of Equation (6) shows the Doppler centroid caused by the motion of MBS, which can be represented as f_{dc}^{m}. The second term on the right side shows the Doppler centroid caused by Earth's rotation, which can be represented as f_{dc}^{e}.

As Earth's rotation and the Moon's revolution are not equal, the specific ground target is always changing and should be renewed. The unit direction of $\mathbf{R}_{me}$ is equal to $\mathbf{A}$, shown in Equation (3) in the Antenna Coordinate System (ACS). The Z-axis in ACS is parallel to the vector from MBS to Earth's centroid, the Y-axis is parallel to the angular momentum of MBS, and the X-axis obeys the right hand rule together with the Y-axis and Z-axis.

2.2 Attitude steering method of MBS

Parameters $\mathbf{A}$ and $\mathbf{B}$ in Equation (3) and (4) are in ACS. If Doppler centroid f_{dc} must be reduced to zero, the track speed should be perpendicular to the direction of the beam center. In the initial orientation of the SAR antenna in ACS, parameter $\mathbf{A}$ can be expressed as:

$$\mathbf{A} = k\cdot\sin\varphi_1\cdot\cos\varphi_2\cdot\mathbf{u}_x + k\cdot\sin\varphi_1\cdot\sin\varphi_2\cdot\mathbf{u}_y + \cos\varphi_1\cdot\mathbf{u}_z \tag{7}$$

In the analysis of this paper, φ_2 is always set to an angle that makes the MBS left-looking (φ_2=90°) or right-looking (φ_2=−90°). If MBS is left-looking, $k=1$, and if MBS is right-looking, $\varphi_2=-1$. The vectors of the X-axis, Y-axis, and Z-axis in ACS are $\mathbf{u}_x$, $\mathbf{u}_y$, and $\mathbf{u}_z$. If parameter $\mathbf{A}$ is perpendicular to $\mathbf{B}$, then the Doppler centroid will be reduced to zero, which meets the purpose of antenna attitude steering. Generally speaking, antenna attitude steering has three methods: the first is no attitude steering, the second is one-dimensional attitude steering, and the third is two-dimensional attitude steering. They consist of three basic methods: roll and then yaw steered, roll and then pitch steered, and yaw and then pitch steered. The calculation of attitude steering angles of the three methods are basically the same.

Here, the pitch steer then yaw steer method is introduced. In this method, the rotation matrix is expressed as:

$$\Xi = R_y(\varpi_2)\cdot R_z(\varpi_1) = \begin{bmatrix}\cos\varpi_2 & 0 & -\sin\varpi_2\\ 0 & 1 & 0\\ \sin\varpi_2 & 0 & \cos\varpi_2\end{bmatrix}\cdot\begin{bmatrix}\cos\varpi_1 & \sin\varpi_1 & 0\\ -\sin\varpi_1 & \cos\varpi_1 & 0\\ 0 & 0 & 1\end{bmatrix} \tag{8}$$

where ϖ_1 and ϖ_2 are the angle of pitch steer and yaw steer, respectively.

After attitude steering, the unit axis vectors in ACS will be changed into the parts below:

$$\begin{aligned}\mathbf{u}'_x &= \cos\varpi_2\cdot\cos\varpi_1\cdot\mathbf{u}_x - \sin\varpi_1\cdot\mathbf{u}_y + \sin\varpi_2\cdot\cos\varpi_1\cdot\mathbf{u}_z\\ \mathbf{u}'_y &= \cos\varpi_2\cdot\sin\varpi_1\cdot\mathbf{u}_x + \cos\varpi_1\cdot\mathbf{u}_y + \sin\varpi_2\cdot\sin\varpi_1\cdot\mathbf{u}_z\\ \mathbf{u}'_z &= -\sin\varpi_2\cdot\mathbf{u}_x + \cos\varpi_2\cdot\mathbf{u}_z\end{aligned} \tag{9}$$

For left-looking or right-looking MBS, parameter A in Equation (8) after attitude steering in ACS can be rewritten as Equation (11):

$$\begin{aligned}\mathbf{A}' &= k\cdot\sin\varphi_1\cdot\mathbf{u}'_y+\cos\varphi_1\cdot\mathbf{u}'_z\\ &= k\cdot\sin\varphi_1\cdot(\cos\varpi_2\cdot\sin\varpi_1\cdot\mathbf{u}_x+\cos\varpi_1\cdot\mathbf{u}_y+\sin\varpi_2\cdot\sin\varpi_1\cdot\mathbf{u}_z)+\mathbf{cos}\varphi_1\cdot(-\sin\varpi_2\cdot\mathbf{u}_x+\mathbf{cos}\varpi_2\cdot\mathbf{u}_z)\\ &= \begin{bmatrix} k\cdot\sin\varphi_1\cdot\mathbf{cos}\varpi_2\cdot\sin\varpi_1-\cos\varphi_1\cdot\sin\varpi_2\\ k\cdot\sin\varphi_1\cdot\cos\varpi_1\\ k\cdot\mathbf{sin}\varphi_1\cdot\sin\varpi_2\cdot\sin\varpi_1+\cos\varphi_1\cdot\cos\varpi_2\end{bmatrix}^T\cdot\begin{bmatrix}\mathbf{u}_x\\ \mathbf{u}_y\\ \mathbf{u}_z\end{bmatrix}\end{aligned} \tag{10}$$

where the upper right subscript 'T' indicates the transpose of a vector. The track speed in ACS is $\mathbf{B}=\xi_x\cdot\mathbf{u}_x+\xi_y\cdot\mathbf{u}_y+\xi_z\cdot\mathbf{u}_z$, and should be perpendicular to the changed unit axis vectors $\mathbf{u}'_y$ and $\mathbf{u}'_z$. That is to say, the Doppler centroid result should be subject to the equations below:

$$\begin{aligned}\mathbf{B}\cdot\mathbf{u}'_y&=0\\ \mathbf{B}\cdot\mathbf{u}'_z&=0\end{aligned} \tag{11}$$

Therefore, the variable equations are changed again as:

$$\begin{cases}\xi_x\cdot\cos\varpi_2\cdot\sin\varpi_1+\xi_y\cdot\cos\varpi_1+\xi_z\cdot\sin\varpi_2\cdot\sin\varpi_1=0\\ -\xi_x\cdot\sin\varpi_2+\xi_z\cdot\cos\varpi_2=0\end{cases} \tag{12}$$

The results are altered again as:

$$\begin{cases}\varpi_1=-\arctan\left(\dfrac{\xi_y}{\xi_x\cdot\cos\varpi_2+\xi_z\cdot\sin\varpi_2}\right)\\ \varpi_2=\arctan\left(\dfrac{\xi_z}{\xi_x}\right)\end{cases} \tag{13}$$

The calculation of rotation angle for the other two steering methods is similar to the equations above.

In the no attitude steering method, the attitude of the antenna will always be consistent, as in ACS. In the one-dimensional attitude steering method, only the first rotation angle of the corresponding two-dimensional attitude steering method is introduced.

2.3 Probability distribution of attitude steering for the near side of the Moon

Theoretically, MBS can be distributed anywhere on the near side of the Moon and as long as the Doppler centroid has at least one intersection on Earth's surface, the intersections on Earth's surface can be called Intersections Perpendicular to Track Speed (ISPT). However, there are reasons why there might not be such an intersection.

The ISPT of MBS in some regions do not exist because of occlusion from the Moon's curvature if we consider the Moon a smooth sphere with an average radius of 1,738 km. In addition to its curvature, the complex topography of the Moon will also cause occlusion on the orientation of MBS after attitude steering, but the influence of lunar topography will be ignored for a simplified calculation in this paper.

While analyzing whether ISPT exists or not, we can devise three situations based on the observable Earth disc. The first situation is when the entire Earth disc can be observed by MBS, the second when part of the Earth disc can be observed by MBS, and the third when no Earth disc can be observed by MBS.

For the region of the Moon where the entire Earth disc can be observed, ISPT exists if Equation (14) is subjected, and the Doppler centroid can be effectively reduced to zero:

$$\alpha <= \frac{\pi}{2} - \beta, \left|\phi - \frac{\pi}{2}\right| \leq \beta \tag{14}$$

where α is the angle between the normal vector of the MBS ground plane and the vector from MBS to Earth's centroid; β is the angle between the edge of the Earth disc and the vector from MBS to Earth's centroid; and ϕ is the angle between the track speed and the vector from MBS to Earth's centroid. The three angles are shown in Figure 1(b).

In Equation (14), formula $\alpha <= \frac{\pi}{2} - \beta$ indicates that the entire Earth disc, about 1.8°, can be observed. Formula $\left|\phi - \frac{\pi}{2}\right| \leq \beta$ indicates MBS always has a zero Doppler line on Earth's surface, and its corresponding frequency of the Doppler centroid can be minimized to zero after attitude steering.

For the region of the Moon where only part of the Earth disc can be observed, there also exists ISPT if Equation (15) and (16) are subjected, and the Doppler centroid can be also effectively reduced to zero.

$$\alpha > \frac{\pi}{2} - \beta \; and \; \alpha <= \frac{\pi}{2} + \beta \tag{15}$$

$$\left|\phi - \frac{\pi}{2}\right| \leq \beta \tag{16}$$

Equation (15) indicates only part of the Earth disc can be observed because occlusion of the Moon's curvature caused by the MBS location far away from the center of the near side of the Moon. Equation (16) indicates MBS has a zero Doppler line on Earth's surface after attitude steering.

For the region of the Moon where no Earth disc can be observed, there is no ISPT, which is to say, no intersection on Earth can make the vector from the MBS center to the intersection perpendicular to the track speed of MBS. We can obtain the distribution of ISPT for all sites on the near side of the Moon, and for Earth observation from MBS, it is better to establish the MBS someplace where more ISPT exists.

3. Results

3.1 Distribution of ISPT for the near side of the Moon

Different sites distributed around the near side of Moon have different rotation and revolution speeds. After unifying the two speeds into the same coordinate system (the ECR) and superimposing them, we noticed that the angle between the vector from the lunar site to the Earth centroid and its corresponding track speed fluctuated slightly around 90°. Table 1 shows maximum, average, and minimum values of the angle between the track speed and the vector from MBS (on the near side of the Moon) to the Earth centroid at different latitude and longitude lines in a sidereal year. In the annual motion of the Moon around Earth, the maximum and minimum values of the angles at different latitude and longitude lines fluctuated slightly around 90°. In this time period, the average apogee of the Moon's orbit was about 4.05×10^5 km, the average perigee of the Moon's orbit was about 3.63×10^5, the average distance from the Moon centroid to the Earth centroid was about 3.8×10^5 km, and the angle between the edge of the Earth disc and the vector from the MBS center to Earth's center fluctuated between about 0.9° and 1° without considering the influence of the irregular ellipsoidal Earth. The maximum value of the angle was 90.02°, and the minimum was 89.98°, as shown in Figure 2. Its fluctuation range does not exceed half the FOV (e.g., 0.9°) of the entire Earth disc, so any site on the near side of the Moon is able to find ISPT of the MBS most of the time without considering the impact of Moon curvature.

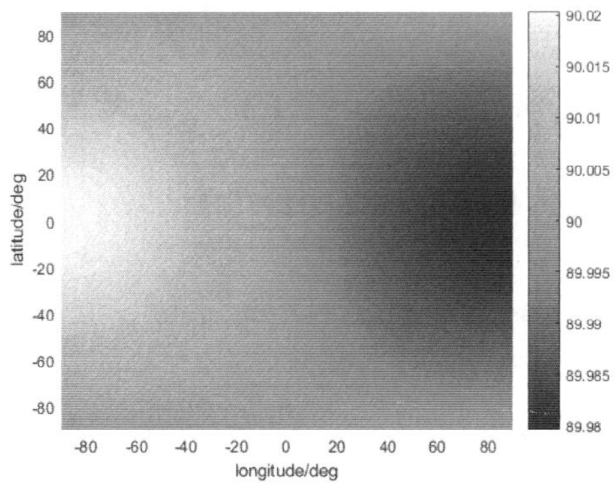

Figure 2. Average degree between the track speed and vector from MBS (distributed around the near side of the Moon) to Earth's centroid in 18.6 years

Table 1. The maximum, average, and minimum values of the angle between the track speed and the vector from MBS (on the near side of the lunar surface) to Earth's centroid at different latitude and longitude lines in a sidereal year

Latitude line	Maximum	Average	Minimum	Longitude line	Maximum	Average	Minimum
80°N	90.0032°	90°	89.9968°	80°W	90.0202°	90.0128°	90°
60°N	90.0099°	90°	89.9901°	60°W	90.0179°	90.0113°	90°
30°N	90.0175°	90°	89.9825°	30°W	90.0106°	90.0067°	90°
0°	90.0204°	90°	89.9796°	0°	90.0004°	90.0002°	90°
30°S	90.0178°	90°	89.9822°	30°E	90°	89.9937°	89.9900°
60°S	90.0105°	90°	89.9895°	60°E	90°	89.9889°	89.9825°
80°S	90.0039°	90°	89.9960°	80°E	90°	89.9873°	89.9800°

However, Moon curvature is an important factor in occlusion of the MBS line of sight after attitude steering. The average probability distribution of ISPT is shown in Figure 2. ISPT will always exist inside the area shown in Figure 3(a), where the ranges from about 80°W to 80°E and 80°S to 80°N during the Moon's nutation period is the region of interest for attitude steering (IRAS). But ISPT does not exist at all during 0.3~0.6 of the Moon's nutation period outside of IRAS, as shown in Figure 3(b). From the point of reducing the Doppler centroid to zero while considering Moon curvature, the area inside IRAS is more suitable than outside.

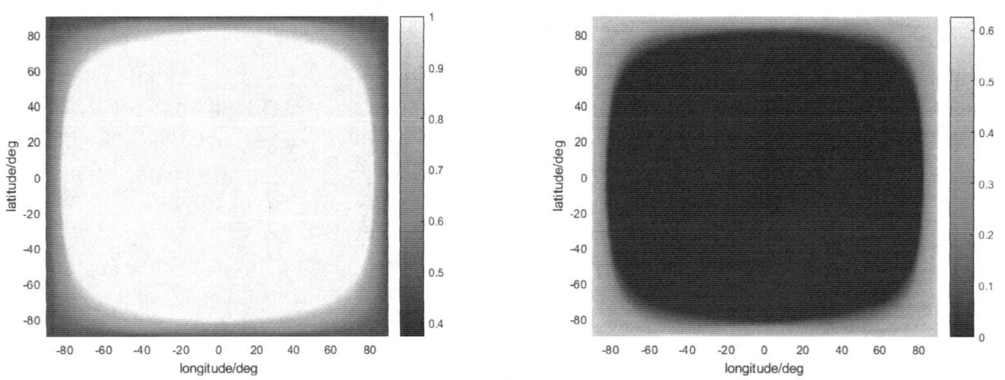

Figure 3. Average probability distribution of ISPT in all observed situations (a), and in no observed situation (b) in 18.6 years

3.2 Attitude steering of selected MBS inside IRAS

Research on the distribution of ISPT in section 3.1 will contribute to attitude steering of specific MBS. Three evenly distributed MBS inside the IRAS will be selected to study attitude steering. The first site, labeled A, is in (30°S, 30°W), B is in (30°S, 30°E), C is in (30°N, 30°W), D is in (30°N, 30°E), and E is in (0°, 0°). As the angle between the edge of Earth's disc and the vector from MBS to Earth's centroid fluctuates between about 0.9° and 1°, the angle φ_1 in Equation (7) of MBS can set a middle number 0.3°.

Under a normal ionosphere environment, the L-band has less influence on MBS imaging [20], so the carrier frequency of MBS can be set at 1.5 GHz. In section 2.2, the angle of attitude steering is mainly related to track speed of MBS when it is left-looking or right-looking. The attitude steering angles for the five evenly distributed MBS in an orbital period are shown in Figure 4.

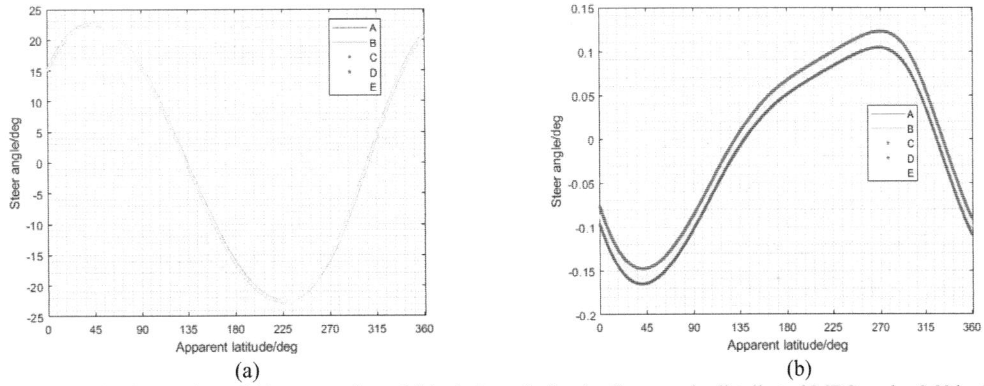

Figure 4. Attitude steering of (a) yaw angle and (b) pitch angle for the five evenly distributed MBS under 0.3° looking angle

In Figure 4, the angle change of attitude steering is continuous and the values of MBS are very similar because the rotation speed of MBS is much smaller than its revolution speed. The revolution speed of the Moon fluctuates between 960 m/s and 1,100 m/s, and the maximum rotation speed of the lunar equatorial region is only about 4.6 m/s, so the influence of Moon rotation on attitude steering is much smaller compared to the Moon's revolution. Figure 4 also indicates that the difference in MBS attitude steering is small in IRAS. This characteristic avoids huge jumps in the process of attitude steering, especially for MBS that contains huge antennas and swivel structures.

Because the angle changes of the five sites are similar, the fifth, labeled E, in (0°, 0°) can be selected to simply analyze the Doppler centroid before steering and after steering. In Figure 5(a), the MBS motion and Earth rotation all contribute to the Doppler centroid. If there were no attitude steering, such a large Doppler centroid would lower the quality of imaging formulation at a great scale, but if there were only one-dimensional attitude steering, the Doppler centroid would be minimized but not to zero, as shown in Figure 5(b), so it is necessary to introduce two-dimensional attitude steering for MBS. After two-dimensional attitude steering, the residual Doppler centroid in (0°, 0°) can be reduced to almost zero, as shown in Figure 6(a).

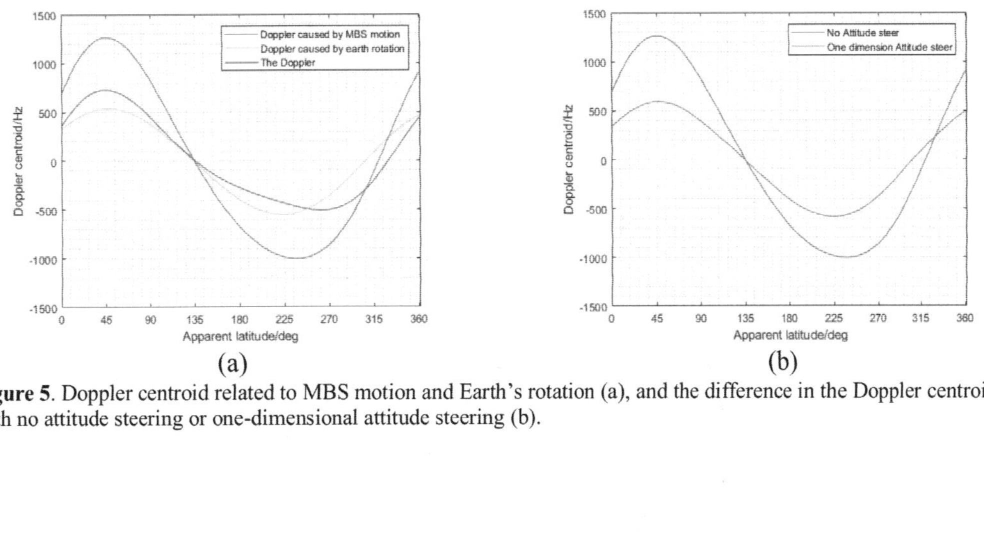

Figure 5. Doppler centroid related to MBS motion and Earth's rotation (a), and the difference in the Doppler centroid with no attitude steering or one-dimensional attitude steering (b).

IOP Conf. Series: Earth and Environmental Science 1004 (2022) 012011 doi:10.1088/1755-1315/1004/1/012011

Parameter θ in Equation (5) is the equivalent squint of MBS. Compare Figure 5(b) with Figure 6(b), though MBS's equivalent squint is far less than 1° under no or one-dimensional attitude steering, its corresponding Doppler centroid is still hundreds of hertz. The pointing accuracy is of great importance along the long distance between MBS and achievable EGT. After two-dimensional attitude steering, the Doppler centroid of MBS can be reduced to a small number very close to zero. In this case, the Doppler centroid is usually regarded as zero, and the equivalent squint angle in Equation (5) can also be reduced to zero as shown in Figure 6(b).

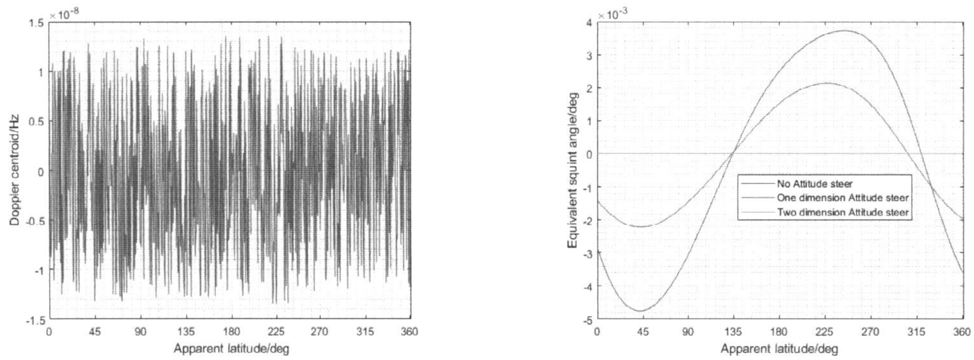

(a) (b)

Figure 6. Doppler centroid of MBS in (0°, 0°) with two-dimensional attitude steering (a) and the equivalent squint angle with no/one-dimensional/two-dimensional attitude steering (b)

4. Conclusions

Not all of the near side of Moon has intersections perpendicular to the track speed (ISPT) of Moon-based SAR (MBS), and the area outside the region of interest for attitude steering (IRAS) is not better for MBS to be established. Because of the complex motion of MBS and Earth, the Doppler centroid is not always zero, which has adverse effects on imaging. Therefore, a method of two-dimensional attitude steering should be introduced to eliminate the effect, and this paper clarifies a two-dimensional attitude steering method and compares it with no attitude steering and one-dimensional attitude steering. The results indicate no attitude steering and one-dimensional attitude steering methods cannot reduce the Doppler centroid to zero, which shows the necessity of a two-dimensional attitude steering method.

References

[1] Ding, Y.X; Guo, H.D; Liu, G.; Han, C.M; Lv, M.Y. Constructing a High-Accuracy Geometric Model for Moon-Based Earth Observation [J]. Remote Sensing. 2019, 11, 2611

[2] Guo, H.D.; Liu, G.; Ding, Y.X. Moon-based Earth observation: scientific concept and potential applications [J]. International Journal of Digital Earth. 2018, 11, 546–557.

[3] G. Fornaro, G. Franceschetti, F. Lombardini, A. Mori, and M. Calamia, Potentials and limitations of Moon-borne SAR imaging [J], IEEE Trans. Geosci. Remote Sens., 2010, 48, 3009-3019.

[4] Xu, W.B.; Zhao, H.B. Deep space exploration of asteroids: the science perspectives [J], Journal of deep exploration, 2005, 11, 31-38.

[5] Shen, G.Z.; Guo, H.D; Liu, G.; Zhang, L.; Huang, J. Geometry Numerical Simulation and Analysis for Moon-Based Earth Observation. IEEE Journal of Selected Topics in Applied Earth Observations and Remote Sensing. 2020, 13, 3381–3393.

[6] Huang, S.P.; Liao, J.J.; Guang, J.; Wu, J.T.; Bi, S.S.; Tang, X.Y.; Kuang, J.J. Towards Moon-based monitoring of energy budget of the Earth climate system. In Proceedings of the IGARSS 2016—2016 IEEE International Geoscience and Remote Sensing. Symposium, Beijing, China, 10–15 July 2016.

[7] Yuan, X.K. Calculation of Doppler Frequency and Total Zero Doppler Centroid Steering for

Space-Borne SAR [J]. Aerospace Shanghai, 2010, 27, 1-8.
[8] Sui, Y.; Guo, H.D; Liu, G.; Ren, Y.Z, Analysis of Long-Term Moon-Based Observation Characteristics for Arctic and Antarctic [J]. *Remote Sensing.* 2019, 11, 2805.
[9] Yao, M. J.; Chen, J. P.; Xu, B.; Ji, J. Z.; Chen, J. Difference Comparison and Evolution Analysis of Lunar Nearside and Farside [J]. Journal of Earth Science and Environment. 2017, 39, 428-438
[10] Xie, H.Y; Zhao, H.Z; Fu, Q.; Tang, G.F. A Doppler parameters estimation technique for squint SAR imaging [C]. 2008 International Conference on Radar. IEEE, 2008.
[11] Moussessian A , Chen C , Edelstein W , et al. System Concepts and Technologies for High Orbit SAR (Invited)[C]. Microwave Symposium Digest, 2005 IEEE MTT-S International. IEEE, 2005.
[12] Xie, H.Y; Zhao, H.Z; Fu, Q.; Tang, G.F. A Doppler parameters estimation technique for squint SAR imaging [C]. 2008 International Conference on Radar. IEEE, 2008.
[13] Li, L.; Hong, J.; Ming, F.; Hu, J, W. Study on Ionospheric Effects Induced by Spatio-temporal Variability on Medium-earth-orbit SAR Imaging Quality [J]. Journal of Electronics & Information Technology, 2014, 36(04), 915-922.
[14] Folkner, W. M.; Williams, J. G.; Boggs, D. H.; et al. The Planetary and Lunar Ephemerides DE430 and DE431 [J]. Interplanetary Network Progress Report, 2014.
[15] Petit. G.; Luzum B. IERS Conventions (2010) [J]. IERS technical note, 2010.
[16] Yao, M. J. Eruption Cycle Division and Evolution Analysis of Lunar Basalts [M]. China University of Geosciences，Beijing, 2020
[17] Wang, J.; Zhou, C, H.; Cheng, W, M. The Spatial Pattern of Lunar Craters on a Global Scale[J]. Geomatics and Information Science of Wuhan University 2017, 42(04):512-519.
[18] Xi, X. N. Orbit Design of Lunar Probe, National Defense Industry Press, 2001
[19] Meeus, J. Astronomical algorithms [J]. Willmann-Bell, Incorporated, 1991.
[20] Iwashita, A.; Baba, H.; Morohoshi, T.; et al. Study of Disaster Information Extraction using C/L-Band SAR Interferometry [M]. IEEE Computer Society, 1970.

Acknowledgments

Thanks the IAU SOFA Collection were used. Copyright © International Astronomical Union Standards of Fundamental Astronomy (http://www.iausofa.org). We also thank the JPL for the free access to the DE430 and the IERS for the free access to EOP data.

IOP Conf. Series: Earth and Environmental Science 1004 (2022) 012012 doi:10.1088/1755-1315/1004/1/012012

Travel-mode classification based on GPS-trajectory data and geographic information using an XGBoost classifier

Huiling Jin[1,*], Hangbin Wu[1,2], Zeran Xu[1], Wei Huang[1,3] and Chun Liu[1,2]

1. College of Surveying and Geo-informatics, Tongji University, Shanghai 200092, China
2. Urban Mobility Institute, Tongji University, Shanghai 201804, China
3. Department of Civil Engineering, Ryerson University, 350 Victoria Street, Toronto, ON M5B 2K3, Canada

E-mail addresses: 1931994@tongji.edu.cn (Huiling Jin)

Abstract. Massive Global Positioning System (GPS) trajectory datasets are being produced owing to the advances in mobile sensors, the Internet, and GPS devices. Accurately inferring travel modes from GPS trajectory data can be helpful in transportation planning and modeling, infrastructure design, etc. However, adverse factors such as data noise, differences in sampling rate, and inadequate features have a negative impact on the results of travel mode classification. In this paper, to address such issues, we first propose a preprocessing workflow, which includes data cleaning, segmentation, and resampling, to preprocess raw trajectories. Then, we add new features related to the road and bus stop information for travel mode classification using an XGBoost (eXtreme Gradient Boosting) classifier, along with various basic features of the trajectories. We conducted a set of experiments on the GeoLife dataset using a group of state-of-the-art methods. The results showed that the proposed methods can improve the classification accuracy by using all the classifiers we compared and the classification accuracy using the XGBoost classifier can reach a maximum of 90.41%.

1. Introduction

Identifying travel modes is an important task in the human behavior research field and is valuable for both individuals and cities [1]. For individuals, the information mined from travel mode classification can reflect on their activity patterns, which can be useful for daily life and work [2]. For cities, accurate travel mode data are of great value in traffic flow prediction [3]. The traditional methods of acquiring travel mode data are mainly through interviews or questionnaires, which are inconvenient, time consuming, and expensive [4]. Recently, ubiquitous computing and advanced Global Positioning System (GPS) devices have enabled the use of GPS datasets as a new way to study travel modes.

Common travel mode classification approaches involve two main steps. First, we extract features or motion attributes for each GPS trajectory and then feed them into the classifiers. Researchers have extracted the statistical values of motion attributes (speed, acceleration direction change, etc.) as features [5,6]. Xiao et al. [7] generated some discriminative global and local features, where a tree-based ensemble method was used to infer the travel modes solely from GPS data, and their work was proved to outperform other traditional methods. A convolutional neural network (CNN) was used in [8,9] to infer travel modes on the basis solely of raw GPS trajectories and achieved a high accuracy. Li et al. [10] used generative adversarial networks (GANs) to create some samples to balance the dataset and then applied a CNN to classify the travel modes.

Trajectory data are a sampling sequence with position and time information, whose quality is affected by the recording equipment and the surrounding environment. The GPS trajectory data from different sources have different scales, which are reflected in the difference in both the sampling rates and the noise levels of the trajectory points. In terms of the noise of GPS data, most studies simply removed GPS points with an abnormal velocity and/or acceleration in the samples within a certain and realistic range of its transportation mode [8,9]. Etemad et al. [11] proposed a method to deal with noise, proving that the noise-removal task affects the performance of the classification models tested. For unlabeled trajectories, owing to a lack of knowledge of transport modes, a rule was designed in [12] to remove the possible outliers according to the statistics of the velocity and acceleration of the trajectory segment. However, GPS trajectory data from different sources have different sampling rates, which can affect the characteristic expression of these data. The higher the sampling rate, the richer the information contained in the data, and the more accurate the features expressed. Through trajectory cleaning, the attributes derived from the trajectories tend to represent the common characteristics of multiscale trajectories more promisingly.

In travel mode recognition, different types of motorized travel modes such as cars and buses are more difficult to recognize; therefore, the features based solely on GPS data may not be sufficient to accurately distinguish similar travel modes. Socioeconomic attribute data were used in [13] to improve the prediction of e-bikes and solve the confusion between bus and car modes. Nawaz et al. [14] proposed a convolutional long short-term memory model to learn traffic patterns in combination with weather datasets, which improved the accuracy of the model. Although the use of additional data can improve the accuracy of pattern recognition, such external data are not always available, whereas open geographic information is relatively easy to obtain.

To address the abovementioned shortcomings, we propose a preprocessing method that is not based on prior knowledge of travel patterns to reduce the impact of GPS data on classification and then use an XGBoost (eXtreme Gradient Boosting) classifier to classify different travel modes on the basis of GPS data and geographic information accessible from open data portals and map application program interfaces (e.g., OpenStreetMap). The main contributions of this study are summarized as follows:

(1) A preprocessing method, which includes data cleaning, segmentation, and resampling, is proposed to preprocess raw trajectories, which can improve the quality of the domain expert features and the classification accuracy.

(2) To further improve the classification accuracy, we propose new features from open geographic information to further distinguish between buses and cars.

(3) A group of widely used classification models was applied to prove that our methods can improve the classification accuracy in each of the compared classification models, which further proves the superiority of the method proposed in this paper.

2. Data

The GPS trajectory dataset used in this study was obtained from the GeoLife project of Microsoft Research Asia. The dataset was collected by 182 users from April 2007 to August 2012, and the majority of the data were generated in Beijing, China [1]. Of the 182 users, 73 labeled their trajectories with travel modes, including airplane, walk, bike, bus, taxi, car, subway, boat, run and motorcycle. These trajectories were recorded by different GPS loggers and GPS-enabled phones and had a variety of sampling rates; 91.5% of the trajectories were recorded in a dense representation (e.g., every 1-5 s or 5-10 m per point). We only considered the trajectories of the common travel modes: walk, bike, bus and car (taxi and car) in the Beijing area. In addition to the GPS trajectories, road network information and bus stop information for Beijing were obtained from OpenStreetMap (https://www.openstreetmap.org) for the study. The road network information included vector files for all levels of roads, and we only considered the main road. The main road includes motorways, primary roads, secondary roads, tertiary roads, trunk roads, and cycleways.

IOP Conf. Series: Earth and Environmental Science 1004 (2022) 012012 doi:10.1088/1755-1315/1004/1/012012

3. Methodology

3.1 Trajectory preprocessing

3.1.1 Trajectory cleaning
In the process of data collection, during the warm start/cold start or in "urban canyons," the GPS signal is most likely to be lost or degraded, leading to a significant fluctuation in the collected data [15]. Thus, the collected data may have a blank area or drift for a short period such as when the user momentarily stops because of traffic congestion, stops by a station, or moves to a place surrounded by high-rise buildings.

To perform trajectory cleaning, first, we remove the stop segments. The specific steps are described as follows:

(1) Use the velocity (0.8 m/s) to distinguish the possible stop points from the non-stop points.

(2) If the length of a segment is composed of consecutive stop points or non-stop points of less than three points, merge the segment into its backward segment.

(3) If the distance between the start and end points of the stop segment is less than the distance that the user can travel during a certain period (the sampling rate of this trajectory) and the average heading change of the stop segment is more than 50°, consider it as a certain stop segment.

(4) Remove all the certain stop segments.

Second, after removing the stop segments, we filter out the points with significant changes in orientation. The specific steps are described as follows.

(1) From the starting point, calculate the heading change every three consecutive points; if the heading change is more than 130° at a distance of 150 m, remove the three points.

(2) Repeat (1) with the filtered data until no points satisfy this condition; then, the loop is broken.

Finally, we apply the Savitzky-Golay filter to smooth all trajectories and set the window size and the polynomial order to 9 and 3, respectively.

3.1.2 Trajectory segmentation
One GPS trajectory sample usually contains multiple trips and travel modes because users cannot label every single travel mode accurately. The most likely labeling error is the inaccurate marking of the transition point of the traffic patterns, and walking is often a transition between different modes, resulting in a mix of walk segments in other non-walk modes. After the trajectory segmentation, the walk and non-walk segments can be distinguished effectively and a trajectory segment that contains only the bus mode can be relabeled. Therefore, the change point-based segmentation method proposed in [1] is used to remove the possible walking segments from the non-walking patterns, ensuring the labeling accuracy of the sample data. Ultimately, every trip contains a single travel mode.

3.1.3 Trajectory resampling
The trajectories have various sampling rates owing to the difference in GPS devices and users' preferences, and this difference has an effect on the characteristic expression. Thus, to maintain the consistency of the trajectory data, we propose a method to standardize the sampling rate of the trajectory. We dilute the dense trajectory points and use the cubic spline interpolation method to reconstruct relatively sparse trajectory points that can be restored to the original trajectory relatively well, and the sampling rate of a sparse trajectory can be standardized to 2 s after the trajectory resampling.

3.2 Feature extraction
The features used for the classification of travel modes are categorized into 39 basic features and 19 new features. The basic features refer to the features associated with the GPS trajectory data only, whereas the new features are the features related to the geographic information.

3.2.1. Extraction of the basic features

First, four motion attributes (i.e., speed, acceleration, jerk, and heading change) are calculated. Then, the trajectory features are calculated using a statistical method. The mean, percentiles (10th, 30th, 50th, 70th, and 95th), standard deviation, skewness, and kurtosis of the four abovementioned motion attributes are calculated. Then, three trajectory features, i.e., the heading change rate (*HCR*), stop rate (*SR*), and velocity change rate (*VCR*), proposed in [16], are used.

The *HCR* reflects the frequency of the change in direction, and it is defined as equation (1):

$$HCR = P_c/N \tag{1}$$

where P_c is the number of points at which the user changes the direction exceeding a certain threshold in a trip and *N* is the number of all points in a trip. The heading change threshold is set to 19°.

SR and *VCR* reflect the characteristics of the velocity when people take different travel modes. *SR* is defined as equation (2):

$$SR = P_s/N \tag{2}$$

where P_s is the number of GPS points with a velocity below a certain threshold in a trip. The threshold in *SR* is set to 1 m/s.

VCR is defined as equation (3), and the threshold in VCR is set to 0.1.

$$VCR = P_v/N \tag{3}$$

where P_v is the number of GPS points whose $VRate$ exceeds a certain threshold during a trip. $VRate$ is defined as equation (4):

$$VRate = |V_2 - V_1|/V_1 \tag{4}$$

3.2.2. Extraction of the new features

New features are proposed to better describe the relationship between the user and the road environment by taking different travel modes. The Euclidean distance (*Dr*) of the trajectory point to the nearest road is calculated; $Dr_i(i = 1, ..., n)$ is first calculated for a trajectory point sample; the mean, percentiles (10th, 30th, 50th, 70th, and 95th), standard deviation, skewness, and kurtosis of the *Dr* value are subsequently calculated.

To realize a better identification of bus trajectories, we integrate bus stop-related features to obtain the characteristics related to bus geographic information. For each trajectory point, the Euclidean distance (*Ds*) to the nearest bus stop is calculated. $Ds_i(i = 1, ..., n)$ is first calculated for a trajectory point sample, and then the mean, percentiles (10th, 30th, 50th, 70th, and 95th), standard deviation, skewness, and kurtosis of the *Ds* value are calculated. Buses have the feature of stopping at a bus stop; therefore, buses are more likely to stop within the buffer zone of a certain distance of the bus stop than other travel modes. The stop rate at the bus stop (SR_{bus}) of a trajectory is calculated using equation (5), which reflects the characteristics of the buses stopping at the bus station.

$$SR_{bus} = \frac{P_b}{P_s} \tag{5}$$

where P_b is the number of GPS points whose velocity is below a certain threshold and the *Ds* value is below a certain threshold during a trip. P_s is the number of GPS points whose velocity is below a certain threshold during a trip. The velocity threshold in the SR_{bus} is 1 m/s, and the *Ds* value threshold is set to 50 m.

4. Results

XGBoost was used to separate the different travel modes. After the preprocessing, the trajectories were divided into trips if the time interval between two consecutive points exceeded 10 min. Then, the trips were converted into fixed-size segments, with the number of GPS points for each segment being equal to 200 to enlarge the sample size. Finally, all samples were randomly divided into a training set and a testing set at a ratio of 4:1. To minimize the effects of unreasonable sample division on the model evaluation, we divided the dataset randomly for 50 times and calculated the average evaluation indicators of the model. The evaluation indicators used in this study included accuracy, precision, recall, F-score, and confusion matrix.

4.1 Comparison of the detection models using raw data and the preprocessed data based on the basic features

In Experiment 1, the raw GPS data and the basic features were used, and the confusion matrix of Experiment 1 is shown in Table 1. Experiment 2 was carried out for comparison with Experiment 1 by using the preprocessed data and the basic features, and the classification results are shown in Table 2. It can be seen clearly that the F-score of each category improved and the overall accuracy increased from 86.45% to 88.87% by performing data preprocessing. In particular, the error of misclassification of the other classes to the walk class decreased, the precision increased by ~7% for walking, and the recall increased by ~4.5%, ~3.5%, and ~2% for the bikes, buses, and cars, respectively. In summary, the results indicate that the classification accuracy improved by preprocessing the raw data.

Table 1. The confusion matrix of Experiment 1

Actual	**Predicted**							
	Walk	*Bike*	*Bus*	*Car*	*Sum*	*Precision (%)*	*Recall (%)*	*F-Score (%)*
Walk	799.34	0.26	0.4	0	800	85.55	99.92	92.18
Bike	51.98	431.92	22.08	4.02	510	94.64	84.69	89.39
Bus	61.2	16.44	566.34	66.02	710	84.27	79.77	81.95
Car	21.86	7.78	83.26	342.1	455	83.01	75.19	78.90
Overall accuracy=86.45%								

Table 2. The confusion matrix of Experiment 2

Actual	**Predicted**							
	Walk	*Bike*	*Bus*	*Car*	*Sum*	*Precision (%)*	*Recall (%)*	*F-Score (%)*
Walk	799.9	0.1	0	0	800	92.50	99.99	96.10
Bike	25.24	456.8	23.4	4.56	510	94.83	89.57	92.12
Bus	31.36	15.48	590.88	72.28	710	84.43	83.22	83.82
Car	8.24	9.32	85.58	351.86	455	82.08	77.33	79.63
Overall accuracy=88.87%								

4.2 Comparison of detection models using the basic features and all features based on the preprocessed data

Experiment 3 was carried out for comparison with Experiment 2 by using not only the preprocessed data but also 19 additional new features. The confusion matrix of Experiment 3 is presented in Table 3. Compared with those in Experiment 2, the F-score of each category improved and the accuracy improved from 88.87% to 90.41% with the addition of new features. Because buses often pull over at bus stops, the distance from the center line of the same direction road to the buses is more likely to be greater than the distance to the cars, and buses are more likely to stop within the buffer zone at a certain distance from the bus stop than cars. The precision of the buses increased by ~3% and the recall of the cars increased by ~4.5% with the addition of road and bus stop information, implying that the new features are effective for the identification of these two classes.

Table 3. The confusion matrix of Experiment 3

Actual	Predicted							
	Walk	*Bike*	*Bus*	*Car*	*Sum*	*Precision (%)*	*Recall (%)*	*F-Score (%)*
Walk	799.94	0.06	0	0	800	92.54	99.99	96.12
Bike	25.32	460.02	21.28	3.38	510	94.97	90.20	92.52
Bus	31.1	14.7	604.7	59.5	710	87.59	85.17	86.36
Car	8.02	9.6	64.42	372.96	455	85.57	81.97	83.73
Overall accuracy=90.41%								

4.3 Comparison results of the three experiments

The overall accuracies of Experiments 1, 2, and 3 were 86.45%, 88.87%, and 90.41%, respectively, which constantly increased. The F-score comparison results of the three experiments are presented in Figure 1. In each experiment, the F-scores of the four travel modes all improved on the basis of the previous ones. The preprocessing operation in Experiment 2 played a significant role in improving the classification accuracy of the four travel modes. Experiment 3 improved the classification accuracy of each category compared with Experiment 2 by adding new features, and it had a greater impact on buses and cars.

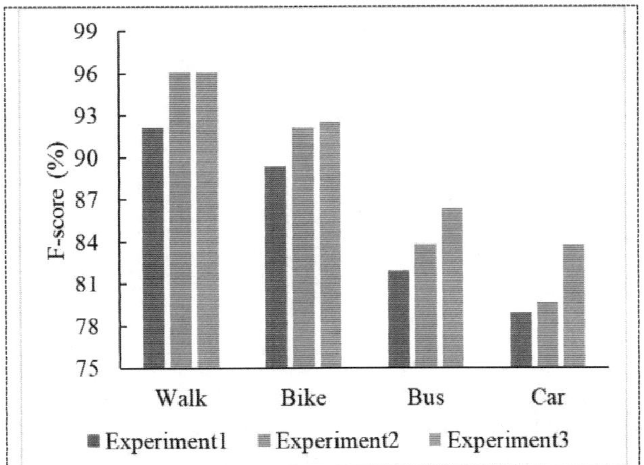

Figure 1. F-Scores of the different experiments.

4.4 Comparison between different classifiers

Several classical machine learning algorithms were used for classification using the same preprocessed data and all the features in Experiment 4. Table 4 shows that XGBoost is capable of achieving better prediction accuracy than those of the other models in each group of experiments. XGBoost achieved a prediction accuracy above 90%, outperforming the other classifiers. For all the classifiers used in Experiment 4, the classification accuracy improved by performing data preprocessing and adding new features. Therefore, the methods proposed in this study can contribute positively to travel mode classification.

Table 4. Overall accuracy (%) of the different models

Model	Raw data + basic features	Preprocessed data + basic features	Preprocessed data + all features
DT	81.65	83.46	84.61
RF	85.02	87.38	88.6
GBDT	83.25	86.19	87.84
XGBoost	86.45	88.87	90.41

5. Conclusion

In this paper, a travel mode classification model that includes a trajectory preprocessing method and a feature extraction method is proposed. First, a preprocessing method, which includes trajectory cleaning, segmentation, and resampling, is proposed to preprocess the raw trajectories and optimize the sample quality. Then, the new features related to road and bus stop information are added to improve the prediction accuracy for buses and cars.

An XGBoost classifier was used for the experiments, which considered the trajectories of the common travel modes: walk, bike, bus and car, the four types of GPS trajectories from the Microsoft GeoLife dataset. The results showed that the classification accuracy was 86.45% without trajectory quality optimization and the addition of new features. After the preprocessing, the sample quality improved and the overall accuracy of the classification improved from 86.45% to 88.87%. With the addition of the new features, the overall accuracy improved from 88.87% to 90.41%, with the precision of buses improving by ～3% and the recall of cars improving by ～4.5%. The comparison with other classifiers performed in this study showed that XGBoost achieved the best accuracy. In the future, more accurate information can be mined using GPS data with better positioning accuracy and a higher sampling rate.

Reference

[1] Zheng Y, Liu L, Wang L, et al. 2008 Learning transportation mode from raw GPS data for geographic applications on the web. International World Wide Web Conference Committee (IW3C2)
[2] Zheng Y and Xie X 2011 Learning travel recommendations from user-generated GPS traces. ACM Transactions on Intelligent Systems and Technology, 2(1)
[3] Duan Z, Yang Y, Zhang K, Ni Y and Bajgain S 2018 Improved deep hybrid networks for urban traffic flow prediction using trajectory data. IEEE Access, 6 31820-7
[4] Xiao G, Juan Z and Zhang C 2015 Travel mode detection based on GPS track data and Bayesian networks. Computers, Environment and Urban Systems, 54 14-22
[5] Feng T and Timmermans H J P 2016 Comparison of advanced imputation algorithms for detection of transportation mode and activity episode using GPS data. Transportation Planning and Technology, 39 180-94
[6] Wang B, Gao L and Juan Z 2018 Travel mode detection using GPS data and socioeconomic attributes based on a random forest classifier. IEEE Transactions on Intelligent Transportation Systems, 19 1547-58
[7] Xiao Z, Wang Y, Fu K and Wu F 2017 Identifying different transportation modes from trajectory data using tree-based ensemble classifiers. ISPRS International Journal of Geo-Information, 6 57
[8] Karatzoglou A, Schnell N and Beigl M 2018 A convolutional neural network approach for modeling semantic trajectories and predicting future locations. Artificial Neural Networks and Machine Learning, 61-72

[9] Dabiri S and Heaslip K 2018 Inferring transportation modes from GPS trajectories using a convolutional neural network. Transportation Research Part C: Emerging Technologies, 86 360-71

[10] Li L, Zhu J, Zhang H, Tan H, Du B and Ran B 2020 Coupled application of generative adversarial networks and conventional neural networks for travel mode detection using GPS data. Transportation Research Part A: Policy and Practice, 136 282-92

[11] Etemad M, Soares Júnior A and Matwin S 2018 Predicting transportation modes of GPS trajectories using feature engineering and noise removal. Canadian Conference on Artificial Intelligence, 259-64

[12] Dabiri S, Lu C, Heaslip K and Reddy C K 2020 Semi-supervised deep learning approach for transportation mode identification using GPS trajectory data. IEEE Transactions on Knowledge and Data Engineering, 32 1010-23

[13] Wang B, Gao L and Juan Z 2017 Travel mode detection using GPS data and socioeconomic attributes based on a random forest classifier. IEEE Transactions on Intelligent Transportation Systems, 19 1547-58

[14] Nawaz A, Huang Z, Wang S, Hussain Y and Khan Z 2020 Convolutional LSTM based transportation mode learning from raw GPS trajectories. IET Intelligent Transport Systems, 14 570-7

[15] Gong H, Chen C, Bialostozky E and Lawson C T 2012 A GPS/GIS method for travel mode detection in New York City. Computers, Environment and Urban Systems, 36 131-9

[16] Zheng Y 2008 Understanding mobility based on GPS data. Proc. 10th International Conference on Ubiquitous Computing

Acknowledgment

We would like to express our appreciation to those who provided suggestions and comments for this paper.

IOP Conf. Series: Earth and Environmental Science 1004 (2022) 012013 doi:10.1088/1755-1315/1004/1/012013

Multi-view imaging simulation of a space born TDI CCD camera and a 3D geometric reconstruction estimate

Yue Qingxing
Baishengcun No1, Baisheng village, Zizhuyuan, Haidian District (Beijing, 100048)
627105426@qq.com

Abstract. This paper proposes a multi-view imaging simulation method for a space-born TDI CCD camera, focusing on the integral time and satellite attitude calculation method of a TDI CCD camera when the imaging target is far away from the nadir point. The accuracy of the calculation method was verified by a target imaging simulation. Furthermore, a multiview stereo imaging simulation of a single camera was realized. Based on the high-precision three-dimensional (3D) building model and the simulation method, the simulation image was generated under various side swing angles and 3D angles. The generated building point cloud data was based on the belief propagation and graph-cuts stereo matching methods. Finally, we evaluated the impact of the side-view angle and stereo angle on the 3D reconstruction effects of the overall area and building façade based on the effective point percentage, root mean square, NMAD, and LE90.

1. Introduction

Presently, Airbus's Pleiades satellite and Maxar's worldview series satellites have flexible multiview stereo image acquisition capabilities, and can automatically mass-produce three-dimensional (3D) data with the help of advanced data processing technology. Their pixel factory systems have had mature satellite 3D modeling technology since 2016, when Maxar launched the satellite 3D product vricon3d. The vricon series products offer unique technology that is a "combination of stereo photogrammetry and big data processing." In terms of uncontrolled geometric accuracy, the 90% spherical error (se90) of the 3D geometric model is as high as 3M, the absolute vertical error (le90) of the 3D point cloud is 3M and the relative vertical error is 1m. Moreover, Airbus and Maxar are planning their next-generation 3D optical satellite constellations, Pleiades Neo and Worldview Legion, whose resolution is approximately 30 cm. With multiple satellites and flexible and efficient imaging characteristics, as well as advanced cloud processing systems and processing technology, they are bound to take the lead in global 3D data acquisition.

Satellite 3D technology is not only a unilateral problem of satellite data processing, but also a system engineering approach closely related to satellite platforms and loads. Domestic stereo mapping satellites such as ZY-3 and TH-1 adopt a three-line-array stereo imaging mode, which has low resolution and poor flexibility, and mainly focuses on topographic mapping. Gaojing-1, Jilin-1, and other satellites have stereo imaging capability and high resolution, but their performance is still inferior to that of Pleiades, Worldview, and other advanced satellites in terms of multi-view stereo imaging quality. Today, with the lack of multi-view stereo image acquisition ability of domestic satellites, generating the required stereo image through simulation technology will undoubtedly greatly reduce the cost and speed up the progress of relevant research. Although researchers in related fields have conducted research on high-resolution satellite imaging simulation [1-7], there have been few simulation studies on stereo mapping applications. This paper presents a technical method for multi-view stereo imaging simulation of a satellite TDI CCD camera, and focuses on solving the problem of multi-level integral imaging blurs caused by image motion and velocity mismatch. Simulation images obtained under multiple side view angles and 3D included angles were generated. The point cloud data were generated based on two global optimization matching technologies [8-12] of confidence propagation (hereinafter referred to as BP) and graph cuts (hereinafter referred to as GC). All effective point cloud data and building facade point cloud data were counted and compared with a variety of accuracy indicators [13-16].

2. Satellite multi view imaging simulation

2.1. Exterior orientation element calculation

The simulated track height was 500000 m, the track inclination was 98.5 °, and the track was circular; that is, the eccentricity was 0. The input includes the following: Two ground coordinates, one is the coordinates of the sub-satellite point, and the other is the center coordinate of the target area. The first is the CCD integration time, the other is the imaging time, and the third is the time offset of the imaging time of the target center relative to the "nearest point time" which will be explained later. Camera geometric parameters, including focal length (7.5m), CCD size (7e-6m), and integral series.

Given the satellite flight trajectory and target point position O(Xo,Yo,Zo), the main step of the external orientation element calculation is obtaining the attitude rotation matrix so that the vector corresponding to the camera main point "aligns" the target point after rotation. The satellite flight trajectory is represented by the satellite position and velocity time series in the geocentric rectangular coordinate system, i.e., $P(t)=\{X_t,Y_t,Z_t\},V(t)=\{V_{X_t},V_{Y_t},V_{Z_t}\}$.

First, we determine the satellite position corresponding to the integer second t_{near} closest to the target O and the satellite position at the time $t_{near}-1$ and $t_{near}+1$, calculate the distance between the three satellite positions and the target point O, respectively, and obtain the time t_0 closest to the target point (hereinafter referred to as the nearest point time) and the corresponding satellite position $P(t_0)$ and speed $V(t_0)$ through the parabola fitting method.

The angle component between the satellite position P(t) of the current time t and the line vector of the sub-satellite point and the target point in the vertical orbit direction is the side angle, while the angle component along the orbit direction is the front and rear angle. Suppose that the line vector between the satellite and the sub-satellite point at the nearest point is v_{00}, the connection vector between the satellite and the target point at the nearest point is v_{01}, and the line vector between the satellite and the sub-satellite point at the current time is v_{10}. The connection vector between the satellite and the nearest point at the current time was v_{11}. The side view θ uses the values of v_{00} and v_{01} calculated by the inverse cosine formula, and the front and rear viewing angles ∅ are v_{10} and v_{11}. They are calculated by the inverse cosine formula.

Because the variable of analog input is the time offset of the imaging time of the target relative to the nearest point, the included angle along the track direction corresponding to the same time offset under different side view angles is slightly different. For comparison, the variables along the track direction were unified as a time offset. Table 1 shows the front and rear view angles corresponding to different time offsets when the side view angles were 21.127 °and 37.152 °. It can be observed that there is little difference between the two. In addition, when the front and rear view offset times are the same, the difference between the corresponding front and rear view angles is small, which can be approximately regarded as time symmetry. The front and rear view angles have an approximately linear relationship with the offset time in a short time. The approximate front and rear viewing angle ∅ can be obtained by multiplying the offset time by –0.7377. However, if you want the satellite to image with a certain front and rear viewing angle ∅, you only need to divide ∅ by –0.7377 to obtain the time offset. Because the real satellite attitude adjustment takes time, the shorter the adjustment time, the higher the flexibility of the representative satellite, and the easier it is to obtain an image of a specific

IOP Conf. Series: Earth and Environmental Science 1004 (2022) 012013 doi:10.1088/1755-1315/1004/1/012013

observation angle. The longer the attitude adjustment, the shorter the imaging time, and even the required image cannot be obtained.

Table 1. Forward and backward viewing angles with different time offsets.

Side view	–40"	–30"	–20"
21.127°	29.423	23.046	15.896
37.152°	29.415	23.039	15.891
21.127°	10"	20"	30"
37.152°	–8.124	–15.897	–23.047

According to the satellite position $P(t_0)$ at the nearest point and speed $V(t_0)$, the rotation matrix $R_{obt2eth}$, $P(t_0)$ from the instantaneous orbit coordinate system to the geocentric rectangular coordinate system corresponding to the sub-satellite point coordinates M(t)(X,Y,Z) can be calculated. The rotation matrix $R_{eth2obj}$ can be obtained according to the P-M and P-O vectors.

It is assumed that the camera installation matrix (such as the rotation matrix from the spatial coordinate system to the satellite body coordinate system) is R_{set}, and the rotation matrix from the satellite body coordinate system to the instantaneous orbit coordinate system is R_{ypr} (determined by the yaw angle, pitch angle, and roll angle). After the observation vector of the image space coordinate system is transferred to the object space, the rotation matrix of the "see" target point O is

$$R = R_{rth2obj} * R_{obt2eth} * R_{ypr} * R_{set} \tag{1}$$

The real satellite imaging of the target does not need to go through three attitude deflections, i.e., R_{ypr}, $R_{obt2eth}$, and $R_{rth2obj}$, to "see" the target, as $R_{rth2obj}*R_{obt2eth}*R_{ypr}$ considers all the three deflections. This paper decomposes this process into three steps, which makes understanding the related concepts easier.

The particularity of TDI CCD camera imaging is that it requires that the image movement direction is consistent with the TDI integration direction, and the charge transfer period is consistent with the integration time Δt of the single-stage CCD. This study converts these two requirements from the perspective of photogrammetric central projection to make them more applicable to the image application field.

The TDI CCD is regarded as a two-dimensional (2D) area array detector with m rows and n columns. The rows represent the integral series and columns represent the actual imaging width. Assuming that the focal length of the camera is f (unit meter), the row and column coordinates of a point P on the TDI CCD are R and C, respectively, and the row and column coordinates of the main point of the image are r0 and c0, respectively. If the size of a single CCD is pix (unit meter), then the vector U of the image space coordinate system corresponding to p is

$$U = [x, y, -f]^T \tag{2}$$

Among them, $x = (r-r_0)*pix$ and $y = (c-c_0)*pix$.

The object direction corresponding to U is

$$U' = R*U \tag{3}$$

The differences between the origin of the instantaneous orbit coordinate system (the specific position of the satellite flight trajectory at a certain time), the origin of the satellite body coordinate system, and the origin of the image space coordinate system are not considered, hereinafter referred to as the origin coordinates, and the origin coordinates at time t are P(t). The corresponding ground coordinates O(t) can be obtained from P(t), u, and the earth ellipsoid elevation or DEM data. After integration time Δt, the origin coordinate becomes P(t+Δt). $R_{eth2obj}$ at this time is calculated by the position and velocity vector at time t+Δt, while $R_{obt2eth}$ at this time is consistent with time t. The rotation matrix R' at time (t+Δt) is calculated using equation (3). By substituting O(t), P(t+Δt), and R' into the collinear equation, the image plane coordinates (x', y ') at time (t+Δt) can be obtained.

At this time, the requirement that "the image shift direction is consistent with the TDI integration direction, and the charge transfer period is consistent with the integration time Δt of single-stage CCD" is equivalent to x – x' '= pix, y – y '= 0.

If x – x' ≠ pix, the integration time Δt is modified to Δt'.

$$\Delta t' = \Delta t * (x - x')/pix \tag{4}$$

If y – y'≠0, under the condition of multi-level integral imaging, the same CCD column cannot aim at the same target at the center of the integration time, resulting in imaging blur. Assuming that Δy = |y – y'|/|x – x'|, after Δt is corrected to Δt', Δy = |y – y'|. The greater the value of Δy, the more serious is the ambiguity. The experiments in this study reveal that the magnitude of Δy is related to the deviation of the target point and subsatellite point along the orbit and vertical orbit. The line vector between the satellite instantaneous position and the target point is v_{p2m}, vector v_{p2m} rotates at an angle α, and Δy = 0.

$$\alpha = arctan\,(\Delta y) * 0.9 \tag{5}$$

The three normalized components of v_{p2m} are $[v_0, v_1, v_2]$, and they correspond to the quaternion $[q_0, q_1, q_2, q_3]$:

$$q_0 = cos\left(\frac{\alpha}{2}\right), q_1 = v_0 * sin\left(\frac{\alpha}{2}\right), q_2 = v_1 * sin\left(\frac{\alpha}{2}\right), q_3 = v_2 * sin\left(\frac{\alpha}{2}\right) \tag{6}$$

The corresponding rotation matrix $R(\alpha)$ is

$$\begin{matrix} 1 - 2 * (q_2^2 + q_3^2), 2 * (q_1 * q_2 - q_0 * q_3), 2 * (q_0 * q_2 + q_1 * q_3) \\ 2 * (q_1 * q_2 + q_0 * q_3), 1 - 2 * (q_1^2 + q_3^2), 2 * (q_2 * q_3 - q_0 * q_1) \\ 2 * (q_1 * q_3 - q_0 * q_2), 2 * (q_2 * q_3 + q_0 * q_1), 1 - 2 * (q_1^2 + q_2^2) \end{matrix} \tag{7}$$

Find $R(\alpha)$. After, the total matrix becomes

$$R = R(\alpha) * R_{rth2obj} * R_{obt2eth} * R_{ypr} * R_{set} \tag{8}$$

Table 2 lists the results under different side views and time offsets α. It can be seen that, when the side angle of view and the front and rear angles of view are large at the same time, α When the yaw angle is close to 0 or the front and rear viewing angle is 0, value of α is very small. As shown in figure 1, the value of α changes gently with the row and column direction, and the required value can be obtained quickly through 2D interpolation. α approximate value.

IOP Conf. Series: Earth and Environmental Science 1004 (2022) 012013 doi:10.1088/1755-1315/1004/1/012013

Table 2. α values of different side view angles and time offsets.

	0.000°	10.993°	21.127°	29.888°
40"	-0.013044	2.577922	4.843275	6.593452
30"	-0.005698	1.992732	3.731206	5.061330
20"	-0.001729	1.360829	2.540581	3.434735
10"	-0.000215	0.692170	1.288696	1.735570
0"	0.00000	0.003112	0.002998	-0.003311
-10"	0.000246	-0.685842	-1.282635	-1.742304
-20"	0.001909	-1.354143	-2.534685	-3.441767
-30"	0.006305	-1.985291	-3.724380	-5.068730
-40"	0.014602	-2.569110	-4.835426	-6.601112

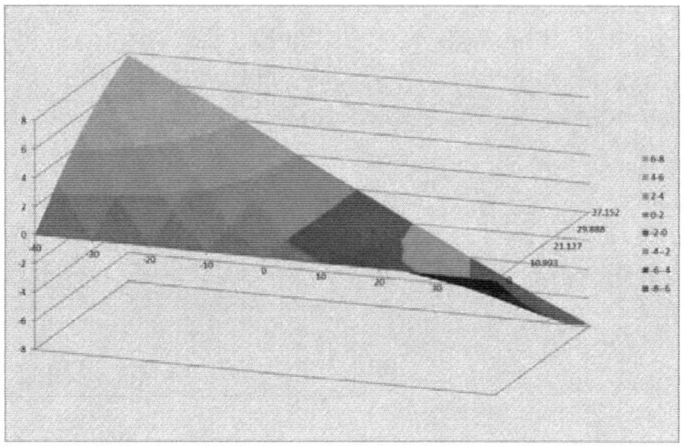

Figure 1. Two-dimensional distributions of α under different side angles and time deviations.

Figure 2 is a partial DEM with an analog input, a DOM with a target, and 3D rendering. Figure 3 shows the simulated images from top to bottom under the following conditions of "side view – time offset": 10.993° – 10"; 21.127°– 20"; 29.888°– 30"; and 37.152°– 40". The first three figures in each line were obtained without adding $R(\alpha)$, and the integral series were 24, 48, and 96, respectively. The fourth figure in each line was obtained by the addition of the $R(\alpha)$ analog image with rotation and an integral series of 96. It can be seen that under the same integral series, the larger the value of α, the more blurred the image. After rotation, Join $R(\alpha)$ was eliminated. In addition, the change in the four-line image resolution and the distortion caused by the terrain and observation angle can also prove the correctness of the simulation method to a certain extent.

Figure 2. Input DEM, target DOM and 3D rendering.

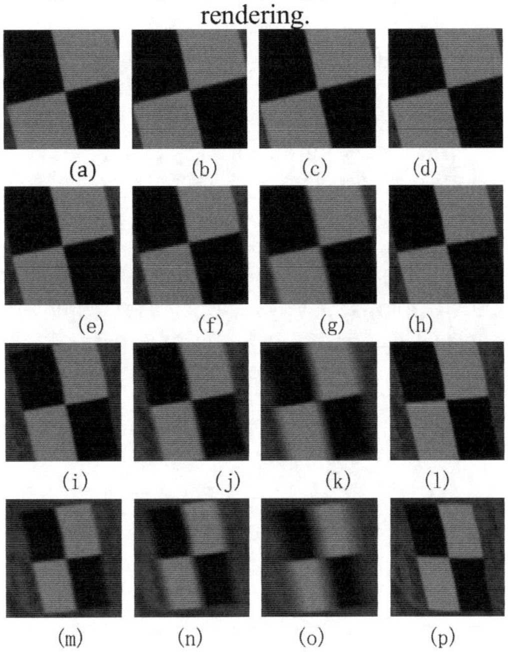

Figure 3. Target imaging under different side views, front and back views, integral series, and deflection angle α.

2.2. Multi view imaging simulation

Only two view stereo images are needed to obtain an ideal matching effect in areas with small terrain slopes, while three view stereo images are needed to solve the problems of sparse matching points and occlusion caused by excessive relative deformation or occlusion in mountainous areas or urban buildings with large slopes. For urban buildings, when the side view angle is small, whether it is two or three dimensions, only the point cloud data at the top of the building can be obtained through stereo matching. Owing to the limitations of the matching algorithm and image quality, the matching results at the top often have some defects, such as missing point clouds or flying points. To better realize 3D reconstruction of buildings, it is necessary to form 3D images on each side of a building, and it is better to use three-view 3D images to solve the problems of occlusion and relative deformation caused by the complex geometric structure or angle of some building sides. This study does not consider the limitations of the number, distribution, and attitude adjustment ability of real satellites; instead, it focuses on the quality of 3D point clouds obtained from stereo images with different side viewing angles and stereo angles from the perspective of application.

2.2.1. Data introduction. The simulation data range was approximately 3 km × 3 km, and the elevation range was 5 m–150 m, including many types of buildings. The geometric and texture modeling accuracies were high, which is representative.

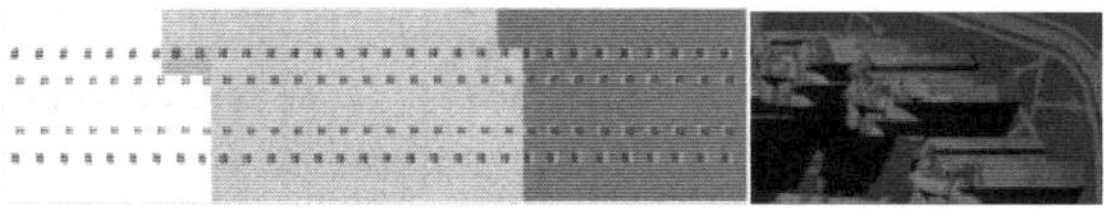

Figure 4. Texture contrast between 3D model and simulated image.

2.2.2. Simulation calculation. The time t for imaging the center of the target area and the rotation matrix R at this time was obtained using equation (8). In the simulation period, $R_{obt2eth}$ and R_{ypr} varied with times. $R_{obt2eth}$ at the center time of each line of the simulated image was calculated by Lagrange interpolation through discrete orbit data with an interval of 1 s. According to the attitude stability parameters, three attitude angle sequences with a certain frequency were obtained using the power spectrum synthesis method. The yaw, pitch, and roll angles of each line of the simulated image center time were calculated by linear interpolation, and then R_{ypr} was obtained. If the line vector between the real-time origin P and the real-time sub-satellite point, and the line vector between the real-time origin P and the real-time target point are approximately regarded as parallel within the imaging time period, then $R(\alpha)$ and $R_{rth2obj}$ are fixed matrices in the imaging time period. Suppose that the multiplication result of two matrices is R_{obj}, and assume R_{set} is the identity matrix, only $R_{obt2eth}$ at the specified time t needs to be obtained and R_{ypr} can be used to obtain the corresponding global rotation matrix R(t):

$$R(t) = R_{obj} * R_{obt2eth} * R_{ypr} \tag{9}$$

After obtaining the projection center and rotation matrix at the center time of each line of the image, the image can be simulated and generated by the method described in the Reference [3], and the corresponding RPC can be calculated simultaneously. The projection center and rotation matrix of the sub-pixel corresponding time required for time-domain integration were obtained by linear interpolation of the integer row time data. Each pixel was divided into 5 × 5 sub-pixels and calculated by intersection with the ground triangulation network.

2.2.3. Ground truth. One advantage of using a simulation to study this problem is that a ground truth value can be used as the evaluation basis. This study focused on the effect of 3D geometric reconstruction; therefore, it was necessary to record the ground geometric information corresponding to each image point. Because each CCD uses a 5 × 5 sub-pixel to intersect with the ground triangulation, the geometric information of each CCD is also the average of the geometric position of the sub-pixel. In addition, this paper provides statistics on the overall regional matching effect and the building facade matching effect. In addition to recording the geometric information of all effective pixels, the geometric information of the building façade should be recorded separately. The building facades were distinguished so that the normal vector of triangular patches could be calculated and normalized to obtain the normalized normal vectors $v[v_0, v_1, v_2]$. If the triangle belongs to the elevation, v is parallel to the horizontal plane, i.e., v_2 equals 0. Simultaneously, the slope s of the triangular patch relative to the horizontal plane can be calculated by v:

$$s = arctan\left(\frac{v_2}{\sqrt{v_0^2+v_1^2}}\right) \tag{10}$$

In the actual simulation, triangles with s less than 0.1 were recorded as elevation triangles.

3. Stereo matching and effect evaluation

3.1. Stereo matching

Obtaining high-quality 3D point cloud data through dense stereo matching is the core step in building a 3D reconstruction. The rapid development and wide application of 3D modeling technology are closely related to the development of stereo matching algorithms. Stereo matching often includes multiple links, among which matching cost calculation and matching cost optimization are two important links. The difference between the different matching methods (classification basis) is mainly reflected in these two links, especially the matching cost optimization link. The difference between the BP and GC matching algorithms used in this study lay in the matching cost optimization.

3.1.1. Matching cost calculation method. The matching cost is the amount of similarity between two image points on a stereo image. Common matching costs include the normalized correlation coefficient, census distance, and mutual information. In this study, the census distance of the 9 × 7 window and the weighted average of mutual information were used to calculate the matching cost. First, the quasi-kernel image pairs are generated by the projection trajectory method, and the length and width were h and w, respectively. The search range was the parallax range corresponding to the minimum and maximum elevation ranges, plus a certain margin. The total search range was d. After calculating the matching cost per parallax, a matching cost cube of w*h*d was formed. Because the calculation of mutual information was related to the accuracy of the initial parallax, the final matching cost was calculated through two iterations. In the first iteration, only the census distance was used to calculate the matching cost. After the matching cost was optimized, the parallax with the lowest matching cost was selected for each image point as the new initial parallax value. Through this process, the mutual information lookup table was calculated, and then the average value of the normalized mutual information and the normalized census distance was multiplied by 100 to become an integer between 0 and 100.

3.1.2. Matching cost optimization. The goal of matching cost optimization is to minimize the global energy. The energy value of each image point consists of two parts: a data item and a smoothing item. The energy value of the data item is the matching cost E_d after the confidence propagation of the BP algorithm or alpha expansion of the GC algorithm. The smoothing term refers to the energy E_s corresponding to the minimum matching cost (optimized) of adjacent pixels and their parallax distance Δd. The calculation of this energy value follows the principle that the greater the parallax distance Δd, the greater the energy value. The model adopted in this study was as follows:

$$E_s = \begin{cases} \Delta d^2 * P_1 & \Delta d \leq 2 \\ P_2 & \Delta d > 2 \end{cases} \tag{11}$$

In addition, the calculation of the smoothing term is related to the selection of neighborhood pixels, and four neighborhoods, eight neighborhoods, or sixteen neighborhoods can be used. Generally, the greater the number of neighborhoods, the better the matching effect, especially in weak texture areas. Both optimization algorithms in this study used eight neighborhoods.

3.1.3. Effective match point filtering. In the simulation link, the geometric information of each image point of the simulation image and the identification information on whether it is a building façade are recorded, and the matching link is carried out on the quasi-nuclear line image. To obtain the geometric information of each image point of the quasi-nuclear line image, it is only necessary to pass the image point coordinates through an average object elevation h_0 that is projected onto the original image from the same angle of view. The corresponding geometric information is obtained through bilinear interpolation, and the elevation identification information is obtained through the nearest point search. The parallax truth value can be obtained by projecting the image points of the quasi-nuclear line image onto the paired quasi-nuclear line image through the elevation truth value h. Compared with the parallax value obtained by matching, if the difference is within a certain threshold (set as 1 pixel in this study), it is considered to be the correct matching result. The matching image points are converted into ground 3D coordinates, which can be compared with the true value to determine the matching effect. This study focused on evaluating the three-view stereo matching effect under different side view angles and takes the middle view image as the main image. To facilitate the display, the parallax value between the main image and the front and rear images was converted into the ground elevation value, which was called the elevation map (including the main front elevation map and the main rear elevation map), and the two elevation maps were combined to obtain the final elevation map. Thus, the 3D geometric accuracy of each image point could be expressed by the accuracy of the elevation value.

3.2. Evaluating indicator

In this study, the effect of building a geometric 3D reconstruction was evaluated by obtaining the traditional root mean square error and the integrity of the reconstructed geometric information, i.e., how many effective points can be obtained through stereo matching. Among them, the effective number and quality of the building façade were counted separately. The statistical indexes of the effective points in this study included the effective point proportion, root mean square, nmad, and le90. The effective point ratio is the ratio of matching effective points to the number of ground points participating in the evaluation multiplied by 100. nmad is the normalized median absolute deviation.

4. Results

A total of 22 groups of three-view stereo images were generated in this study. It was assumed that the orbit of the sub satellite point passing through the center of the survey area was the front view orbit, the orbit offset to the West relative to the front view orbit was the left view orbit (hereinafter referred to as the left orbit), and the orbit offset to the East was the right view orbit (hereinafter referred to as the right orbit). Each track generated images from three perspectives. The image from the middle perspective was the main image, and the absolute value of the time offset of the front and rear views was the same as that of the main image (hereinafter referred to as the front-view image; note that it is not the front view track). The BP and GC methods described above were used for stereo matching. The evaluation of the matching results is shown in Tables 3–6. The first number in the "Order" column in the table is the table serial number, the second number uses 1 and 2 to distinguish the BP and GC algorithms, and the third number is the row serial number.

Tables 3 and 4 correspond to the left visual track, and Tables 5 and 6 correspond to the right visual track. The left half of each table shows the statistical results of the BP matching algorithm, and the right half shows the statistical results of the GC matching algorithm. The only difference between Table 3 and Table 4 is that the former counts the results of all image points in the entire region, while the latter only counts the elevation results. This difference is shown in Figure 5, as is the difference between Tables 5 and 6. The results in rows 1–3 of each table correspond to a side view angle of 29.888 °, rows 4–6 correspond to a side view angle of 21.127 °, and rows 7–11 correspond to a side view angle of 10.993°. According to the previous definitions of left and right visual tracks, the absolute values of the side viewing angles in the same rows of Tables 3 and 4 and Tables 5 and 6 are the same, and the symbols are opposite to indicate that opposite sides of the building were observed, as shown in Figure 6. It can be seen that the illumination of the left and right track images is very different, and the left track has more shadows. Lines 1–3 of each table correspond to the front and rear viewing angles, which are described by the time offset relative to the nearest point time, which are offset by 20 s, 15 s, and 10 s. Lines 4–6 are offset by 20 s, 15 s, and 10 s, respectively, and lines 7–11 are offset by 20 s, 15 s, 10 s, 7 s, and 5 s, respectively.

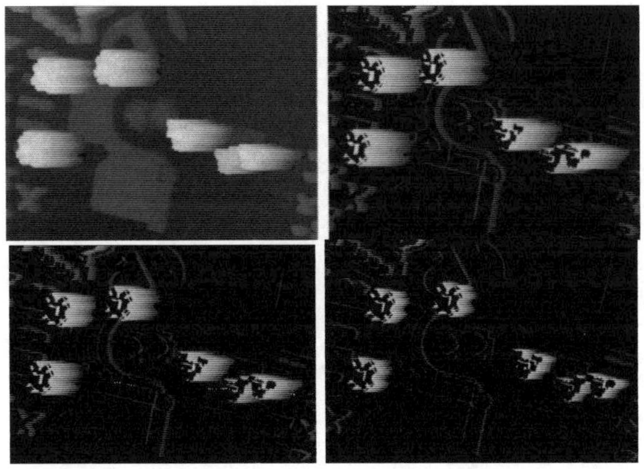

Figure 5. Overall evaluation area and the elevation evaluation area corresponding to the three side views.

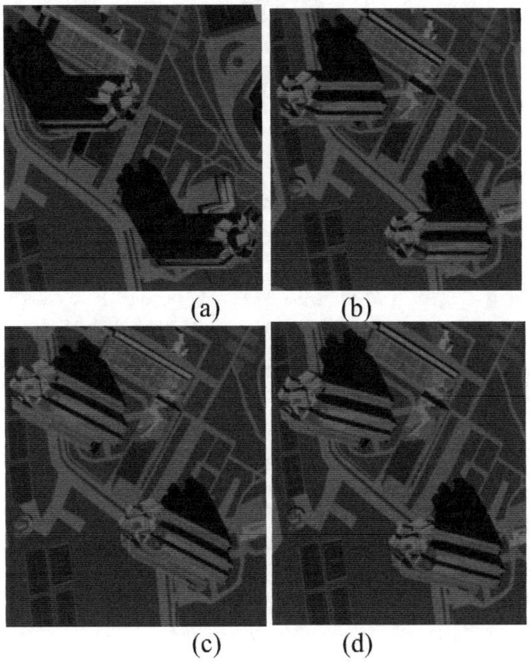

Figure 6. Left track "front view" image, right track "front view" image (side view), right track “rear view offset 20," and right track "rear view offset 10."

The statistical index changes caused by the differences in observation conditions and processing algorithms in each table were then compared and further analyzed in combination with the images.

4.1. Comparison of the same side angle and different stereo angles

According to the traditional photogrammetry theory, the ground elevation measurement accuracy is higher when the base height ratio of the stereo image is between 0.8–1.0. However, in modern urban areas, a smaller base height ratio is required because of building density and significant shielding. From the statistical indicators in Tables 3–6, as the included 3D angle reduces, the proportion of effective points increases, and the accuracy index generally decreases. Compared with the overall indicators, the indicators of building façades change more violently. Figure 7 shows a comparison of the local matching results of 4-1-1, 4-1-2, and 4-1-3 (also the matching results of 3-1-1, 3-1-2, and 3-1-3. The difference is shown in Table 4. Only the elevation was counted). It can be seen that, with a reduction in the included angle, the elevation recovery becomes increasingly complete, and the statistical index is reflected in the increase in the proportion of effective points.

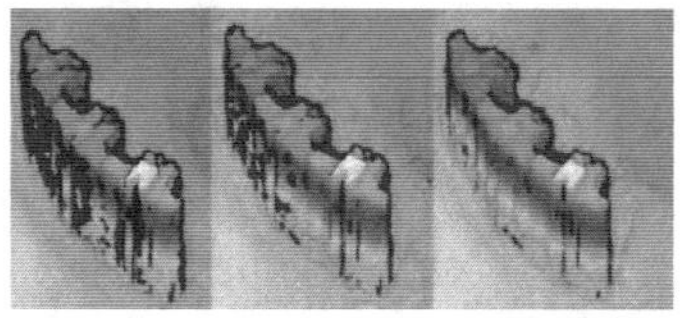

Figure 7. 4-1-1, 4-1-2, 4-1-3 local matching results.

4.2. Comparison of the same stereo angle and different side angles
This study focused not only on the matching effect of the top of the building, but also on the matching effect of the building facade. Owing to the constraints of the track itself, the included angle between the building façade and the observation line of sight (i.e., the included angle between the façade normal vector and the observation line of sight vector; the smaller the included angle, the closer it is to the "front" wall) was large. The larger the angle, the more significant the image shrinkage of the façade, and it is a "inclined plane" with a gradually changing parallax on the parallax map. The implicit "fronto-parallel" assumption of the BP and GC algorithms makes the parallax change sharply, which worsens the regional matching effect. Figure 8 shows the local matching results of 4-1-7, 4-1-8, 4-1-9, 4-1-10 and 4-1-11. Under the condition that the included angle is consistent with that in Figure 7, the elevation matching effect is very different. Although the effective points are improved with further reduction of the included angle, the facade is obviously roughened, and the statistical indicators are obviously worse except for the effective points. Even if the accuracy requirements are reduced, the density of the elevation points is much lower than that in Figure 7 because of the large difference in the number of image points corresponding to the elevation. In addition, such a small time offset also has high requirements for satellite imaging capability.

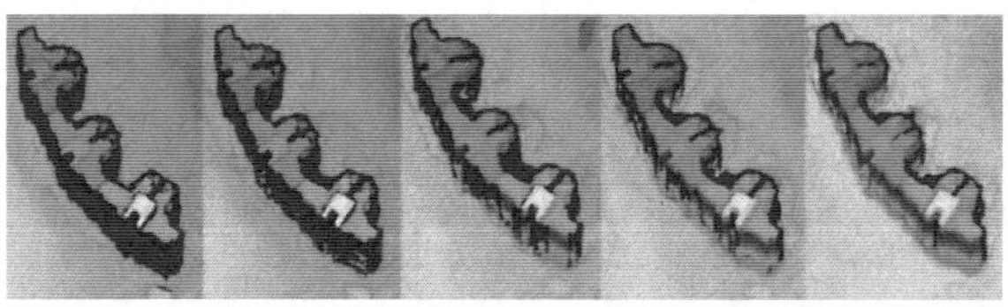

Figure 8. 4-1-7, 4-1-8, 4-1-9, 4-1-10, 4-1-11 local matching results.

4.3. Different matching methods under the same observation conditions
The GC algorithm is a robust matching algorithm. It can be seen from the proportion of effective points that the proportion of effective points of the GC matching results is generally higher than that of the BP algorithm. However, the classical GC algorithm returns an optimal parallax position corresponding to the matching cost. Generally, the sub-pixel matching accuracy is improved by increasing the matching cost sampling density. For example, in addition to calculating the matching cost of the integer points, the matching cost of the sub-pixel points is also calculated. In this study, only the matching cost of the integer points was calculated, and the matched integer parallax position was fitted parabolically with the original matching cost to obtain the sub-pixel position. The BP algorithm can fit the parabola with the aggregated matching cost. From the statistical index, the BP algorithm is superior to the GC algorithm in terms of overall accuracy.

4.4. Comparison of the left and right sides

From the perspective of the proportion of effective points, when the solid angle is large, the left track (shadow surface) is lower than the right track (light surface). With a decrease in the solid angle, the proportion of effective points on the shadow surface gradually increases. Here, the change in the statistical index cannot be simply attributed to the change in the solid angle because, regardless of whether the BP or GC algorithm is applied, the amplification ratio of the normalized matching cost should be directly proportional to the smoothing parameters. Because of the low discrimination of the shadow area, the matching cost often becomes smaller; therefore, it is more advantageous to reduce the smoothing parameters. If the matching cost scaling coefficient (or smoothing parameter) is dynamically adjusted, the variation range of various indicators of the shadow surface will be greater than that of the illumination surface. The matching scaling coefficient and smoothing parameters used in Tables 3–6 were fixed. Therefore, the reason for the index change also includes the adaptability of the matching parameters.

4.5. Comparison of the change trends reflected by different statistical indicators

Because the matching parallax eliminated the error greater than one pixel based on the true value, the final point elevation participating in the evaluation was considered to be the "correct" elevation. The statistical results in Tables 3–6 indicate that the larger the effective point percentage, the better the effect. The other indicators are the accuracy indicators. The smaller the value, the better the effect. In this study, the regional overall index and elevation index were counted. The corresponding numerical change curves are shown in Figures 9–12. The effective point proportion and le90 change trend of the building façade are close to the regional overall index, while the elevation mean square deviation is quite different from the overall index, and some show the opposite change trend. Therefore, it is not recommended to use the mean square deviation to evaluate the geometric accuracy of the building. In addition, the nmad index of the façade is slightly different from the overall index. Figures 13 and 14 compare and analyze the nmad and le90 indexes of the BP and GC algorithms on the left and right tracks. It can be observed that the trends of the two reactions are very different. Owing to the good illumination of the right track, the result of the right track should be better.

The proportion of effective points is dominated by the right track in the left part of the curve with a large solid angle. However, as the solid angle decreases, the proportion of effective points on the right track is gradually lower than that of the left track. In this case, the adaptability of the matching parameters should be considered. The le90 index also reveals that, when the included 3D angle is large, the right track is better than the left track. With a reduction in the included angle, the left and right results gradually become close. For the nmad index, the smaller the side angle, the smaller the 3D included angle, and the greater the nmad value (the lower the accuracy), which clearly reflects the trend. However, under the same side angle and solid angle, the Nmad index indicates that the left track is generally better than (less than) the right track. The advantage of nmad is that it is robust to gross errors; however, as the matching results in this study are based on the ground truth value and eliminate gross errors, they cannot reflect this advantage. In addition, simply eliminating the error value according to a pixel threshold may also be related to the "anomaly" of the root nmad statistical index.

IOP Conf. Series: Earth and Environmental Science 1004 (2022) 012013 doi:10.1088/1755-1315/1004/1/012013

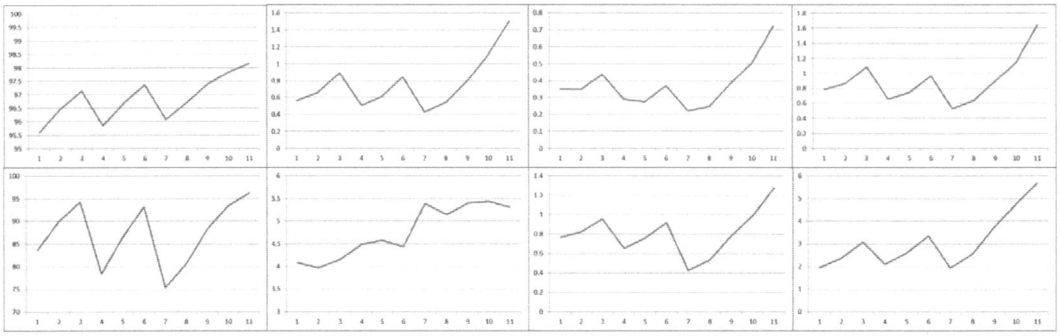

Figure 9. Comparison of the overall index using the BP algorithm (top) of the left track and the elevation index (bottom). (a) the effective point proportion, (b) mean square deviation, (c) nmad, and (d) le90.

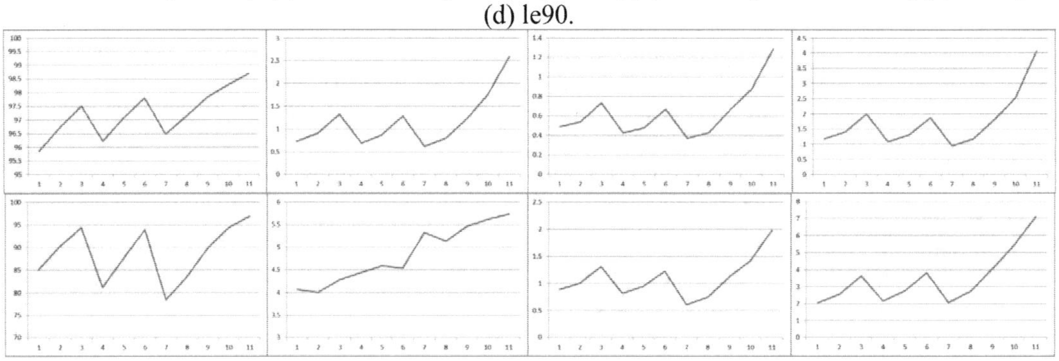

Figure 10. Comparison of the overall index using the GC algorithm (top) of the left track and the elevation index (bottom). (a) the effective point proportion, (b) mean square deviation, (c) nmad and (d) le90.

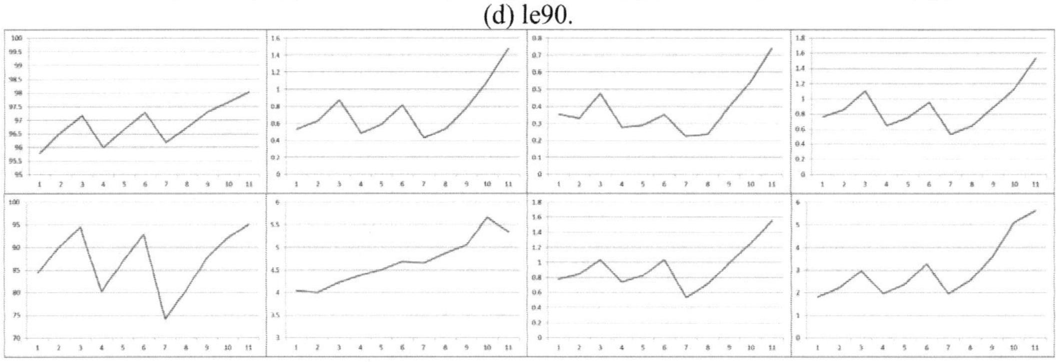

Figure 11. Comparison of the overall index using the BP algorithm (top) of the right track and the elevation index (bottom). (a) the effective point proportion, (b) mean square deviation, (c) nmad and (d) le90.

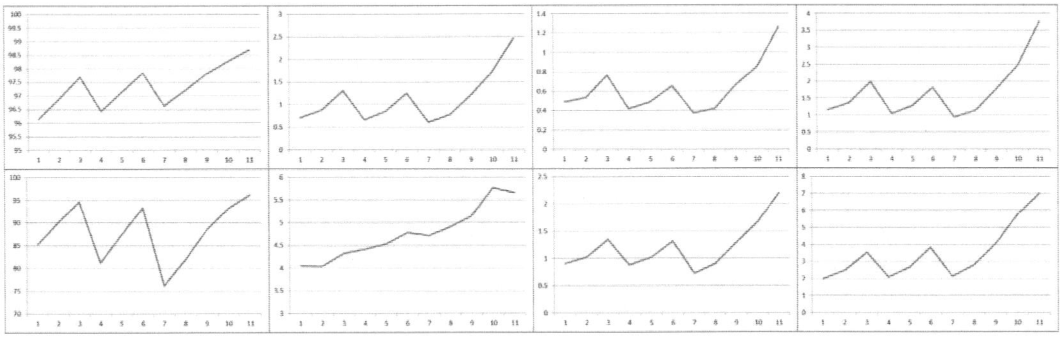

Figure 12. Comparison of the overall index using the GC algorithm (top) of the right track and the elevation index (bottom). (a) the effective point proportion, (b) mean square deviation, (c) nmad, and (d) le90.

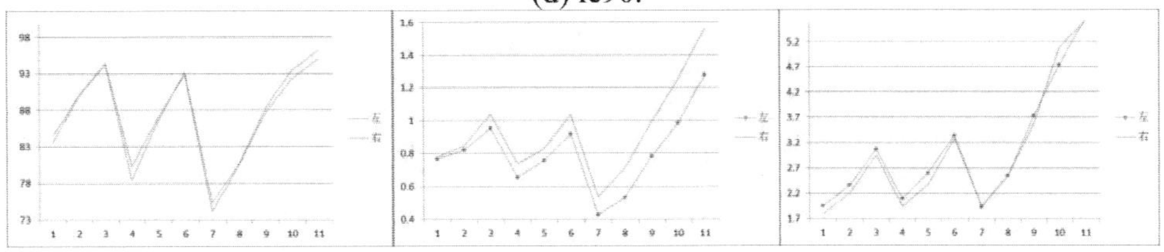

Figure 13. Comparison of the (a) effective proportion, (b) nmad, and (c) le90 indexes of the BP algorithm for the left and right tracks of the facade.

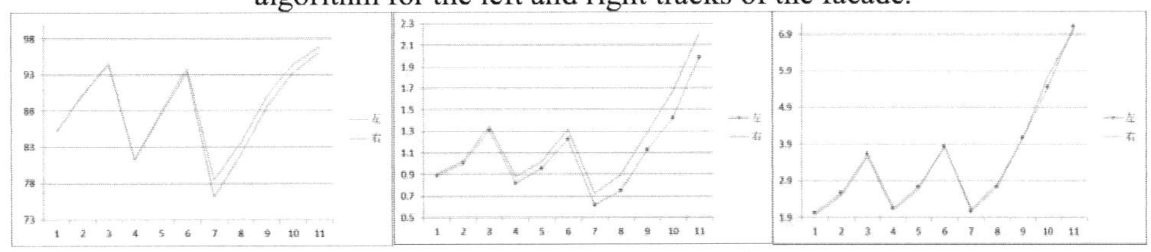

Figure 14. Comparison of the (a) effective proportion, (b) nmad, and (c) le90 indexes of the GC algorithm for the left and right tracks of the facade.

Table 3. Statistical indicators of the left side view in all the zones.

Order	Efficient	rmse	nmad	LE90	Order	Efficient	rmse	nmad	LE90
3-1-1	95.60025	0.565	0.349	0.780	3-2-1	95.85292	0.730	0.487	1.173
3-1-2	96.47140	0.659	0.348	0.866	3-2-2	96.73524	0.900	0.539	1.402
3-1-3	97.13161	0.891	0.438	1.083	3-2-3	97.50568	1.317	0.731	1.989
3-1-4	95.84843	0.507	0.290	0.651	3-2-4	96.22512	0.687	0.425	1.082
3-1-5	96.68576	0.610	0.274	0.744	3-2-5	97.08588	0.864	0.479	1.309
3-1-6	97.36503	0.846	0.368	0.963	3-2-6	97.80026	1.281	0.669	1.874
3-1-7	96.07835	0.431	0.221	0.527	3-2-7	96.48332	0.618	0.368	0.946
3-1-8	96.70343	0.546	0.247	0.637	3-2-8	97.14061	0.798	0.429	1.175
3-1-9	97.38601	0.799	0.384	0.891	3-2-9	97.82261	1.235	0.663	1.813
3-1-10	97.82948	1.102	0.506	1.145	3-2-10	98.27384	1.752	0.872	2.536
3-1-11	98.16395	1.505	0.724	1.640	3-2-11	98.68981	2.578	1.284	4.074

IOP Conf. Series: Earth and Environmental Science 1004 (2022) 012013 doi:10.1088/1755-1315/1004/1/012013

Table 4. Statistical indicators of the left side view in the vertical section.

Order	Efficient	rmse	nmad	LE90	Order	Efficient	rmse	nmad	LE90
4-1-1	83.54893	4.086	0.765	1.954	4-2-1	85.09523	4.070	0.891	2.038
4-1-2	89.82664	3.965	0.820	2.360	4-2-2	90.25046	4.005	1.008	2.563
4-1-3	94.28577	4.151	0.953	3.064	4-2-3	94.39004	4.286	1.310	3.633
4-1-4	78.39868	4.489	0.654	2.096	4-2-4	81.21793	4.451	0.819	2.161
4-1-5	86.54463	4.576	0.758	2.596	4-2-5	87.79643	4.594	0.956	2.746
4-1-6	93.18845	4.431	0.916	3.339	4-2-6	93.94100	4.543	1.228	3.827
4-1-7	75.38460	5.392	0.427	1.928	4-2-7	78.46575	5.327	0.615	2.075
4-1-8	80.55541	5.148	0.533	2.551	4-2-8	83.50082	5.136	0.752	2.736
4-1-9	88.25593	5.402	0.782	3.734	4-2-9	89.94826	5.462	1.126	4.062
4-1-10	93.44560	5.442	0.985	4.747	4-2-10	94.38880	5.614	1.424	5.449
4-1-11	96.26189	5.315	1.276	5.674	4-2-11	96.89782	5.733	1.987	7.111

Table 5. Statistical indicators of right side view in all zones.

Order	Efficient	rmse	nmad	LE90	Order	Efficient	rmse	nmad	LE90
5-1-1	95.78443	0.531	0.354	0.761	5-2-1	96.14631	0.709	0.487	1.148
5-1-2	96.54933	0.633	0.331	0.857	5-2-2	96.88382	0.876	0.533	1.363
5-1-3	97.17692	0.875	0.476	1.104	5-2-3	97.69765	1.305	0.765	1.986
5-1-4	95.99117	0.487	0.278	0.649	5-2-4	96.42720	0.661	0.416	1.039
5-1-5	96.65966	0.590	0.290	0.751	5-2-5	97.14015	0.844	0.487	1.282
5-1-6	97.28397	0.814	0.351	0.954	5-2-6	97.84272	1.247	0.655	1.814
5-1-7	96.17546	0.430	0.223	0.533	5-2-7	96.61968	0.609	0.373	0.936
5-1-8	96.73331	0.539	0.237	0.647	5-2-8	97.20088	0.773	0.417	1.136
5-1-9	97.30219	0.785	0.397	0.889	5-2-9	97.81943	1.208	0.660	1.765
5-1-10	97.64840	1.092	0.541	1.135	5-2-10	98.27113	1.728	0.856	2.480
5-1-11	98.02264	1.476	0.740	1.534	5-2-11	98.69732	2.476	1.264	3.769

Table 6. Statistical indicators of right side view in vertical section.

Order	Efficient	rmse	nmad	LE90	Order	Efficient	rmse	nmad	LE90
6-1-1	84.39633	4.043	0.778	1.802	6-2-1	85.18805	4.044	0.906	1.976
6-1-2	89.98239	4.002	0.847	2.218	6-2-2	90.11892	4.036	1.028	2.495
6-1-3	94.49623	4.220	1.038	2.957	6-2-3	94.62807	4.318	1.345	3.539
6-1-4	80.26621	4.381	0.737	1.943	6-2-4	81.21332	4.417	0.880	2.099
6-1-5	86.98521	4.503	0.827	2.372	6-2-5	87.41877	4.526	1.021	2.668
6-1-6	92.92323	4.689	1.035	3.279	6-2-6	93.30919	4.772	1.316	3.834
6-1-7	74.16342	4.653	0.534	1.949	6-2-7	76.14360	4.709	0.724	2.137
6-1-8	80.43826	4.873	0.713	2.551	6-2-8	81.82402	4.906	0.898	2.820
6-1-9	87.66495	5.057	0.993	3.573	6-2-9	88.54829	5.144	1.288	4.060
6-1-10	92.19004	5.663	1.252	5.091	6-2-10	93.19166	5.768	1.672	5.759
6-1-11	95.07916	5.333	1.557	5.619	6-2-11	96.17059	5.654	2.200	7.008

5. Conclusion

The satellite multiview imaging simulation method proposed in this paper can realize stereo image simulation of a TDI CCD camera for different side views and stereo angles.

For the building façade, both the side view and the included 3D angle should be controlled within a reasonable range. This paper suggests a side view angle between 20–40° and an included angle between the front view and front and rear views between 10–15°.

The stereo-matching algorithm used in this study utilizes two representative classical algorithms. For stereo images with small included angles, some stereo matching algorithms with a small base height ratio may offer a better matching effect. Therefore, it is necessary to comprehensively consider the image shooting angle, matching algorithm, calculation efficiency, and other factors to build an economical and feasible processing flow.

To evaluate the effect of the 3D geometric reconstruction of urban buildings, a single statistical index should not be used. The integrity of each side of the building should be a primary consideration. For building façades, more attention should be paid to the density of the effective points. In terms of accuracy, le90 is recommended as the preferred evaluation index. In actual satellite image evaluations, there is often a certain gross error distribution.

References

[1] Zhang L 2012 *Analysis and Simulation of a Dynamic Image Captured by a Satellite Based TDI CCD Camera* (Harbin: Harbin Institute of Technology)

[2] Li B 2013 *Analysis of Random Vibration on Image Quality Degradation of a TDICCD Camera* (Harbin: Harbin Institute of Technology)

[3] Yue Q, Tang X, and Gao X 2015 Imaging simulation of sub-meter satellite TDI CCD camera for surveying and mapping *Geomatics Inf. Sci. Wuhan Univ.*

[4] Fan B and Yu Q 2017 Simulation analysis of platform vibration on image quality of satellite-based high-resolution optical system. *Chinese Space Sci. Technol.*

[5] Su H 2018 *Image Restoration Method and Vibration Degradation Simulation of TDI Remote Sensing Image* (Zhejiang: Zhejiang University)

[6] Ye P 2017 *Optical Remote Sensing Motion Deration Simulation and Image Restoration Technology* (Zhejiang: Zhejiang University)

[7] Guanter L, Segl K, and Kaufmann H 2009 Simulation of optical remote-sensing scenes with application to the EnMAP hyperspectral mission. *IEEE Trans. Geosci. Remote Sens.* **47** 2340–51

[8] Felzenszwalb P F and Huttenlocher D P 2004 Efficient belief propagation for early vision. In: *CVPR* pp. 261–8

[9] Marshall F, William T, Freeman T 2003 Comparison of graph cuts with belief propagation for stereo using identical MRF parameters *Proc. Ninth IEEE Int. Conf. Comp. Vision* pp. 900–07

[10] Sun J, Zheng N, Shum H 2003 Stereo matching using belief propagation *IEEE PAMI* **25** 787–800

[11] Boykov Y, Veksler O, and Zabih R 2001 Fast approximate energy minimization via graph cuts. *IEEE PAMI* **23** 1222–39

[12] Boykov Y and Kolmogorov V 2004 An experimental comparison of min-cut/max-flow algorithms for energy minimization in vision. *IEEE PAMI* **26** 1124–37

[13] Höhle J and Höhle M 2009 Accuracy assessment of digital elevation models by means of robust statistical methods *ISPRS J. Photogramm. Remote Sens.* **64** 398–406

[14] Wang B, Shi W, Liu E. 2015 Robust methods for assessing the accuracy of linear interpolated DEM *Int. J. Appl Earth Obs. Geoinf.* **34** 198–206

[15] Kuschk G, D'Angelo P, Qin R, et al. 2014 DSM accuracy evaluation for the ISPRS commission I image matching benchmark. *ISPRS–Int. Arch. Photogramm. Remote Sens. Spatial Inf. Sci.* **XL-1** 195–200

[16] Wang B, Shi W, and Liu E 2015 Robust methods for assessing the accuracy of linear interpolated DEM *Int. J. Appl. Earth Obs. Geoinf.* **34** 198–206

Image enhancement method for visual localization and mapping in complex light scenarios

Shoujun Jia, Chun Liu*, Hangbin Wu, Chen Chen

College of Surveying and Geo-informatics, Tongji University, Shanghai, P.R. China, 200092
Correspondence should be addressed to Chun Liu; liuchun@tongji.edu.cn

Abstract. Visual localization and mapping have received considerable attention in the fields of computer vision, photogrammetry, and remote sensing. Image matching is key to visual localization and mapping. However, light is complex and uneven in real scenarios, creating difficulties in feature extraction and matching. Hence, feature mismatch and loss reduce the efficiency, accuracy, and robustness of visual localization and mapping. We developed a visual localization and mapping method for complex light scenarios based on image enhancement. Starting with initial images, the irradiance and reflectance components were separated based on logarithmic transformation. Our method strengthened high-frequency components and restrained low-frequency components with improved homomorphic filtering, restraining the light component and enhancing the important reflection component. The SIFT algorithm was used for feature detection and matching. The proposed method was tested on images with uneven light captured using a stereo vision camera in an indoor environment, focusing on visual localization and mapping. The experimental results emphasized that the method improved the rate of image localization and number of point clouds, as well as the reprojection error, which ranged from 0.85 to 0.82 on average. Thus, the proposed method is robust rather than probabilistic for improving visual localization and mapping under complex light conditions.

Keywords: Visual Localization and Mapping, Complex Light, Homomorphic Filtering, Image Enhance, Image Matching

1 Introduction

Localization and mapping have been applied in geographic information systems, computer vision, robotics, and other fields. These applications include indoor localization, unmanned ground vehicles, simultaneous localization, and mapping, among others. Image matching is a fundamental issue in image localization and mapping and comprises feature detection, descriptor assignment, and feature matching [1]. However, a limitation to this approach is that image matching algorithms are sensitive to complex light in practical situations. Particularly, complex light leads to a loss of important features and an increase in false features, and thus significantly degrades the performance of image matching [2].

Various techniques have been proposed for addressing the problem of complex illumination in image matching. Numerous feature detectors and descriptors have been described in the literature, most of which can be used for illumination-invariant image matching but show low applicability in practical situations [3]. An illumination robust feature descriptor was proposed based on the local patch clustering distribution [4], which performs well but the input reference images must be under good illumination. Feature matching algorithms have been suggested for handling the illumination variation issue [5], which consist of three main steps: local region matching, overlapping area enhancement, and feature matching; however, these factors are subject to the parameter setting. A robust visual localization algorithm under varying lighting conditions was developed [6] which performs well in the real world but at the cost of appropriate or harsh camera exposure time. Moreover, a histogram equalization and sharpen (HES) approach was adopted to enhance the image [7, 8]. Although these solutions can mitigate the effects of uneven lighting, their effectiveness and applicability are limited.

Here, a novel method is proposed for addressing the problem of localization and mapping under complex light conditions based on image feature enhancement. First, the original images were processed based on an improved homomorphic filtering function, in which light components were restrained and important reflection components were enhanced. Using the enhanced images, the SIFT algorithm was used to obtain preliminary image matching before filtering with the random sample consensus (RANSAC). Finally, the image relative orientation and forward intersection algorithms were used for localization and mapping.

2 Methodology

This paper presents an improved visual localization and mapping theory under complex light conditions based on an image feature enhancement approach that reduces the irradiation components and enhances the reflection components of an image. As shown in Figure 1, the proposed approach is easily employed in existing visual localization and mapping theory. The proposed method can be applied not only in indoor environments but also in scenarios with complex light conditions.

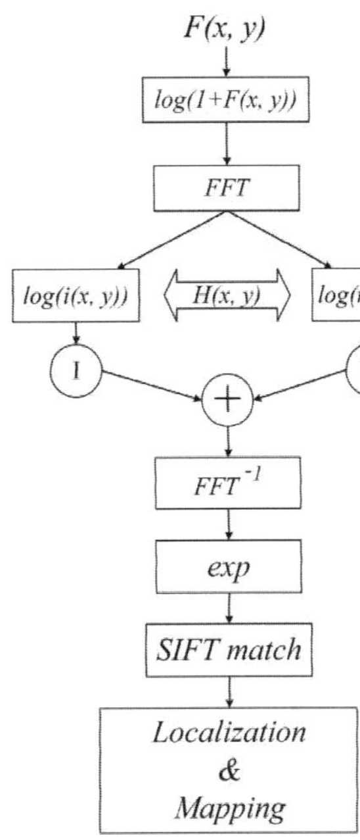

Figure 1. Framework of proposed method inserted into the localization and mapping system

2.1 Light-robust feature enhancement

We used homomorphic filtering theory to separate the irradiation and reflection components of the image based on logarithmic transformation, for example, $F: * \rightarrow +$, which are related to low-frequency components and high-frequency components in the frequency domain, respectively, after the fast Fourier transform [9, 10]:

$$F(x, y) = i(x, y) * r(x, y) \tag{1}$$

$$ln\left[F(x, y)\right] = ln\left[i(x, y)\right] + ln\left[r(x, y)\right] \tag{2}$$

$$FFT \rightarrow \{ln\left[F(x, y)\right]\} \tag{3}$$

$$Z(x, y) = I(x, y) + R(x, y) \tag{4}$$

where FFT is the fast Fourier transform and $F(x, y)$, $i(x, y)$, and $r(x, y)$ denote the original image, irradiation, and reflection components, respectively.

The improved homomorphic filtering function model $H(x, y)$ was constructed to strengthen the high-frequency components and restrain the low-frequency components, strengthening the reflection components and restraining the irradiation components correspondingly after inverse fast Fourier transform, which improves visual image feature extraction and matching under uneven light conditions. The homomorphic filtering function $H(x, y)$ and image enhancement model are as follows [11, 12]:

$$H(x, y) = \frac{h-l}{\left[1+c*\left(\frac{D_0}{D(x, y)}\right)^n\right]} + l \tag{5}$$

$$FFT^{-1} \rightarrow \{H * Z = H * I + H * R\} \tag{6}$$

$$Z^{-1}(x, y) = I^{-1}(x, y) + R^{-1}(x, y) \tag{7}$$

$$F'(x, y) = exp\,(Z^{-1}) = i'(x, y) * r'(x, y) \tag{8}$$

where $IFFT$ is the inverse fast Fourier transform and $F'(x, y)$, $i'(x, y)$, and $r'(x, y)$ denote the image, irradiation, and reflection components after fast Fourier transform. Additionally, $D(x, y)$ is the Euclidean distance from the current point in the frequency domain to the center of the frequency domain, c is the adjusting parameters of the function shape, n is the iteration order of the filter, and l and h are the inhibition and enhancement coefficients of the filter, respectively. D_0 is the cutoff frequency radius, whose optimal value is obtained using a traditional low-pass filter. The model of the low-pass filter is as follows [13]:

$$low(x, y) = 1/\left[1 + c * \left(D(x, y)/D_0\right)^n\right] \tag{9}$$

Note that the improved homomorphic filtering function model $H(x, y)$ divides the frequency domain range of the image according to the value of D_0. The range of low-frequency components relevant to the irradiation components affected by light is from 0 to D_0. The remaining frequency components reflect the properties of the object in scenarios that are unrelated to light components or rarely. Therefore, the different ranges of frequency components can be controlled by $H(x, y)$ constructed at varying levels, where the reflection components are intensive and irradiation components can be decreased, reducing the adverse impact on feature extraction from bright light and increasing the identification and extractability of image features under uneven light conditions.

2.2 Feature extracting and matching

Considering that the SIFT algorithm is representative in image matching, tolerant and robust to scaling changing, spatial rotation, and affine transformation, it was applied for feature extraction and matching in this study [14]. The SIFT descriptors are computed first by constructing the scale space and Gaussian difference pyramid at each image, and then by determining the location and direction of key points with a 128 element feature vector, generating SIFT feature descriptors [15]. Preliminary matching of the SIFT feature descriptor was conducted based on the nearest neighbor distance ratio. Finally, the random sample consensus (RANSAC) algorithm was adopted to remove outliers from the preliminary matching results [16].

2.3 Image localization and mapping

Image localization and mapping was performed by using image relative orientation and forward intersection theory with feature matches as stereopairs, which has been thoroughly studied in photogrammetry [17, 18] and computer vision [19]. First, the relative position of images can be computed based on image matching with the polar geometry relationship between image frames. Based on this information, the three-dimensional mapping was conducted using forward intersection theory to calculate the three-dimensional coordinates of key points. Subsequently, the bundle adjustment algorithm was utilized to decrease the accumulated error from the process of relative orientation and forward intersection, optimizing the image position and coordinates of three-dimensional points simultaneously and globally, reducing reprojection error and improving the precision visual localization and mapping [20].

3 Experimental results and analysis

3.1 Experimental dataset

The improved visual localization and mapping methodology proposed in this paper was tested on a dataset containing 623 image pairs acquired using a mobile camera system for an indoor scenario with complex and uneven light conditions. The images were captured with a stereo vision camera system comprised of two Canon 600D SLR cameras with 18 million effective pixels and a synchronizer trigger. Furthermore, the images in this dataset cover some planar objects under uneven light conditions, including walls, floods, and roofs. Therefore, the dataset can provide several overlapping image pairs with uneven light components for testing the proposed methodology. Note that the stereo vision cameras were calibrated before data collection, and the calibration error was controlled within 1.5 pixels. Table 1 lists the image pairs used and their relevant information as a sample.

3.2 Evaluation criteria

The proposed feature enhancement method was evaluated with numerical and visual indicators and compared with other image enhancement methods after preliminary matches were filtered by RANSAC and in image localization and mapping. Two numerical indicators were adopted after the matches were filtered. The first numerical indicator was the rate of increasing correct matches (RICM), which is based on the number of correct matches (NCM) obtained by filtering the preliminary matches with RANSAC. RICM is defined as $RICM = (NCM2 - NCM1)/NCM1$, where $NCM1$ is the number of correct matches of the original image pair and $NCM2$ is the number of correct matches of the image after the image feature is enhanced. The second is the matching precision (MP), defined as $MP = NCM/NPM$, where NPM is the number of preliminary matches. Additionally, to evaluate the

IOP Conf. Series: Earth and Environmental Science 1004 (2022) 012014 doi:10.1088/1755-1315/1004/1/012014

effectiveness of the image feature enhancement approach proposed in image localization and mapping, the success rate of image localization (SRIL), number of point clouds (NPC), and projection error (RE) are calculated based on the dataset before and after enhancing the image feature, and SRIL is defined as $SRIL = (NSL/NL) \times 100\%$, where NSL is the number of successful image localizations and NL is the number of all images calculated.

3.3 Results of feature matching

The proposed image enhancement approach and existing HES methods were tested on the image pairs explained in Section 3.1. The original and enhanced images were obtained for image matching based on the SIFT algorithm, and then the numerical indicators were compared to evaluate the performance of the image enhancement method. Note that 10 image pairs were selected randomly from the dataset to avoid randomness in the results, whose performance is shown numerically in Figures 2–5.

Table 1. Sample images from the datasets are listed with their information

Image Information	Left Image	Right Image
Bright light condition Image type: RGB Image size: 2592 × 1728		
Weak light condition Image type: RGB Image size: 2592 × 1728		

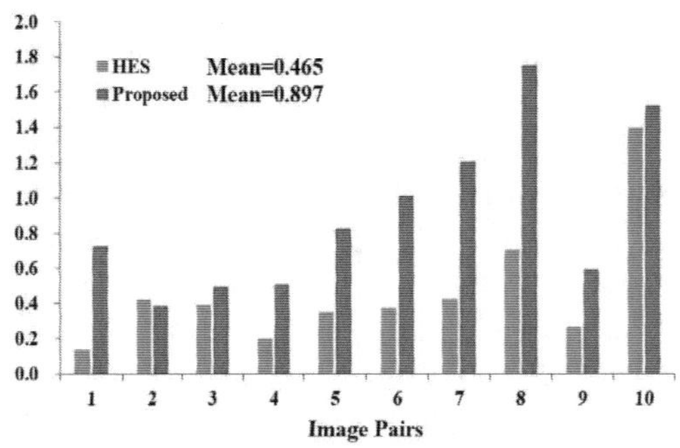

Figure 2. RICM of 10 image pairs with different methods under bright light

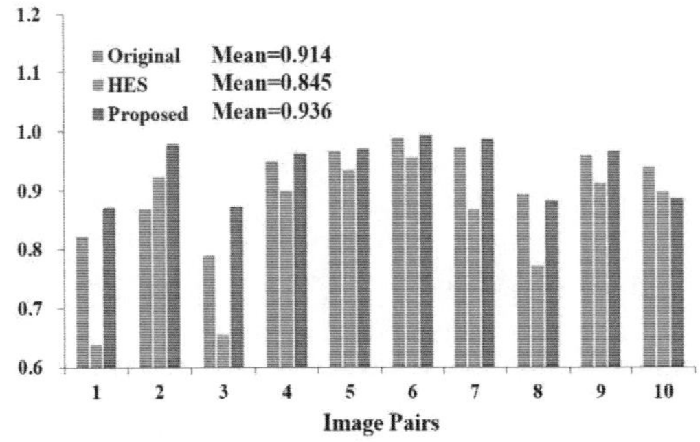

Figure 3. Matching precision of 10 image pairs with different methods under bright light

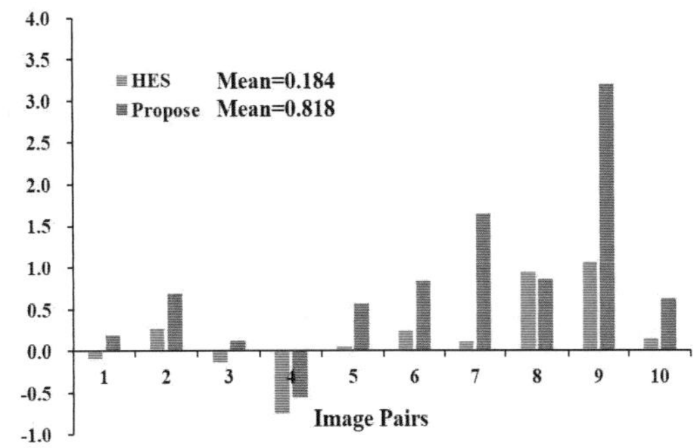

Figure 4. Rate of increasing correct matches of 10 image pairs with different methods under weak light

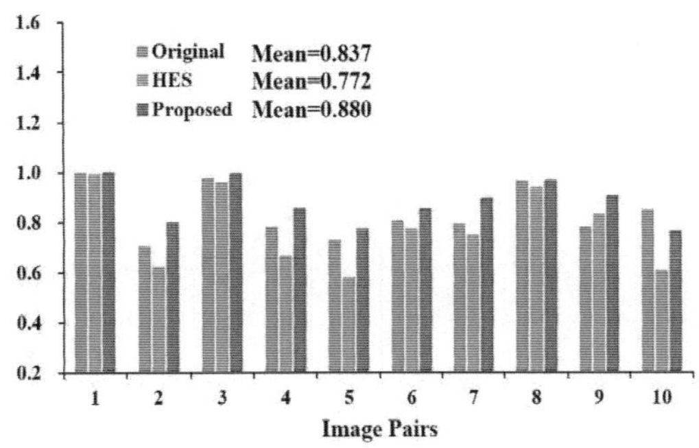

Figure 5. Matching precision of 10 image pairs with different methods under weak light

As shown by the numerical results in Figure 4–7, in terms of image feature matches without image enhancement, the mean MP was lower at only 83% under the weak light condition. The HES method improved the RICM, particularly in bright light, but showed poor performance in the MP under weak light conditions. However, the proposed approach can significantly improve the uneven light and shows better performance than HES under both light conditions, which enhances visual features in similar areas of images with weak texture. Furthermore, the RICM and MP of the proposed method remained at a high level in both cases, where the mean value of MP remained above 88%. Therefore, the proposed approach can perform well for feature matching and is superior to the HES method.

3.4 Results of localization and mapping

All images from the dataset were processed based on the proposed approach and HES method, which were then used for image localization and mapping. Subsequently, quantification criteria, including SRIL, NPC, and RE, were computed to evaluate the effect of the proposed approach on localization and mapping, which are listed in Table 2.

Table 2. Indicators in localization and mapping

Indicators	SRIL	NPC	RE
Original	89.4%	27359	0.85
HES	91.2%	62984	0.97
Ours	92.9%	60054	0.82

The higher SRIL, less RE, and better NPC of the proposed approach compared with the original and HES methods are shown in Table 4. The experimental results showed that the HES improved the density of mapping but not precision. Compared with existing methods, which are subject to complex and uneven light conditions, the proposed approach can achieve high robustness and applicability.

4 Conclusions

Light conditions create difficulties in image matching; therefore, we proposed a feature enhancement method for improving the precision and efficiency of image matching. In the proposed method, the irradiation components of the image are reduced and reflection components are enhanced in the frequency domain based on homomorphic filtering. Moreover, the proposed approach was tested in image matching under complex light and employed for visual localization and mapping. The experimental results showed that the proposed method achieves higher precision in image matching and more robustness in image localization and mapping than the current HES method. Further studies are needed to extend the applicability of the proposed method to more complex scenarios under rapidly changing lighting conditions.

Acknowledgment

This work was supported by the National Science and Technology Major Program (Grant No. 2018YFB1305003) and Major Program of the National Natural Science Foundation of China (Grant No. 42130106).

References

[1] Cao T, Cai H, Fang D, Huang H and Liu C 2017 Keyframes global map establishing method for robot localization through content-based image matching *J. Robot.* 2017 1–16 [DOI: 10.1155/2017/1646095]

[2] Park S, Schops T, Pollefeys M, et al 2017 Illumination change robustness in direct visual SLAM. international conference on robotics and automation p 4523–30.

[3] Yu Y, Huang K, Chen W and Tan T 2012 A novel algorithm for view and illumination invariant image matching *IEEE Trans. Image Process.* **21** 229–40 [DOI: 10.1109/TIP.2011.2160271]

[4] Wang H, Yoon S M, Han D K and Ko H 2017 A feature descriptor based on the local patch clustering distribution for illumination-robust image matching *Pattern Recognit. Lett.* 94 46–54 [DOI: 10.1016/j.patrec.2017.05.010]

[5] Shao Z, Chen M, Liu C, et al 2015 Feature matching for illumination variation images *J. Electron. Imaging* **24**: 033011

[6] Kim P, Coltin B, Alexandrov O, et al 2017 Robust visual localization in changing lighting conditions. international conference on robotics and automation p 5447–52.

[7] Qiu Y M, Chen S H, Zhou Y and Wang Y 2013 3-D video Enhancemnt based on histogram equalization and edge sharpening *Appl. Mech. Mater.* 333–335 1129–33 [DOI: 10.4028/www.scientific.net/AMM.333-335.1129]

[8] Jafar I and Ying H 2007 Image contrast enhancement by constrained variational histogram equalization *Electro Information Technology* p 120–5.

[9] Madhav B T P, Pardhasaradhi P, Manepalli R K N R and Pisipati V G K M 2015 Homomorphic filtering textural analysis technique to reduce multiplicative noise in the 11Oba nano-doped liquid crystalline compounds *Phase Transit.* **88** 735–44 [DOI: 10.1080/01411594.2015.1025781]

[10] Zhang C, Liu W and Xing W 2018 Color image enhancement based on local spatial homomorphic filtering and gradient domain variance guided image filtering *J. Electron. Imaging* **27** [DOI: 10.1117/1.JEI.27.6.063026]

[11] Plichoski G F, Chidambaram C and Parpinelli R S 2017 Optimizing a homomorphic filter for illumination compensation in face recognition using population-based algorithms Workshop of Computer Vision (WVC), Natal, 2017 p 78–83. doi: 10.1109/WVC.2017.00021

[12] Tseng C and Lee S 2017 A weak-illumation image enhancement method uisng homomorphic filter and image fusion. ieee global conference on consumer electronics p 1–2.

[13] Agarwal M and Mahajan R 2017 Medical images contrast enhancement using quad weighted histogram equalization with adaptive Gama correction and homomorphic filtering *Procedia Comput. Sci.* 115 509–17 [DOI: 10.1016/j.procs.2017.09.107]

[14] Liu J 2016 Feature matching of fuzzy multimedia image based on improved SIFT matching *Adv. Electr. Electron. Eng.* **9** 34–8.

[15] Sun Y, Zhao L, Huang S, Yan L and Dissanayake G 2014 -SIFT: SIFT feature extraction and matching for large images in large-scale aerial photogrammetry *ISPRS J. Photogramm.* **91** 1–16 [DOI: 10.1016/j.isprsjprs.2014.02.001]

[16] Hosseinnejad Z and Nasri M 2017 An adaptive image registration method based on SIFT features and RANSAC transform *Comput. Electr. Eng.* 524–37.

[17] Stewenius H, Engels C and Nister D 2006 Recent developments on direct relative orientation *ISPRS J. Photogramm.* **60** 284–94 [DOI: 10.1016/j.isprsjprs.2006.03.005]

[18] Kubik K 1991 Relative and absolute orientation based on linear features *ISPRS J. Photogramm.* **46** 199–204 [DOI: 10.1016/0924-2716(91)90053-X]

[19] Hartley R and Zisserman A 2003 *Multiple View Geometry in Computer Vision* (Cambridge University Press).

[20] Khatiwada B and Budge S E 2016 Three-dimensional image reconstruction using bundle adjustment applied to multiple Texel images *Proc. SPIE.*

Active Contour Building Segmentation Model based on Convolution Neural Network

Mengjia Liu[1,2], Peng Liu[1], Bingze Song[1,2], Yuwei Zhang[1,2], Luo Zhang[1,2]

1. Aerospace Information Research Institute, Chinese Academy of Sciences, Beijing, China
2. School of Electronic, Electrical and Communication Engineering, University of Chinese Academy of Sciences, Beijing, China

liupeng@radi.ac.cn

Abstract. In high-resolution remote sensing images, artificial features on the surface account for a large proportion. In artificial features, buildings, as special artificial features, buildings have different shapes. They are easily affected by light, so it takes a long time to extract using traditional image segmentation methods. It can't effectively design feature engineering to depict the high-dimensional features of the target building. We propose an active contour model based on a convolution neural network, which integrates the prior knowledge and constraints of active contour model, such as continuity of boundary, smooth edge, and geometric characteristics of buildings, into the learning process of convolution neural network to realize the close unity of ACM and CNN. According to our work, a fundamental end-to-end trainable image segmentation framework which is composed of convolution neural network (CNN) and ACM with learnable parameters is implemented, the problem of semantic segmentation of buildings in aerial images was dealt with, the model was evaluated on the publicly available dataset called Vaihingen, and some parameters were explained. In building semantics, the active contour model based on a convolution neural network has good performance.

Keywords: convolution neural network（CNN）; active contour model（ACM）; semantic segmentation

1. Introduction

In remote sensing, the development of image acquisition technology has always been ahead of the development of image processing technology. As remote sensing technology grows, obtaining remote sensing images is becoming more and more diversified. All kinds of military, civilian and commercial remote sensing satellites provide high-resolution, hyperspectral, and all-weather remote sensing data services for global users. In addition, the application field of remote sensing is also expanding, for instance, applied in the areas of land and resources exploration, agriculture production, forest protection, environment monitoring. At present, the amount of remote sensing image data is surging, the variety of remote sensing images is more abundant, and the remote sensing image acquisition technology is developing rapidly. However, the corresponding remote sensing image processing and key information extraction technology are relatively backward, slowing down the pace of remote sensing in people's daily lives [1].

Remote sensing image segmentation technology plays a vital role in remote sensing image digital processing, and it is the first step of many remote sensing image processing and analysis technologies. Effective segmentation guarantees deep feature extraction, such as remote sensing image analysis, recognition, information extraction, etc. The target detection of remote sensing images is easily affected by their multiplicative speckle noise. The distribution of ground objects in some areas is complex, the contrast between target and background is low, and the target boundary is blurred in remote sensing images. The existence of these problems increases the difficulty of remote sensing image segmentation.

According to the research of Kim[12] et al., the CV model is used as a loss function, the parameter f is replaced by ground truth, tested in PASCAL VOC 2012 and Cityscapes, and FCN and DeepLab are used as a baseline, which ultimately increases the mIOU by 3% to 4%. Although the implementation of the model is simple, the reproduction effect is not good. The research of Chen[13] et al. has improved using the Convexified CV model, adding the regularization term related to length and achieving the effect of a 1% increase in Hausdorff distance. ACM is only designed as a loss function in the above research rather than part of an end-to-end model.

After 2010, artificial intelligence is becoming more and more popular. The neural network model is applied to computer vision and shines brilliantly in the field of computer vision. The new generation of image segmentation models generated by deep neural networks usually performs better than traditional image segmentation algorithms in undertaking and processing effects.

According to our work, an active contour model based on a convolution neural network is proposed, which combines the learning ability of a convolution neural network with people's prior knowledge of the target to be segmented (description of target shape, empirical statistics of brightness and color). Taking the aerial remote sensing image of urban architecture as an example, a semantic segmentation model for extracting dense urban buildings is constructed.

2. Related Work

Kass[2] et al. proposed an active contour model called Snake in 1987, based on global energy minimization. The model is guided by external constraints and pulls the initial contour to the target boundary under external rules. The internal energy term maintains the shape of the active contour.

Snake does not divide the areas used in the image, and it only divides specific regions or parts of particular areas. Snake relies on other mechanisms, such as interaction with users, higher-level image understanding processes, and information from image data adjacent to each other in time or space. This interaction must specify an approximate shape and initial profile of the starting position for the Snake somewhere near the target profile. Then, a priori and image-based information push the Snake to an appropriate energy minimization function. The minimized energy is a weighted combination of internal and external forces[3].

The active contour model depends on the deformable function controlled by the energy minimization function. The energy of the active contour relies not only on the image itself but also on the parameters of the deformable function and different prior knowledge. The energy of the active contour depends not only on the image itself but also on the parameters of the deformable function and additional prior knowledge:

The first and most important constraint is that because the goal of the active contour model is to perform segmentation based on object and shape detection, the active contour is required to be affected by force to make the active contour approach to the edge of the target object.

The second constraint depends on the profile model we want. We want to get a deformable model, which means that we want to have a continuous curve whose curvature will match the curvature of the target profile. Therefore, this means that there should be a theoretical model that we can use to describe the behavior of the active contour itself.

The energy of the active contour model is the sum of the internal energy term and the external energy term of the image[5]. Therefore, the energy function formula of the active contour model is obtained:

$$E_{snake} = E_{int} + E_{ext} = \oint_C \alpha \parallel \gamma'(s) \parallel^2 + \beta \parallel \gamma''(s) \parallel^2 ds - \delta \oint_C \parallel \nabla I \parallel^2 \big(\gamma(s)\big) ds \quad (1)$$

As mentioned earlier, this active contour method depends on the minimization of the energy function. Now that the energy function is defined, the following work builds a way to minimize the energy function. The goal at this stage is to find the best parameters that reduce the defined energy

IOP Conf. Series: Earth and Environmental Science 1004 (2022) 012015 doi:10.1088/1755-1315/1004/1/012015

function. These parameters are position vectors that define the location of the active contour line, thus minimizing the energy function. The purpose of minimizing the energy function is to find the value of $\gamma(s) = (\gamma_x, \gamma_y)(s)$ when E_{snake} is minimal.

Therefore, the optimization formula for this problem is:

$$S_{optimal} = \underset{\gamma \in F}{\operatorname{argmin}} \, \mathrm{E}(\gamma(s)) \tag{2}$$

Infinite dimensions, one of the most used methods, is the gradient descent method. This method comes from function analysis because if the function's gradient is equal to zero, the function has reached the local extremum. The gradient descent method is a first-order optimization technique based on first-order Taylor series decomposition. Other higher-order decomposition methods, such as Newton's method, can also be used.

The active contour can be expressed as a discrete contour region y = (upright v), which contains L control nodes, $y = (up\ pence\ v) \in \mathbb{R}^2$, where $s \in (1, L)$ each s represents a control node of the discretized profile. Change the shape of the discrete contour region y to minimize the following energy functions:

$$\mathrm{E}(y, x) = \sum_{s=1}^{L} \left[D(x, (y_s)) + \alpha(x, (y_s)) \left| \frac{\partial y}{\partial s} \right|^2 + \beta(x, (y_s)) \left| \frac{\partial^2 y}{\partial s^2} \right|^2 \right] + \sum_{u,v \in \Omega(y)} k\,(x, (u, v)) \tag{3}$$

Where $\mathrm{D}(\mathrm{x}) \in \mathrm{R}^{\mathrm{UxV}}$is called data correlation, depending on the input image of $U \times V$ size, then $\mathrm{x} \in \mathrm{R}^{\mathrm{UxVxd}}$, where d is the number of image channels). $\alpha(\mathrm{x}), \beta(\mathrm{x}) \in \mathrm{R}^{\mathrm{UxV}}$according to the introduction of the snake model in the second chapter, these two parameters make the discrete contour region tend to converge. Smooth, respectively., $k(x)$ is a balloon force term. The introduction of the balloon force term [6]can effectively prevent the discrete contour region from collapsing into a point where Ω (y) refers to the area surrounding the contour line. $D\ (x, (u_s, v_s))$ represents the value of D (x) at the control point yard $\mathrm{y_S} = (\mathrm{u_s}, \mathrm{v_s})$. In the design of this model, D, β, and k are $U \times V$ vectors, and α is the scalar in our experiment.

2.1. *Data Item*

Data dependencies define the area where the active outline should be located. D (x) is usually a predefined function on an image, usually related to the image gradient. A suitable D (x) should correspond to a relatively low value at the edge of the target object and a higher value in other areas. In the active contour model, we generally use the direction in which the gradient decreases the fastest, so we will

$$-D(x) = -\left[\frac{\partial D(x)}{\partial u}, \frac{\partial D(x)}{\partial v}\right] \tag{4}$$

As a data correlation, the contour tends to the boundary of the target object.

2.2. *Internal energy term*

In the traditional active contour model, the values of α and β are usually scalar, which means that there is the same feedback intensity in all image pixels, so we need to weigh the value of β in the traditional active contour method. Changing the dimension of β to assign different β punishments to each pixel to avoid this tradeoff allows the precise edge information to be extracted effectively.
The function of the internal energy term is:

$$E_{int} = \alpha(x, (y_s))|y' \;|^2 + \beta(x, (y_s))|y'' \;|^2 \tag{5}$$

furthermore, we get the discretized form of the internal energy term:

$$E_{int} = \sum_{s=0}^{L} \alpha(y_s) \left| \frac{y_{s+1} - y_s}{\Delta\, s} \right|^2 + \beta(y_s) \left| \frac{y_{s+1} - 2y_s + y_{s-1}}{\Delta\, s^2} \right|^2 \tag{6}$$

thus the derivative of the internal energy term to the contour control point is calculated:

$$\frac{\partial E_{int}}{\partial y_s} = \frac{2}{\Delta s}[-\alpha_{s-1}, \alpha_{s-1} + \alpha_s, -\alpha_s] \cdot [y_{s-1}, y_s, y_{s+1}]^{\top}$$
$$+\frac{2}{\Delta s^2}[\beta_{s-1}, -2\beta_s - 2\beta_{s-1}, \beta_{s-1} + 4\beta_s + \beta_{s+1}, -2\beta_{s+1} - 2\beta_s, \beta_{s+1}] \cdot$$
$$[y_{s-2}, y_{s-1}, y_s, y_{s+1}, y_{s+2}]^{\top} \tag{7}$$

transforming it into the form of an elegance-bit matrix:

$$\frac{\partial E_{int}}{\partial y} = (A + B)\mathrm{y} \tag{8}$$

Where A (α) is a tridiagonal matrix, B (β) is a pentagonal matrix.

2.3. *Balloon force term*

The balloon force term refers to the expansion of the active contour by adding a constant external force in the average direction of each contour control node. Like the coefficient β, we increase its flexibility by taking a different value at each image location.

In the LD Cohen et al. [6], the balloon term is only considered the force increased after calculating the minima of other energy terms. In the model proposed in this paper, the balloon force term is regarded as part of the energy function related to the loss function we define.

Represent all balloon forces as a vector:

$$n_s = [y_{s+1} - y_{s-1}]_{+90^\circ} = [v_{n+1} - v_{n-1}, u_{n+1} - u_{n-1}] \tag{9}$$

further expressed as:

$$\mathrm{n} = [Cv, u^{\top}C]$$

Where C is a tridiagonal matrix, the main diagonal is 0, the upper diagonal is 1, and the lower diagonal is -1. Therefore, we can express the energy function of the balloon force term as:

$$E_b = u^{\top}C\mathrm{v} = \iint_{u,v\in\Omega(y)} d\, udv \tag{10}$$

to flexibly control the balloon force, the coefficient k\inR$^{M\times N}$ similar to β is added and discretized based on the original function:

$$E_k = \sum_{u,v\in\Omega(y)} k\,(u, v) \tag{11}$$

to obtain the derivative of the balloon force energy $E_k$to the vertical balloon force u_s, we first obtain the relationship between\mathrm{ Δ } *E_k* and\mathrm{ Δ } *u_s*:

$$\Delta E_k = \frac{\Delta u_s}{v_{s-1}-v_s}\int_{h=0}^{v_{s-1}-v_s} h\mathrm{k}(h)\mathrm{d}h + \frac{\Delta u_s}{v_{s+1}-v_s}\int_{h=0}^{v_{s+1}-v_s} h\mathrm{k}(h)\mathrm{d}h \tag{12}$$

therefore, the derivative of the balloon force term energy function concerning the vertical balloon force is:

$$\frac{\partial E_k}{\partial u_s} = \frac{1}{v_{s-1}-v_s}\int_{h=0}^{v_{s-1}-v_s} h\mathrm{k}(h)\mathrm{d}h + \frac{1}{v_{s+1}-v_s}\int_{h=0}^{v_{s+1}-v_s} h\mathrm{k}(h)\mathrm{d}h \tag{13}$$

When we exchange u and v positions, we get the balloon force energy function derivative concerning the horizontal balloon force.

Finally, we obtain the iterative formula of the edge contour control points in the active contour model:

$$y^{t+1} = y^t - \frac{dE_{ext}}{dy^t} - (A + B)y^{t+1} \tag{14}$$

further simplification to get:

IOP Conf. Series: Earth and Environmental Science 1004 (2022) 012015 doi:10.1088/1755-1315/1004/1/012015

$$y^{t+1} = (I + A + B)^{-1}\left(y^t - \frac{dE_{ext}}{dy^t}\right) \tag{15}$$

3. Network Architecture

3.1. *Loss function*

For each pair of input images and their corresponding ground truth images, it is expressed as $(y^i, x^i) \in y \times x, \mathrm{i} = 1 \ldots N$, our goal is to minimize the deviation $\Delta\,(y, \hat{y})$ between the predicted contour image and ground truth.

From the energy function in section 2, we can infer the expression of y times:

$$\hat{y}^i = \underset{y \in \mathcal{Y}}{\mathrm{argminE}}(y, x, w) \tag{16}$$

to minimize the $\Delta\,(y, \hat{y})$, we turn the problem into:

$$\widehat{w} = \underset{w}{argmin} \textstyle\sum_i \Delta\left(y^i, \underset{y \in \mathcal{Y}}{argmin} E(y, x, w)\right) \tag{17}$$

because of the discontinuity of $\Delta\,(y, \hat{y})$, we can replace it with a continuous convex function, such as the hinge loss function [7]. By adding L2 regularization and summing all the training samples, the maximum margin can be obtained. The formula is as follows:

$$\mathcal{L}(\mathcal{Y}, x, w) = \frac{1}{2} \parallel w \parallel^2 + \mathrm{C} \textstyle\sum_i \left(\underset{y \in \mathcal{Y}}{\max}\left[0, \Delta(y, y^i) - E(y, x^i; w) + E(y^i, x^i; w)\right]\right) \tag{18}$$

Since $\mathcal{L}(\mathcal{Y}, x, w)$ is a non-differentiable convex function, we need to calculate the sub-gradient [8], which requires us to determine that the current ω is worthy of the maximum penalty constraint. In this case, the expressions of $\hat{y}^i$ becomes:

$$\hat{y}^i = \underset{y \in \mathcal{Y}}{\mathrm{argmax}}\left[\Delta\,(y, y^i) - E(y, x^i; w)\right] \tag{19}$$

the subgradient of w is expressed as:

$$\frac{\partial \mathcal{L}(\mathcal{Y}, x, w)}{\partial w} = \mathrm{w} + \mathrm{C} \textstyle\sum_i \left(\frac{\partial E(y^i, x^i; w)}{\partial w} - \frac{\partial E(\hat{y}^i, x^i; w)}{\partial w}\right) \tag{20}$$

therefore, we get the subgradients of D, α, β, k.:

$$\frac{\partial \mathcal{L}(y^i, x^i, w)}{\partial D_w(x^i)} = \left[(u, v) \in y^i\right] - \left[(u, v) \in \hat{y}^i\right] \tag{21}$$

$$\frac{\partial \mathcal{L}(y^i, x^i, w)}{\partial \alpha_w(x^i)} = \left|\frac{\partial y^i(u,v)}{\partial s}\right|^2 \left[(u, v) \in y^i\right] - \left|\frac{\partial \hat{y}(u,v)}{\partial s}\right|^2 \left[(u, v) \in \hat{y}^i\right] \tag{22}$$

$$\frac{\partial \mathcal{L}(y^i, x^i, w)}{\partial \beta_w(x^i)} = \left|\frac{\partial^2 y^i(u,v)}{\partial s^2}\right|^2 \left[(u, v) \in y^i\right] - \left|\frac{\partial^2 \hat{y}(u,v)}{\partial s^2}\right|^2 \left[(u, v) \in \hat{y}^i\right] \tag{23}$$

$$\frac{\partial \mathcal{L}(y^i, x^i, w)}{\partial k_w(x^i)} = \left[(u, v) \in \Omega(y^i)\right] - (u, v)\left[\Omega \in (\hat{y}^i)\right] \tag{24}$$

Where [] is Iverson's parenthesis, which means that if the condition in the square bracket is met, it is 1, and if it is not satisfied, it is 0.

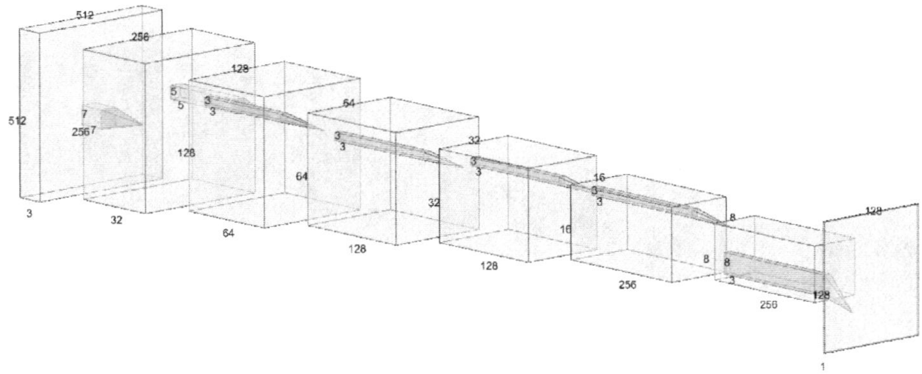

Figure 1. The structure of CNN

3.2. *Network model*

As shown in **Figure 1**, the CNN-based active model consists of two parts: the CNN module[4] and the ACM module. The CNN module learns four parameters of the energy function of the active contour model through the training set: the coefficient α related to the discreteness of the control points, the coefficient D of the data correlation term, the coefficient β associated with the smoothness of the discrete contour, and the coefficient κ related to the balloon force. The ACM part iterates continuously according to the initial contour to minimize the total energy and get the predicted contour. The structured loss function connects the CNN part with the ACM part and updates four main parameters by backpropagation. As shown in figure X. During each iteration, CNN is followed by ACM, which predicts a profile to calculate the structured loss, which is then sent back to CNN for backpropagation to update the parameters for the next iteration.

In the CNN module we designed, the input size of CNN is fixed, and the first six layers of the network are FCN networks, in which the convolution kernel size of the first layer is 7x7, the convolution core size of the second layer is 5x5. The convolution core size of the remaining four layers is 3x3. Each convolution layer is followed by ReLu, Batch normalization, and 2x2 pooling layer, and the output of each layer is sampled (or deconvolution) to the output size and spliced together. The number of convolution cores in the first six layers of the network is 32, 64, 128, 128, 256, 256, respectively. Figure 5.2 shows the network structure of the FCN part of the CNN module 错误!未找到引用源。. Note that the pooling process and upsampling process are not shown in the diagram.

The FCN output is a three-dimensional vector, flattened into a one-dimensional vector, and input to a double-layer MLP with 256 and 64 hidden units to predict four output maps: $\mathrm{D}(x), \alpha(x), \beta(x), \mathrm{k}(x)$.

The ADAM optimizer with a learning rate of 10-4 is adopted in the CNN-based ACM model. We use geometric transformation for data enhancement. The number of iterations of the ACM module is set to 50.

4. Data and Experiment

In the experiment, we used the Vaihingen buildings dataset, which consists of 168 buildings extracted from the "2Dsemanticlabelingcontest" training sets[9][10][11]. The image has three bands corresponding to near-infrared, red and green wavelengths, and the resolution is 9cm. The Vaihingen data set uses high-resolution orthogonal photos and corresponding dense image matching technology to generate the digital terrain model (DSM). Vaihingen is a relatively small village with many separate buildings and small multi-story buildings. Each dataset has been manually classified into the six most common land cover categories. Vaihingen is a multi-category semantic segmentation data set, but we only need building categories in this experiment, so we need to process the original ground truth reference image.

IOP Conf. Series: Earth and Environmental Science 1004 (2022) 012015 doi:10.1088/1755-1315/1004/1/012015

As shown in **Figure 2**, the left is the original ground-truth reference image, and the right is the processed single building category image.

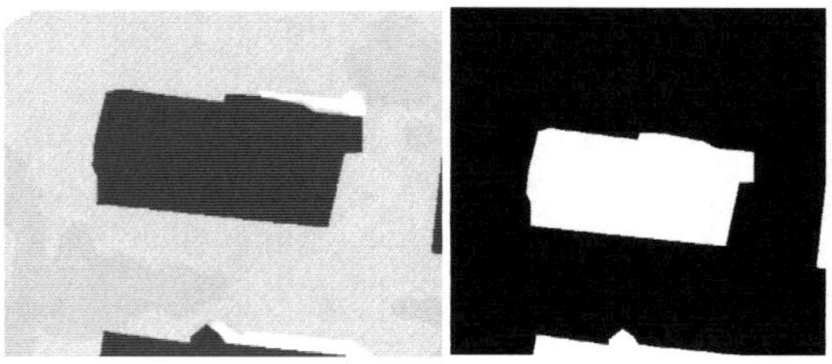

Figure 2. multi-labels to single label

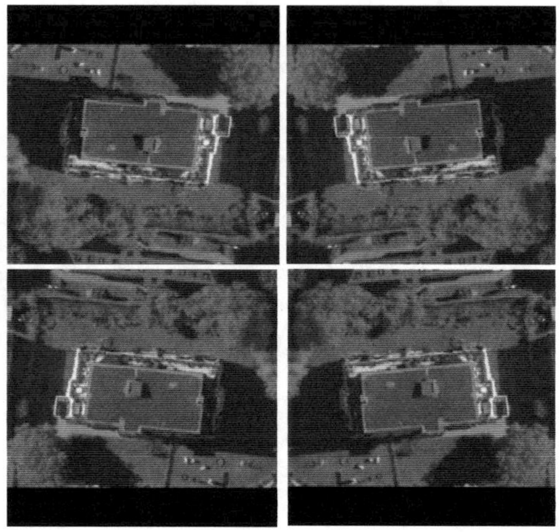

Figure 3. Data Augmentation

For remote sensing images, in the imaging process, the sensor will produce different images of different positions and shapes when the same object is photographed at different angles. Therefore, the image after geometric transformation can make the neural network model better learn the rotation invariance characteristics of ground objects to better adapt to varying forms of images. Therefore, we need to perform data enhancement operations on the dataset. Figure 3 shows the four states of an image: original image, horizontal flip, vertical flip, and diagonal flip.

The bing huts data set and NWPU VHR-10-car data set are selected for training to test the model's performance in low-resolution images. The Bing huts data set consists of 605 individual huts, and aerial images with 30cm resolution are selected from rural Tanzania on Bing maps. Among them, 335 shots of 80 × 80 pixels are used to train the model, and the remaining 270 images are used for testing. Low spatial resolution, low contrast between buildings and surrounding soil, and high label noise make Bing huts a very challenging data set. NWPU VHR-10-car data set is an aerial remote sensing image set extracted from the NWPU VHR-10 data set, which only contains automobile class. Northwestern Polytechnic University released the NWPU VHR-10 data set in 2014. the color images of the data set were obtained from Google Earth with a spatial resolution range of 0.5m -2m. The extracted NWPU VHR-10-car has 339 images with a size of 128x128 pixel, of which 170 are used for training models and the other 169are used for testing.

5. Result and Discussion

Table 1. Evaluation index of the model in each data set [a]

	mIoU		
	Vaihingen	NWPU VHR-10-car	Bing huts
Basic CNN	0.77	0.55	0.56
ACM based on CNN	**0.83**	**0.63**	**0.65**
ACM based on CNN (β, k as scalar)	0.64	-	-
ACM based on CNN (no k)	0.63	-	-
ACM based on CNN (α as a vector)	**0.83**	-	-

[a] the symbol - means this value below 0.5

Table 1 shows the average intersection and merges ratio ((mIoU)) of different models in different data sets. The ACM model based on CNN significantly improves the IOU benchmark. At the same time, the above other models confirm the necessity of κ and β non-scalar, while α can be treated as a scalar without affecting the performance. It also emphasizes the importance of the balloon force term for contour convergence. The CNN-based ACM model is also improved by more than 5% on low-resolution data sets.

Figure 4. model process result

In **Figure 4**, the ground truth is a solid green line, the active outline is a yellow dashed line, and the initial outline is a blue dashed line. The example of the segmentation result of the Vaihingen data set shows that the parameters learned by the convolution neural network do produce smooth and straight edges while allowing sharp corners to existing. By looking at the energy terms predicted in **Figure 5**, we observe the following situations: The energy of the data correlation term D (X) is low on the boundary of the building, and the balloon force term κ (X) only produces a positive value in the interior of the building, especially near the corner of the building. In the curvature-related β (X), the weight near the corner of the building is low or even has no curvature.

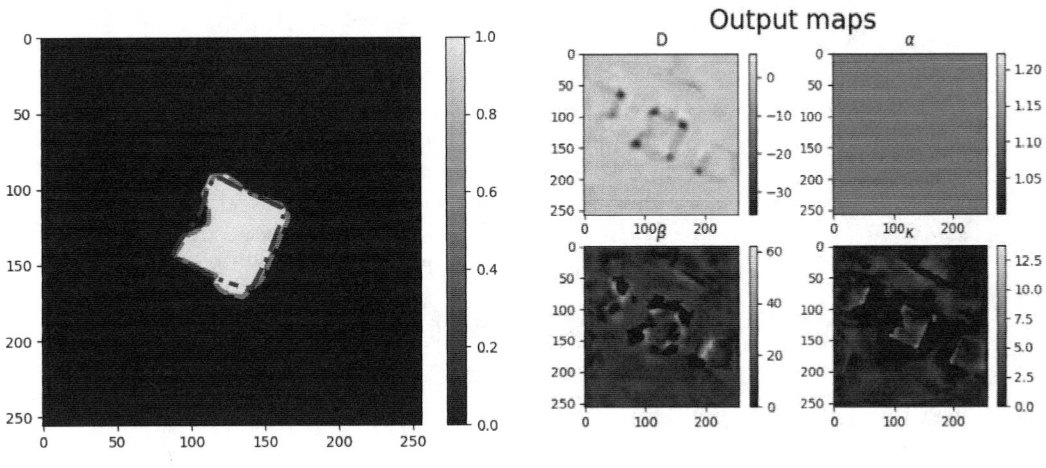

Figure 5. The map of $D(x), \alpha(x), \beta(x), k(x)$

In NWPUVHR-10-car and Bing huts datasets with low resolution and few pixels, the accuracy of the ACM model based on convolution neural network is 5% higher than that of the benchmark model.

The above situation shows the potential of embedding advanced geometric features in the deep learning framework. The ACM model based on convolution neural network uses CNN to determine the parameters of the energy function of the active contour model (ACM) so that its output approaches the closed contour of the edge of the target object. By incorporating ACM reasoning into CNN training and using the creation of ACM and Ground Truth images to assess the structural loss, and then using backpropagation to update the CNN parameters for end-to-end training.

The ACM model based on a convolution neural network uses many energy item sets to match different data sets because the neural network automatically learns the balance between them. The main drawback of this model is that initialization is given through some external method, so it is not included in the learning process.

The ACM model based on a convolution neural network uses many energy item sets to match different data sets because the neural network automatically learns the balance between them. The main drawback of this model is that initialization is given through some external method, so it is not included in the learning process.

6. Conclusion

With the popularity of deep learning frameworks such as TensorFlow and PyTorch, neural network models have become more convenient and fast. The traditional image processing algorithms gradually fade out of the mainstream field of vision under the impact of neural networks represented by CNN. The conventional image processing algorithm is based on a solid theory. It can be explained by a perfect thesis, while the neural network model learns the characteristics of the image from the data set, which is abstract and difficult to explain. The interpretability of the model is theory-centered, and the model's defects are known and further improved through the complete explanation of the model so that human beings can trust the model. The lack of an interpretable network model is like a black box, which can only be found and patched through various black-box tests.

By combining the traditional image processing algorithm with the neural network, the universal standard image processing algorithm can be optimized in a specific field, and the significance of the neural network parameters can be inferred from the perfect theoretical basis of the traditional image processing algorithm.

References

[1] Haralick R M, Shapiro L G. Image segmentation techniques[J]. Computer vision, graphics, and image processing, 1985, 29(1): 100-132.
[2] Kass M, Witkin A, Terzopoulos D. Snakes: Active contour models[J]. International journal of computer vision, 1988, 1(4): 321-331.
[3] Choi S, Kim C. Automatic initialization active contour model for the segmentation of the chest wall on chest CT[J]. Healthcare informatics research, 2010, 16(1): 36.
[4] LeCun Y, Bengio Y, Hinton G. Deep learning[J]. nature, 2015, 521(7553): 436-444.
[5] Kichenassamy S, Kumar A, Olver P, et al. gradient flows and geometric active contour models[C]//Proceedings of IEEE International Conference on Computer Vision. IEEE, 1995: 810-815.
[6] Cohen L D. On active contour models and balloons[J]. CVGIP: Image understanding, 1991, 53(2): 211-218.
[7] Vishwanathan S V M, Murty M N. SSVM: a simple SVM algorithm[C]//Proceedings of the 2002 International Joint Conference on Neural Networks. IJCNN'02 (Cat. No. 02CH37290). IEEE, 2002, 3: 2393-2398.
[8] Boyd S, Xiao L, Mutapcic A. Subgradient methods[J]. Lecture notes of EE392o, Stanford University, Autumn Quarter, 2003, 2004: 2004-2005.
[9] Gong Cheng, Junwei Han, Peicheng Zhou, Lei Guo. Multi-class geospatial object detection and geographic image classification based on collection of part detectors. ISPRS Journal of Photogrammetry and Remote Sensing, 98: 119-132, 2014.
[10] Gong Cheng, Junwei Han. A survey on object detection in optical remote sensing images. ISPRS Journal of Photogrammetry and Remote Sensing, 117: 11-28, 2016.
[11] Gong Cheng, Peicheng Zhou, Junwei Han. Learning rotation-invariant convolutional neural networks for object detection in VHR optical remote sensing images. IEEE Transactions on Geoscience and Remote Sensing, 54(12): 7405-7415, 2016.
[12] Kim Y, Kim S, Kim T, et al. CNN-Based Semantic Segmentation Using Level Set Loss[C]//2019 IEEE Winter Conference on Applications of Computer Vision (WACV). IEEE, 2019: 1752-1760.
[13] Chen X, Williams B M, Vallabhaneni S R, et al. Learning Active Contour Models for Medical

Image Segmentation[C]//Proceedings of the IEEE Conference on Computer Vision and Pattern Recognition. 2019: 11632-11640.

Acknowledgment
This work was supported by the National Natural Science Foundation of China under Grant 61731022, Grant 41971397.

A Standardized GIS Web Service Oriented Shared-nothing Architecture - Taking Species Distribution Modeling as an Example

X Y Liu[1], J S D Yu[1*], D D Wei[1], R J Tong[2], D Y Li[1] and S Luo[1]

[1] The Academy of Digital China(Fujian), Fuzhou University, Fuzhou, China
[2] Department of Transportation, Fujian University of Technology, Fuzhou, China

*Corresponding author, e-mail: yyx350@126.com

Abstract. The rapid development of big data analytics technologies and tools, particularly the related distributed computing technologies, which divide large problems into smaller ones, provides important technical support for developing big earth data research. To realize the effective utilization of big data computing resources, it is important to overcome the potential incompatibility and inconsistency among the distributed computing nodes. Using standardized GIS web service interfaces is a promising approach to link existing heterogeneous GIS platforms through standardized protocols and establish a universal, platform-independent computing layer to improve the efficiency of computing resources for big earth data analysis. This research analyzed the requirements of distributed computing, and further proposed a standardized GIS web service-oriented shared-nothing architecture by comparing various mainstream distributed computing architectures and related web service specifications in the industry. For example, the prediction of *Lophelia pertusa* coral distribution by random forest is the application of the method discussed in this paper.

Keywords- Big Data; parallel computing; GIS Web services

1. Introduction

The rapid development of emerging big data analytics technologies and tools, especially the related distributed computing technologies, which divide large problems into smaller ones, provides important technical support for the development of big earth data research. Big data analytics always refers to processing datasets of a size that cannot be done on a single computing unit and for traditional data processing and algorithms that are inadequate.

Distributed computing always refers to a technology that carries out many calculations or processes simultaneously by multiple computing devices. There are several different types of distributed computing architectures, mainly including shared-memory architecture, shared-disk architecture, shared-nothing architecture, etc. In a shared-memory architecture, memory may be simultaneously accessed by multiple programs without redundant copies. It may also improve performance for passing data between programs. However, when several processors try to access the same memory location simultaneously, the memory bus is impeded due to memory contention. In a shared-disk architecture, each node has its own private memory, and the disks are accessible from all of the computing nodes. In a shared-nothing architecture, all nodes have sole access to distinct memory or storage. Hence, single points of failure and resource contention among nodes will be avoided. As nothing is shared in each node under a shared-nothing architecture, computing nodes of different communities can have their own processing framework without knowing each other. Hence, consistent functionality of interface compatibility for communication cannot always be guaranteed.

Normally, distributed computing technologies are not particularly designed for use over a wide-area network, such as the global internet, while web services provide the potential for creating and deploying loosely coupled applications globally. However, web services are different from traditional distributed computing architectures. Different services tend to be heterogeneous when the applied interface specifications are neither compatible nor consistent.

Therefore, to realize the effective coordination of big earth data computing resources, it is important to overcome the potential incompatibility and inconsistency among nodes. Standardized interfaces are a promising approach to link existing heterogeneous GIS platforms through standardized protocols and establish a universal, platform-independent computing layer to improve the efficiency of big earth data analytics resources. Additionally, a series of related standardized works that have been delivered by international standardization organizations, including ISO, IETF, W3C, etc., where ISO TC211, in collaboration with Open Geospatial Consortium (OGC), are currently the main driving forces in open standards for interoperable GIS web services [1-4].

Regarding standardized GIS web service-oriented distribution architecture, Owonibi proposed a related D-WCPS framework to support several WCPS servers to share a service request for parallel computing [5]. The approach may still be enhanced by introducing ad hoc computing algorithms under unified standardized web interfaces. Yu proposed a standard-based collaborative analysis method based on Web Coverage Processing Service (WCPS) and R to effectively improve the interoperability and processing of spatial data infrastructure [6]. However, the method does not cover the distribution issues. Efforts have been investigated on the improvement of GIS parallel computing [7][8] without concerning the standardization issue. Some studies have realized the parallel management and visualization of multi-source heterogeneous spatial data based on standardization, but have not further completed the analysis and calculation [9][10]. This research proposes a standardized GIS web service-oriented shared-nothing architecture to address the issue. A study case based on species distribution modeling is examined to explore the application realization method.

The rest of the paper is organized as follows: Section 2 presents the study case. Section 3 presents the configuration of the proposed architecture. Section 4 describes the experiment and includes the discussion. Section 5 analyzes acceleration performance. Section 6 concludes the paper.

2. Case study

Marine species distribution models are often applied to predict the potential distributions of species across geographic space or time using environmental variables. The environmental variables may include temperature, precipitation or water depth. When modeling a large area with dozens of high-resolution environmental variables, a large amount of computing resources is needed to process the big data and build the model. When modeling with terabyte or petabyte data, a series of circumstances occur, such as having to take several hours or even more time for data retrieval, having a large amount of model calculation, having insufficient memory, and having to wait a long calculation time for prediction models. Species distribution prediction can be divided into a model training process and a model prediction process. Thus, parallel computing can be used in the above two processes to improve the computational efficiency, but this study only focuses on the model prediction process for parallel computing acceleration processing.

In this study, modeling the potential distribution of *Lophelia pertusa* coral in the North Atlantic [11] was taken as an example to develop the application framework. Biomod2 is an R language package for species distribution modeling [12]. Based on the environmental variables (EVas) and training samples (TSse), a random forest (RF) model was built through calculation and further evaluated based on testing samples. The RF model, together with EVas, was used to complete the projection for the prediction of species distribution. The serial calculation data flow chart is shown in Figure 1.

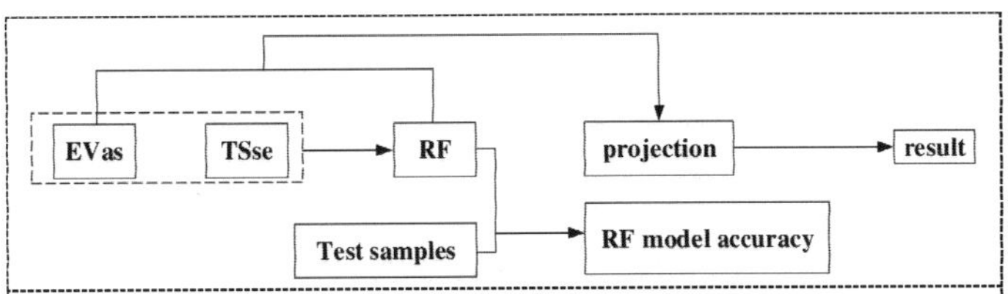

Figure 1. Serical calculation data flow

Species distribution prediction algorithms involve two processes, including RF model establishment and projection. With reference to the "divide-and-conquer" concept of big data processing [13]. The approach is to build an RF model for EVas and TSse, split EVas, project each node in parallel based on the RF model and EVas partition data, and finally combine the partitioned results. The parallel computing data flow diagram is shown in Figure 2.

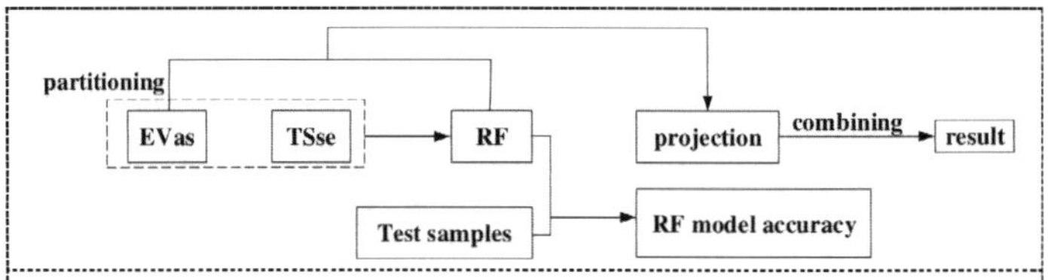

Figure 2. Parallel computing data flow

3. Configuration of the sharded-nothing architectural

Based on a Rasdaman and R parallel approach, the proposed method is improved to support distributed computing of the RF model by using submarine habitat variables and *Lophelia pertusa* coral distributed/non-distributed data.

Rasdaman [14] is a reference implementation of OGC WCS 2.0. In this study, the RF model was trained at the master node, which was delivered to the slave nodes based on the SCP protocol. Then, the shell script of distributed computing was activated through the SSH protocol. Slave nodes retrieved the partitioned EVas based on OGC WCS 2.0. The parallel projection was completed by R scripts, and then the partitioned results were sent back to the master node based on the SCP protocol. Finally, the complete result was combined by GDAL. The architectural is shown in Figure 3.

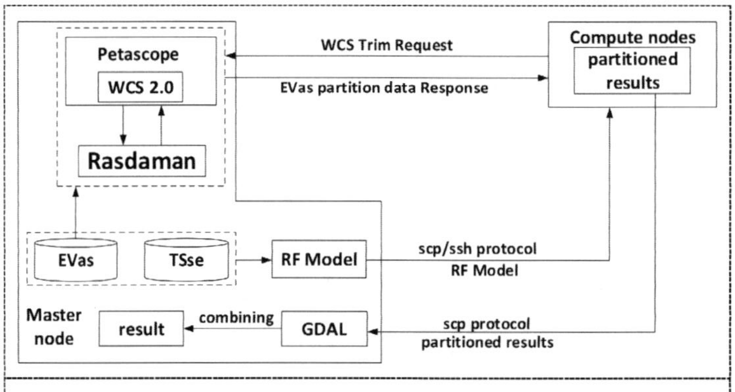

Figure 3. Distributed parallel computing service framework

4. Experiment

This experiment completes the *Lophelia Pertusa* coral distribution prediction based on submarine habitat variables [15], such as water depth, chlorophyll, alkalinity, etc. The computing node configuration is listed in Table 1.

Table 1. Node configuration information

CPU	RAM	Operating System	Mechanical hard disk
Intel(R) Core (TM) i7-9700	16G	Ubuntu 18.04 LTS	1T

Based on the same submarine habitat variables, different sizes of data (as shown in Table 2) were used to assess the performance of a single computing node in the prototyped parallel architecture. The performance is illustrated in Figure 4.

IOP Conf. Series: Earth and Environmental Science 1004 (2022) 012016 doi:10.1088/1755-1315/1004/1/012016

Table 2. Data size table

Experiment NO.	Data Size(M)
1	9216
2	4608
3	2355.2
4	1126.4
5	563.3
6	281.1
7	141.7
8	69.7
9	35.4
10	18.3
11	9.1

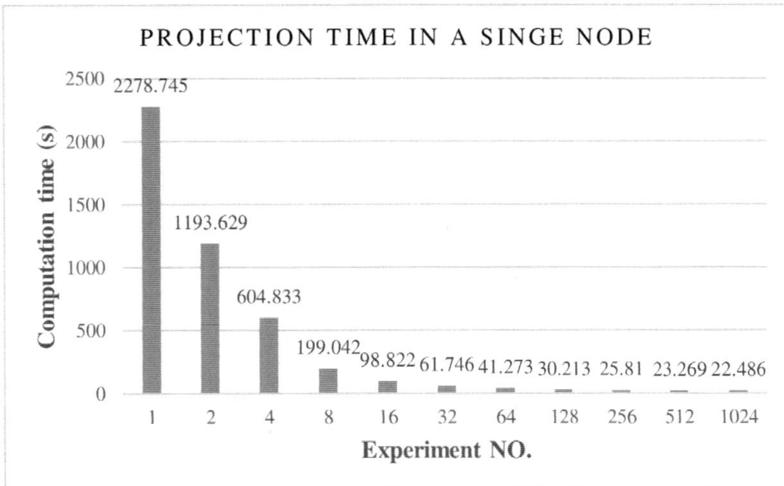

Figure 4. Node computing time

5. Analysis

To explore the theoretical speedup of this study, modeling the speedup function of a single computing node is required. The node computing time in Figure 4 was used to calculate the theoretical speedup ratio of a single computing node under the proposed parallel architecture. The curve is consistent with the exponential semivariogram curve; refer to Figure 5. Its mathematical expression is as follows:

$$y(x)=\begin{cases} 0 & ,x=0 \\ C_0+C(1-e^{-\frac{x}{a}}), & x>0 \end{cases} \quad (1)$$

where C_0 denotes nugget value; $C+C_0$ represents sill.

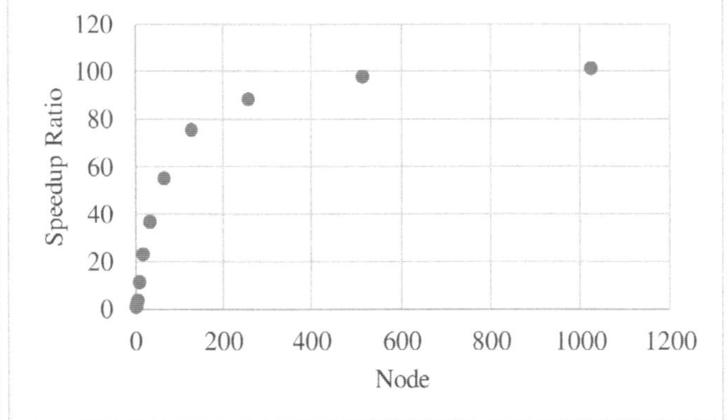

Figure 5. Theoretical speedup ratio

The calculated fitting function is as follows:

$$y(x)=\begin{cases}0 & ,x=0\\ 1.326+95.66*(1-e^{-\frac{x}{77.94}}), & x>0\end{cases} \quad (2)$$

The fitting curve is depicted in Figure 6, and the characteristic parameters are listed in Table 3. R-square is 0.9941. The result reveals that the mathematical function is well fitted with the experiment data. When more than 234 nodes are theoretically adopted for parallel computing, the speedup effect will mostly remain constant when approaching the maximum value of 96.986.

Table 3. Characteristic parameters

Nugget	Sill	Range	R-square
1.326	96.986	234	0.9941

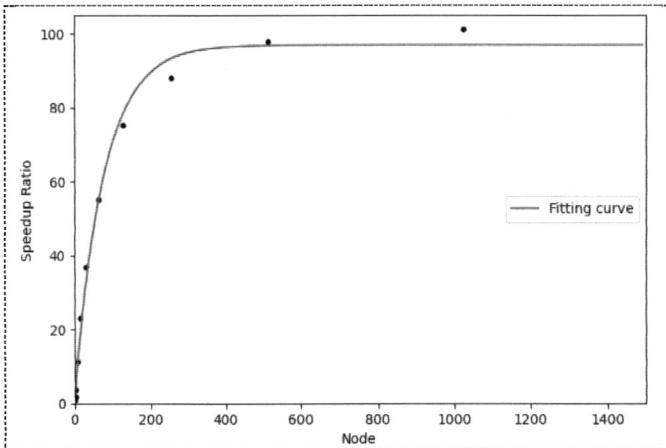

Figure 6. Fitting curve

6. Conclusion

This study delved into the random forest prediction of the distribution of *Lophelia Pertusa* coral in the North Atlantic Ocean. A standardized GIS web service framework was proposed based on the Rasdaman, shared-nothing architecture, and the distributed computing method. The study involved the decomposition of the prediction model considering operator dependence, splitting the heterogeneous data of multiple sources, realizing interoperation and parallel processing, and exploring the theoretical speedup model of a single computing node under the proposed parallel architecture. In this framework, incompatibility and inconsistency among computing nodes are obviated, the problem of insufficient memory is addressed, and the prediction time of the *Lophelia Pertusa* coral distribution is reduced. Therefore, it can potentially provide a faster solution for cold-water coral prediction on a global scale among heterogeneous GIS platforms. However, this research does not split both RF model establishment and projection processes. The next step will focus on both process distribution and optimization to achieve better performance by considering the accuracy tolerance. This study is part of developing a standardized GIS web service-oriented distributed computing architecture for species distribution modeling. In this way, an improved species distribution modeling performance under evolving implementations is expected.

References

[1] International Organization for Standardization 2005 ISO 19123:2005 Geographic information -- Schema for coverage geometry and functions[CrossRef]

[2] Open Geospatial Consortium 2012 OGC 09-110r4 OGC® WCS 2.0 Interface Standard- Core: Corrigendum[CrossRef]

[3] Open Geospatial Consortium 2012 OGC 09-146r2 OGC® Implementation Schema for Coverages[CrossRef]

[4] Open Geospatial Consortium 2021 OGC 08-068r3 Web Coverage Processing Service (WCPS) Language Interface Standard[CrossRef]

[5] Michael O Resource-Aware Decomposition of Geoprocessing Services Based on Declarative Request Languages(Jena: Friedrich Schiller University Jena)[CrossRef]

[6] Yu J S D, Li J X, Peter B, Tong R J and Chen Y J 2018-9-22 A Standard-Based Collaborative Analytics Service of Remote Sensing Imagery *2018 Inter. Workshop on Big Geo. Data and Data Sci. (BGDDS)* (Wuhan) pp 1-4[CrossRef]

[7] Yu J, Wu J X and Mohamed S 2015-11-3 GeoSpark: a cluster computing framework for processing large-scale spatial data *Proc. of the 23rd ACM SIGSPATIAL Int. Conf. on Advances in Geo. Info. Sys.* (Washington) pp 1-4[CrossRef]

[8] Yue P, Wu Z Y and ShangGuan B Y 2018 Design and Implementation of a Distributed Geospatial Data Storage Structure Based on Spark *Geomatics and Info. Sci. of Wuhan University* **43** pp 2295-2302[CrossRef]

[9] Liu Z W, Li S Y, Yu H J and Hao Z W 2016 Key Technology Research of massive multi-source heterogeneous spatial data visualization and management system based on 3D digital earth *2016 Int. Conf. on Info. Sys. and Artificial Intelligence* (Hong Kong) pp 241-246[CrossRef]

[10] Huang Z C, Zhong A R and Li G Q 2016-6-24 One-stop Service for Multi-Source Heterogeneous Remote Sensing Data *7th Int. Conf. on Mechatronics, Control and Materials (ICMCM 2016)* pp 417-423[CrossRef]

[11] Tong R J, Autun p, Janine G, Vikram U, Yu J S D and Zhang C C 2016-3-15 Quantifying relationships between abundances of cold-water coral *Lophelia pertusa* and terrain features: A case study on the Norwegian margin *Continental Shelf Research* **116** pp 13-26[CrossRef]

[12] Wilfried T, Damien G, Maya G, Robin E and Frank B 2021 Package 'biomod2'--Ensemble Platform for Species Distribution Modeling[CrossRef]

[13] Hong S 2016 Big data analytics: six techniques *Geo-spatial Info. Sci.* **19** pp 119-128[CrossRef]

[14] Baumann P, Misev D, Merticariu V, Huu B P, Bell B 2018-11-6 rasdaman: Spatio-Temporal Datacubes on Steroids *Proc. of the 26th ACM SIGSPATIAL Int. Conf. on Advances in Geo. Info. Sys.* (Washington) pp 604-607[CrossRef]

[15] Davies A J and Guinotte J M 2019 Global Habitat Suitability for Framework-Forming Cold-Water Corals - Environmental Data[CrossRef]

Acknowledgments

The authors would like to express appreciations to Large-Scale Scientific Information Systems research group at Jacobs University Bremen for their valuable contributions on relatead standards and reference implementations. This research was supported by the National Key R&D Program of China (NO. 2019YFE0127100), and

Natural Science Foundation of Fujian Province of China (No.2019I0006). The RF model training in this paper has been done on the supercomputing system in the Supercomputing Center of Fujian of China.

GeoMapViz: a framework for distributed management and geospatial data visualization based on massive spatiotemporal data streams

Qi Xu, Longgang Xiang*, Haocheng Wang, Xuefeng Guan, Huayi Wu

State Key Laboratory of Information Engineering in Surveying, Mapping and Remote Sensing, Wuhan University, Wuhan 430079, China.

* Correspondence: geoxlg@whu.edu.cn; Tel.: + 027-6877-8751

Abstract. Spatiotemporal big data have multisource, heterogeneous, high-dimensional and spatiotemporal associations. Due to the limited computing and network resources, while the spatiotemporal data to be rendered are large and dynamic, efficient visual analysis has always been a popular topic and has had difficulty in the research of spatiotemporal big data. As one of the important means of big data visualization, thermal maps play an important role in expressing data flow, information flow, and trajectory flow. At the same time, the development of a distributed computing framework also provides technical support for the online calculation and visualization of spatiotemporal data streams. In response to the above problems, this paper designs and implements GeoMapViz, a distributed management based on massive spatiotemporal data streams and a multiscale geographic spatial visualization framework, which is oriented by the expression of thermal maps of massive point datasets. First, based on the concept of the tile pyramid model and spatiotemporal cube, we propose a thermal map sequential tile pyramid (TS_Tile) model, which realizes scalable storage and efficient retrieval of data flow. GeoMapViz adopts a high-performance Flink stream computing cluster to implement the large-scale parallel construction of hierarchical tile pyramids, implements distributed storage and index construction of data based on HBase and Geomesa, and uses Geoserver to manage the map service to provide a spatiotemporal range query interface. Finally, through using an open dataset as a system simulation test, the results show that the TS_Tile model can effectively organize large-scale, time-space and multidimensional thermal map data, and the query and visualization of the heatmap can reach a subsecond response. Furthermore, GeoMapViz supports the integration of the thermal map and original flow and provides a feasible solution for the visual analysis of large-scale spatiotemporal data.

Keywords: thermal map; tile pyramid; spatiotemporal query; distributed storage and index; data visualization

1. Introduction

With the rapid development of the global internet, the Internet of Things, cloud computing and mobile communication technology, ubiquitous human social behaviour is always accompanied by a large amount of data. These data show the complexity of multidimensional, semantic and spatiotemporal dynamic correlation because they contain temporal, spatial and attribute dimensional information [1]. Each data flow model can be defined by $Object(S,T,V_n)$, where S represents the spatial location generated by the data flow, T is the temporal information of the data flow, and V_n is the n attribute mapping of the $Object$.

Currently, spatiotemporal data streams tend to diversify, and associative computation of spatiotemporal data streams is prevalent in many applications with high concurrent requests and large-scale data production. The traditional relational database gradually has bottlenecks in high concurrent reading and writing, efficient data storage, and scalability, which cannot meet the storage and management requirements of massive spatiotemporal data [2, 3]. Two schemes, file-level data management and nonrelational NoSQL database-level management, are often used in the storage management of data today. File-level data management [4-7] extends the local file system into a file system network with multiple data nodes, which achieves load balancing at the data storage level and relieves the pressure on computer storage by scaling horizontally. Many application studies have also been conducted by domestic and foreign scholars in data storage schemes for distributed file systems [8-10]. Unfortunately, file system-based storage solutions are usually suitable for offline data storage and computation and do not support real-time thermal map visualization with high temporal resolution and frequent updates.

A distributed nonrelational database represented by HBase shows great performance in practice because of its high reliability, high performance and scalability. In the practice of storing massive data based on HBase, there is a lack of research on the storage of temporal dimensions, while the storage of multiresolution tile data is mostly based on remote sensing images. For example, Liu et al. [11] used HBase to store massive image data and used MapReduce for scalable parallel processing. The data stream with spatiotemporal multidimensional information not only has spatiotemporal characteristics but also has multispatial resolution characteristics, which makes the storage scheme based on native HBase difficult to expand storage. Based on frequent updates of the hot map query requirements, HBase does not provide a secondary index mechanism, causing the multicondition application query requirements to have problems.

The storage scheme of data is also closely related to the spatiotemporal dissection scheme of data, and the use of a reasonable spatial data division method can solve the unified management and multiscale representation of massive spatiotemporal data at the global scale. The data-partitioning scheme includes the following: (1) spherical mesh subdivision based on polyhedron subdivision; and (2) spherical mesh generation based on geographic coordinates.

The spherical grid method based on polyhedron subdivision uses a regular polyhedron for data partitioning. For example, the quaternary triangular mesh (QTM, [12]) first divides the spherical surface into eight spherical triangles and then uses the form of a triangular quadtree for recursive subdivision to realize the discretization of data. Kimerling et al. [13] also extended the planar hexagonal grid to the spherical grid system to realize the hexagonal grid subdivision of the Earth projection plane. However, after the spherical subdivision of regular polyhedrons, there are inevitably deformations in size and shape, and the division of polyhedrons and the transformation of geographic coordinates are more complex.

Spherical grid subdivision based on geographic coordinate systems, such as Geohash, divides latitude and longitude information into two dimensions by dichotomy. This grid division method conforms to the law of people's understanding of the Earth [14, 15]. The tile pyramid model is a multiresolution hierarchical structure, and the adopted idea is similar to the level of details (LOD) model, but its organizational object is map tiles. The division method of tiles is based on spatial position, and the division of the time dimension is ignored.

In contrast, in the study of visualization of data flow, data visualization technology is an important method in the field of big data analysis and data mining. It can show the original data flow through calculation and processing in a visual form and can help scholars uncover the laws hidden behind the data. For example, the point of interest (POI) of the taxi in New York can help the taxi company locate the hot area of the taxi. Traditional visualization technology stores and renders map resources through tile technology, which greatly improves the efficiency of webport access. However, its application is limited to image data and does not perform well in unstructured data applications. In the face of a big data environment, classical geospatial map visualization (such as thermal maps) requires considerable time and cost to generate large-scale geospatial data maps, which is inefficient [16]. At the same time, after data preprocessing, most of them are deployed in different data platforms for decentralized storage. Due to the diversification of data types, the data standards of all walks of life are not uniform, which is not conducive to data sharing and publishing, and the efficiency of data integration is low. In the visualization of maps, such as Tableau, ArcGIS and Apache Zeppelin visualization tools, they work well on small- and medium-sized data, but they are difficult to extend to large-scale datasets. Tabula [17] used the sampling cube method to store the sample space and served the user-defined visualization tasks. MapD [18] used a GPU to parallelize, thus speeding up the query and pixelization in visualization, but it was limited to a single machine and had poor scalability. There are also many extended distributed visualization-processing frameworks, such as GeoSparkViz [19], HadoopViz [20], SHAHED [21], MAP_Vis [22], parallelized scatter maps, heatmaps and other map image-rendering pipelines, to realize the organization and visualization of spatiotemporal big data. Most implementations are concentrated in the distributed data-processing platform, using the batch-processing method of data, which has large data delay and does not support the visualization requirements of high real-time data.

In view of the above problems, this paper designs and implements GeoMapViz, a distributed management and multiscale geospatial visualization framework based on massive spatiotemporal data streams. Through distributed parallel computing on large-scale POI data (taxi data, check-in data, and traffic information) to achieve rapid online organization, distributed storage to multiple nodes establishes a spatiotemporal data index and thermal map-oriented visualization of data stream rendering. Based on the concept of the tile pyramid model and spatiotemporal cube, GeoMapViz proposes a thermal map sequential tile pyramid (TS_Tile) model, which realizes scalable storage and efficient retrieval of data flow. GeoMapViz adopts a high-performance Flink stream computing cluster to implement the large-scale parallel construction of hierarchical tile pyramids, implements distributed storage and index construction of data based on HBase and Geomesa, and uses Geoserver to manage the map service to provide a spatiotemporal range query interface. Finally, this paper forms a set of multiscale massive data visualization frameworks, GeoMapViz, which integrates data aggregation, computing, storage, service and visualization. GeoMapViz provides an open source visualization interface that integrates dynamic visualization rendering of thermal maps, visualization of geofences and original flow, and a spatiotemporal query association module.

The main contributions of GeoMapViz are as follows:

- Based on the concept of the tile pyramid model and spatiotemporal cube, GeoMapViz proposes a thermal map sequential tile pyramid model, which is called the TS_Tile model. The organization method of vector tile maps is used to save massive spatiotemporal big data into block data with multiple resolutions. The minimum unit of the spatiotemporal cube in logic is used as the thermal map mapping of the temporal tile, which can effectively organize large-scale thermal map tiles with time information.
- GeoMapViz encapsulates the map visualization process of spatiotemporal data streams, including rasterized spatial objects, pixel aggregation and other operations. Through the distributed ability of clusters, a large-scale parallel-processing process of thermal tiles is quickly constructed, and a multilevel pyramid model of thermal tiles is constructed in the form of TileDataStream without hierarchical aggregation of thermal tiles, which improves the computational efficiency and real-time support of the system.
- GeoMapViz forms a data-processing pipeline that publishes the preprocessing results of data flow as a map service to achieve data integration and sharing. The Geoserver platform is used to perform multiclass queries, including spatial queries, attribute queries, and spatiotemporal queries, with higher readability and applicability.
- GeoMapViz forms a set of multiscale massive data visualization schemes that integrate data aggregation, calculation, storage, service and visualization. The system integrates the modules of dynamic visualization rendering of thermal maps, visualization of geographic fences and

original flow, and spatiotemporal query association and provides an open source user interface to realize the interactive experience of spatiotemporal data flow.

In view of this prospect, the rest of this paper is organized as follows: Section 2 describes the construction method for organizing large-scale spatiotemporal data streams through the TS_Tile model and illustrates the system framework of GeoMapViz. Section 3 tests and evaluates the system through taxi datasets as use case scenarios. Section 4 provides a summary.

2. Functions and framework of GeoMapViz

2.1. Thermal Map Sequential Tile Pyramid (TS_Tile) Model

Considering the structure of data flow (S, T, and V), it is necessary to effectively organize and manage its spatiotemporal information and attribute information. Here, through two classical graphic concepts, the tile pyramid model and spatiotemporal cube are effectively combined to quickly build the TS_Tile model of the thermal map. The tile pyramid model provides a spatial information reference for the TS_Tile structure and constructs spatial divisions at different levels by LoD. The spatiotemporal cube provides a multidimensional extension for the TS_Tile structure and maps and abstracts the time dimension and attribute dimension information through the cube structure, which can effectively store and apply the data flow information.

2.1.1. Tile Pyramid Model

The tile pyramid is a multiresolution hierarchical structure that can organize the area in the same spatial range according to the decreasing resolution from the bottom to the top of the pyramid and the decreasing number of tiles. In the field of geographic information, the tile pyramid model was initially applied to the image pyramid. Through tile-slicing technology, the world map was formed into a multilevel image dataset according to the recursive subdivision principle of a quadtree.

The tile pyramid model adopts a global unified subdivision framework and realizes the transformation of spherical coordinates to a two-dimensional plane through Web Mercator projection. Because of the equiangular projection property of Web Mercator, its spatial range is [85.01°S, 85.01°N], [180°W, 180°E]. When the spatial level = 0, it is a square with the geographic scope of the world map. On this basis, quadtree recursive subdivision is carried out. Each low-level tile corresponds to four high-level tiles, forming a stepped tile map dataset, as shown in Figure 1.

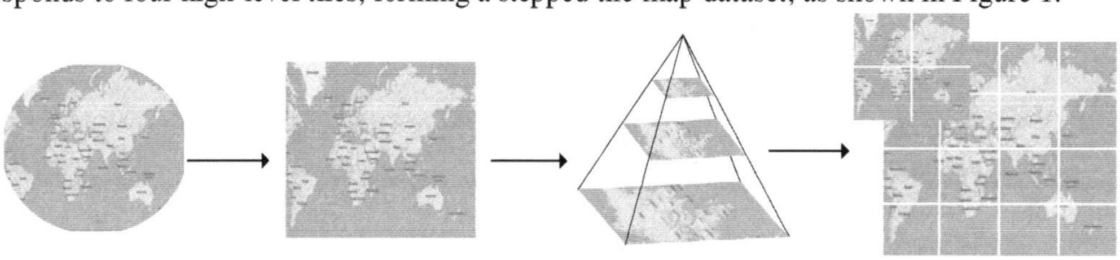

Figure 1. Map projection and the tile pyramid model.

The tile coordinate system needs to be defined in the tile pyramid model to uniquely identify each tile in the pyramid model. The upper left corner is used as the coordinate origin, the longitude direction is the X axis, the latitude direction is the Y axis, and the scaling level is Z. Therefore, each tile can be uniquely identified by (X, Y, and Z), as shown in Figure 2, where each tile contains 256*256 pixels.

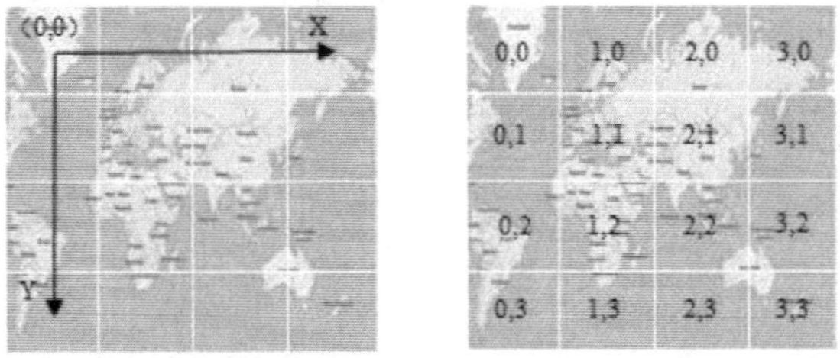

Figure 2. Tile coordinate system.

The image pyramid slice based on the grid slice is the image format after completion: (1) the style is fixed, and the flexibility is poor. (2) There is also a lack of real time and cannot be updated in real time, and grid tiles need to be regenerated when data information changes. (3) Attribute representation information is limited, and grid slicing cannot obtain attribute information. Due to the needs of zooming and roaming, the thermal map can be organized by tile pyramid technology, and users can schedule and render on demand according to needs.

Based on the above problems, vector tile data are more scalable, which is defined as:

- Pixel

Each tile in the pyramid model is composed of 256 × 256 pixels, and the pixel is the smallest construction unit of the pyramid model, which represents a spatial partition area of the finest grain. Z is the spatial level, and X and Y are the positions in the tile coordinate system, where Z-X-Y constitutes the parent node key of the pixel set. Value is the attribute mapping of pixel units (aggregation expression of multiple attributes, such as sum, average, count, max, min, and exist).

$$Pixel = \{Z, X, Y, Value\} \tag{1}$$

- Tile

Tile represents each map tile in the pyramid model, where Z, X, and Y represent the position of the tile in the pyramid model, which can be used as the only specified tile area of the index, including a set of 256 × 256 pixels (n = 256).

$$Tile = \left\{ Z, X, Y, \bigcup_{i=1}^{n\times n} Pixel \right\} \tag{2}$$

- TileMap

TileMap represents a map tile at a spatial level in a pyramid structure, which determines the map range at the Z level, and the tile collection when Z is the key.

$$TileMap_i = \left\{ Z_i, \sum_{m=1}^{2^i} \sum_{n=1}^{2^i} \left(X_m, Y_n, \bigcup_{j=1}^{256\times256} Pixel \right) \right\} \tag{3}$$

- Pyramid

Pyramid is the overall structure of the tile pyramid, which is a set of TileMaps at multiple spatial levels.

$$Pyramid = \left\{ \sum_{i=0}^{Z} TileMap_i \right\} \tag{4}$$

2.1.2. Data Cube

The data cube [23] is a multidimensional data model that allows users to analyse and query data from multiple perspectives and levels and is widely used in online analytical processing (OLAP) for big data management and organization. The data cube is built on top of a multidimensional data model, with each dimensional variable being able to be indexed for data access. The data cube has no heterogeneity, and the dimension range expressed in each cube is consistent. However, due to the uneven distribution of data, the data cube has a problem of data tilt. Many cube spaces are sparse, occupying a large amount of storage space. Therefore, the unit size of the cube dimension also

IOP Conf. Series: Earth and Environmental Science 1004 (2022) 012017 doi:10.1088/1755-1315/1004/1/012017

determines the storage efficiency of the cube. The data cube was first proposed by Grey et al. [24] to provide multidimensional analytical operations: drill-down, roll-up, slice, dice and pivot. Through the integration and subdivision of data, data can be divided from different angles and different levels to meet the analytical needs at different levels.

When the spatial and temporal dimensions are applied to the data cube, it becomes a spatiotemporal cube model. The three branches of the data cube are replaced by longitude, latitude and time, and the entire data space is divided into the same spatiotemporal unit. Each spatiotemporal unit has different partition granularity in each direction, and the obtained spatiotemporal unit size is also different. The dividing granularity of the spatiotemporal cube determines the dividing line position of the data in a certain dimension. The dividing position of the latitude and longitude grid of the thermal map object is closely related to the level of the scaling level, and the time dividing granularity is consistent among the levels. Each spatiotemporal unit represents the dataset with the same spatial and temporal range.

Since there is no upper bound on the time axis, the interval partition on the time dimension can be realized by the time bin and used as a coarse-grained time window. In a spatiotemporal cube, the time information of all data comes from the same time window (as shown in Figure 3); therefore, the multilevel temporal tile cube can be constructed.

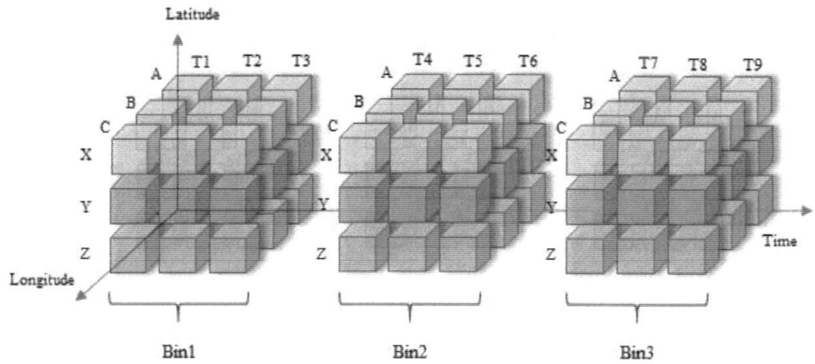

Figure 3. Spatiotemporal cube model.

After organizing the thermal map object through the spatiotemporal cube model, the thermal map tile corresponds to each spatiotemporal unit, but the Lat × Lon × T cube is still a multidimensional (nD) structure. This is because space-time objects have different attribute information, and each cube represents a vector dataset within the same space-time range.

2.1.3. *TS_Tile Model*

Based on the concept of space division and hierarchy of tile pyramids and the multidimensional expansion of the data cube, the TS_Tile model can be constructed, which can effectively store and manage thermal tiles with multidimensional time and space information.

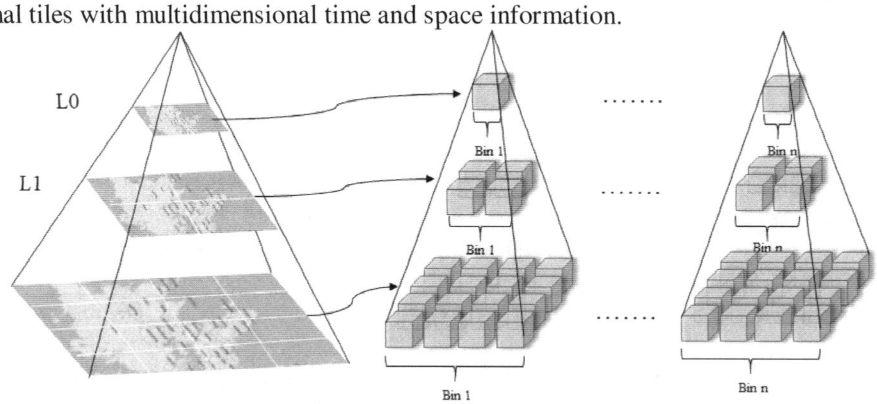

Figure 4. TS_Tile model.

As shown in Figure 4, the TS_Tile model divides the data stream information with spatial and temporal multidimensional information according to the global unified subdivision framework, forming a two-dimensional grid area on the plane and stretching the attribute dimension information longitudinally. At each level, combined with the time window mechanism of Flink stream processing, by designing the predetermined time window size, the spatial data are extended horizontally to form a single level of the spatiotemporal cube model. Between multiple levels, the space-time cubes still form a pyramid model. In the organization of the data model, the time information is divided into interval forms to form a multidimensional spatial dataset. However, in the data visualization, the deadline point of the time window is used. The time dimension and attribute dimension of the spatiotemporal cube need to be mapped to the two-dimensional data slice to show the visualization effect of the sequential continuous thermal map.

Similarly, the definition of the TS_Tile model is as follows:

- ts_Pixel

We define the minimum construction unit in the TS_Tile model as a space-time pixel (ts_Pixel), which represents a space partition region with the finest granularity. It is the pixel unit of a pyramid structure existing in the window sequence bin at a certain time window length.

$$ts\_Pixel = \left\{Bin_n\left(Z, X, Y, Value\right)\right\} \tag{5}$$

- ts_Tile

Spatiotemporal tiles (ts_Tile) represent the pyramid map tiles with time series identification, including the ts_Pixel set with the same time mark and attribute value.

$$ts\_Tile = \left\{Bin_n\left(Z, X, Y, \bigcup_{i=1}^{n\times n} ts\_Pixel\right)\right\} \tag{6}$$

- ts_TileMap

The spatiotemporal map (ts_TileMap) denotes the map under a certain spatial level in the pyramid structure with a certain time window.

$$ts\_TileMap_i = \left\{Bin_n\left[Z_i, \sum_{m=1}^{2^i}\sum_{n=1}^{2^i}\left(X_m, Y_n, \bigcup_{j=1}^{256\times 256} ts\_Pixel\right)\right]\right\} \tag{7}$$

- ts_Pyramid

The spatiotemporal pyramid (ts_Pyramid) is an extension of the tile pyramid in the attribute dimension. It is a set of ts_TileMap under multiple spatial levels and a time window unit in the TS_Tile model.

$$ts\_Pyramid = \left\{Bin_n\left(\sum_{i=0}^{Z} ts\_TileMap_i\right)\right\} \tag{8}$$

- TS_Tile

TS_Tile represents a pyramid set under multiple time windows and constitutes a pyramid sequence in the time dimension.

$$TS\_Tile = \left\{\sum_{i=0}^{n} Bin_i\left(ts\_Pyramid\right)\right\} \tag{9}$$

The advantage of the TS_Tile model is that it can effectively combine Flink's time window-processing mechanism to construct the pyramid model of map tiles in parallel in a single time window by grouping the time window bin as the key. The traditional tile pyramid needs to be aggregated by high-level tiles to generate lower-level map tiles. This requires considerable time, affects the real-time visualization efficiency of the data stream, and enhances the delay of data. Through the distribution mechanism, the data stream is carried in parallel to the map network at all levels, which can realize the parallel construction of map tiles at all levels.

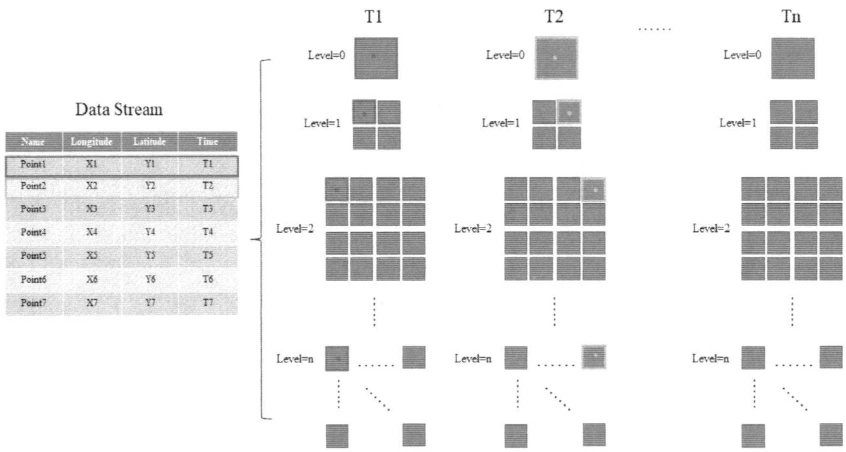

Figure 5. Parallel construction of the TS_Tile model.

As shown in Figure 5, when the data stream information enters the tile pyramid construction window, it is assigned to the tile area of the information on the level map from Level = 0 to Level = n according to its geographic location information. Time information determines the tile map of the data stream on the spatiotemporal cube. The tiles determined by the space dimension and the time dimension are used to encapsulate the thermodynamic data in this concept to determine the tile position of the data flow between different levels.

2.2. Distributed TS_Tile storage structure and index encoding

2.2.1. Distributed Storage in HBase

GeoMapViz data storage is based on a distributed HBase database. The basic unit stored in HBase is CELL, representing a set composed of RowKey, Column Family, Qualifier and Version, namely:

$$CELL=\{RowKey, Column\ Family, Qualifier, Version\} \quad (10)$$

Since the RowKey screening ability is the strongest in HBase, the design of RowKey in HBase directly affects the data retrieval efficiency. There are two types of data tables in HBase; namely, high table and wide table. Considering the frequent updating of the tiles of the sequential thermal map, the time information is integrated into the design of RowKey, and the data-filtering function can be realized by sliding the time window. Therefore, the data structure presents the high table property, and a ts_Tile is represented by a row.

Table 1. Tiles Storage structure for the TS_Tile model.

RowKey	Column Family			
	PK	Tile_id	WindowEndTime	Tile_result
RowKey_1	PK_1	Tile_id_1	WindowEndTime_1	Tile_result_1
RowKey_2	PK_2	Tile_id_2	WindowEndTime_2	Tile_result_2
RowKey_3	PK_3	Tile_id_3	WindowEndTime_3	Tile_result_3

Table 1 shows the logical structure of tile data storage in HBase, and the key value is still used in physical storage. PK is the main key of the data table, which is consistent with RowKey in composition. PK adopts a space-time (T-S) serial-coding scheme, and the main key design method can quickly filter out large-scale time series through time information to improve the retrieval efficiency of time series tiles. The T-S code can explain the temporal and spatial information of tiles so that only one thermal map tile is identified.

Tile_id appears as tile attribute information, representing tile spatial information coding (layer, row, and column). Geomesa can establish attribute indices for tiles through spatial attribute fields to improve the efficiency of data indices. WindowEndTime also appears as an attribute field because the thermal map tile is a spatiotemporal cube and represents section information in the time dimension. In

IOP Conf. Series: Earth and Environmental Science 1004 (2022) 012017 doi:10.1088/1755-1315/1004/1/012017

2D map visualization, it is necessary to map the time dimension and identify the time window in this range with the end time, representing the data falling in this time window.

Tile_result is the sum of ts_Pixel recorded in the spatiotemporal cube, which is divided into 256* 256 pixel planes after mapping. Since the thermal map tiles have data aggregation and obvious data tilts in space, in many thermal map tiles, the pixel matrix shows sparse matrix properties. GeoMapViz only stores pixels with thermal value data to improve the spatial utilization of storage. Finally, the Avro data serialization tool is used to compress the pixel matrix, and Base64 is carried out, which can effectively improve the data transmission efficiency.

2.2.2. Space-time Coding and Index Construction

To map the abstract concept of map tiles into one-dimensional data coding, we need to introduce the space-filling curve to realize the conversion of space-time units to one-dimensional binary coding and store it as RowKey of HBase. The spatiotemporal unit contains the information of the time dimension and spatial dimension; therefore, the information of the time dimension and spatial dimension is mapped to the spatial-filling curve to form the RowKey of the data table.

(1) Space coding: The space-filling curve is used to connect the space-time units at each level to realize the mapping from high-dimensional spatial information to one-dimensional binary coding. The geohash-coding method is introduced in the space-coding part, and the geohash-coding method is applied in the tile coordinate system to form space coding.

(2) Time coding: Time granularity also affects the length of the time-coding sequence. Common time granularity includes year (Y), month (M), week (W), day (D), time (H), minute (M), and second (S). Different application requirements correspond to different time granularities, which accurately depict the time accuracy that you want to give attention to. Due to the characteristics of frequent updating of the thermal map, we define the time window as the time granularity by dividing (M), taking the end time of the first tile time window of the dataset as the starting point, compressing the time stamp information in the original YYYY-MM-DD HH: mm: ss format, and using the offset (s) of the first tile time stamp of the thermal map as the time encoding.

Different concatenations of space coding and time coding also affect the storage structure of data in the database. There is also spatiotemporal coding, such as Z3, but it is found that in most cases, due to the difference in spatiotemporal scale, the query efficiency of the Z3 index strategy is much lower in practice. GeoMapViz uses a T-S series connection, uses the filtering ability of T, can quickly filter the data in a non-time window, and improves the efficiency of data retrieval. The serial encoding method is shown in Figure 6.

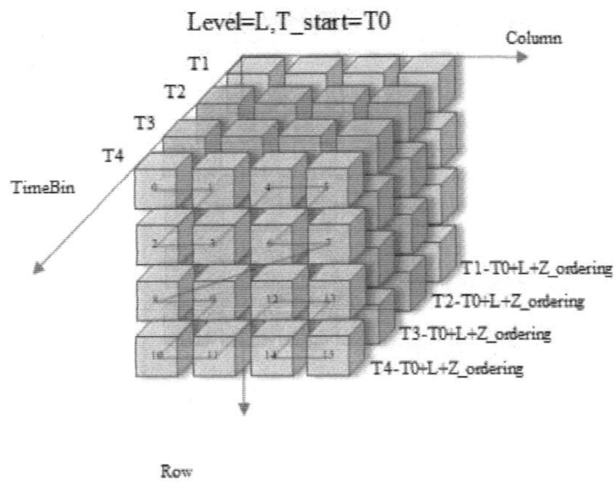

Figure 6. Space-time coding.

Combined with the multiple indices provided by Geomesa, it can support the rapid retrieval of a dataset and return the spatiotemporal query results in a spatiotemporal query.

2.3. Thermal Visualization Algorithm

The thermal map data organization maps the spatiotemporal data flow into the data tiles in the spatiotemporal cube and logically forms the data partition. Thermal map visualization includes three processes: rasterizing the original data flow, pixel aggregation and visual rendering to form a map visualization pipeline, as shown in Figure 7. Its core lies in the process management and parallel processing of data flow processing, which can quickly assemble and render the spatiotemporal data flow in clusters and can present the distribution of data in the form of a thermal map.

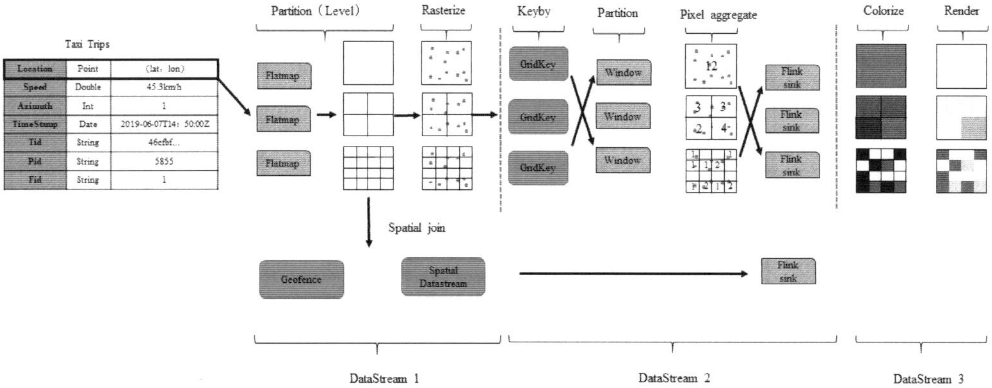

Figure 7. Thermal map visualization process.

2.3.1. Data Flow Mapping

The rasterization of spatiotemporal data flow to pixels: this rasterization process is a mapping process that contains a large number of trajectory data flows and time windows as input and converts spatiotemporal objects into pixel space under multimap pixel resolution in parallel. In the tile coordinate system, there are not only tile coordinates but also pixel coordinates in the tile. The purpose of rasterization is to map the longitude and latitude of the data flow to the pixel of the corresponding tile so that each pixel mapped by the point object has a coordinate/position in the screen coordinate system. In this paper, the tile coordinates and pixel coordinates from the WGS1984 geographic coordinate system to the Web Mercator projection coordinate system are realized by Equation (11):

$$\begin{cases} Tile\_X = Math.floor(\dfrac{Longitude+180}{360} \times 2^{Level}) \\ Tile\_Y = Math.floor\left[\left(\dfrac{1}{2} - \dfrac{ln(\tan(latitude \times \pi/180) + \sec(latitude \times \pi/180))}{2 \times \pi}\right) \times 2^{Level}\right] \\ Pixel\_X = \dfrac{Longitude+180}{360} \times 2^{Level} \times 256\%256 \\ Pixel\_Y = \left(\dfrac{1}{2} - \dfrac{ln(\tan(latitude \times \pi/180) + \sec(latitude \times \pi/180))}{2 \times \pi}\right) \times 2^{Level} \times 256\%256 \end{cases} \quad (11)$$

Through rasterization, massive data streams are distributed in parallel to the corresponding time window, map tiles under each resolution, and further mapped to pixel units. The pixel unit logically presents the key-value pair of position information with data flow, and the tile stores the set of pixel pairs <pixel, value> within its tile range. Because the data stream used in this paper is point information, we only need to consider the point-to-pixel mapping. If the data stream is linear and planar, we can map the pixel trajectory of the line through the Bresenham line-scanning algorithm [25].

2.3.2. Data flow convergence

The aggregation of spatiotemporal data streams: through the rasterization operation of data streams, many overlapping pixels can be mapped at the same position in the screen coordinate system. The reasons for this are as follows. 1. Some spatial objects essentially overlap/intersect with each other and have spatial proximity characteristics. 2. Maps have different resolutions at different zoom levels; therefore, many objects overlap/intersect at low resolutions. However, in map visualization, each pixel can only correspond to an exact thermal value and display the colour of the pixel. Therefore, this paper aggregates the spatiotemporal data flow within the pixel and calculates the weight of overlapping pixels to determine the final weight of the pixel.

The aggregation strategy of pixels is different due to the different application scenarios of data. In the visualization requirements of the thermal map, the data flow is mapped to the pixel through the location. Since there is no numerical difference in thermal values between data, the cumulative strategy is used to record the number of data streams in pixels. After Flink's FlatMap conversion operation, the tiles (Z, X, and Y) are grouped by the Keyby operation, and each pixel is placed in a key value pair of the HashMap. The key is the pixel, and the value is the current weight of this pixel. When traversing a pixel, if a pixel already exists in the hash map, the aggregation strategy is used to update the weight stored in the HashMap.

The calculation methods of topics can be divided into two types: full calculation and incremental calculation. In some complex calculation logics, a full calculation method is needed. In some simple logical calculations, such as maximum, minimum and count, the incremental calculation method can effectively save the processing time cost of data, and the cost is to require additional storage space to save the intermediate results of data. In the pixel aggregation operation of the thermal map, the incremental calculation method can be used to calculate the thermal value in each pixel of the tile.

2.3.3. *Visualization of thermal values*

Thermal map visualization algorithm: After determining the final weight of each pixel by a pixel aggregation operation, the colouring operation can render each pixel according to its weight value. The relationship between colour and weight is usually defined by mathematical equations. After the thermal value is linearly stretched to the unidirectional value of [0-255], the linear Equation (12) and the piecewise colour Equation (13) are often used for pixel visualization.

$$Color(pixel, value) = Color(value, 255, 255) \tag{12}$$

$$Color = \begin{cases} Red & value \in [0, v_0) \\ Green & value \in [v_0, v_1) \\ Blue & value \in [v_1, 255] \end{cases} \tag{13}$$

Through the experiment, it is found that due to the sparse data of tiles in a single time sequence, the visualization effect achieved by pixel colour mapping is not good; therefore, the gradual circle method is used for visualization. The effects achieved include:

(1) Each pixel presents a radial gradual circle, and this gradual circle represents the radiation effect of the thermal value data from strong to weak. The central data are concentrated, and the data in the radial circular direction are dispersed.

(2) Multiple circles can be superposed with each other, and they are linearly superposed, which shows the superposition of data strength in the radial range of data.

(3) Colour mapping with intensity.

It is worth mentioning that when the circularity is filled with intensity chromatography, because the colour obtained in this way is three-dimensional, it is nonlinear when superimposed, and it is prone to data incompatibilities such as colour mutation. Therefore, the single colour channel in the RGBA colour channel is selected as the dimension representing the strength, the data value in the pixel is linearly mapped to a vector, and alpha is used as the measurement index of the outward radiation from the centre of the circle. After colouring all pixel values in the current view, the vector is mapped to the RGBA value through Equation (14). The colouring method is shown in Figure 8.

$$limit = ln(value+1)/\ln(v_0)$$

$$Color(r,g,b,a) = \begin{cases} r = Math.floor(255 * \min(1, limit)) \\ g = Math.floor(255 * \min(1, \max(0, limit-1))) \\ b = Math.floor(255 * \min(1, \max(0, limit-2))) \\ a = Math.\min(1, limit) \end{cases} \quad (14)$$

where v_0 is the controllable parameter, which can adjust the truncation value *limit* of colouring, and *value* is the thermal value reflected by the vector.

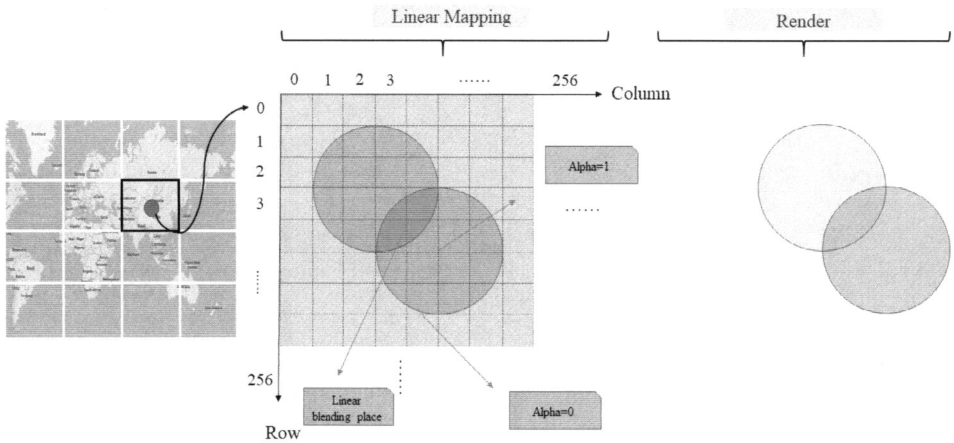

Figure 8. Gradient circle visualization algorithm.

The original data stream is rasterized mapping, pixel aggregation and visualization rendering to form a map visualization pipeline and to finally present the form of a thermal diagram. The method is illustrated in Algorithm 1.

Algorithm 1: Visualization of heatmap

```
Input: <coordinate,datetime>dataFlow
  highLevel = L0
  lowLevel = L1
Output: <pixel,rgba> tileList
Function Vizheatmap(dataFlow, highLevel, lowLevel)
  Create an empty tileList;
  Create an empty <Pixel, weight> HashMap HM;
   for data in dataFlow do
     for lowLevel < Level < highLevel do
       datebin = getBin(data.datetime);
       Tile = createTile(Level,datebin);
       pixel= rasterize(data.coordinate);
       If pixel in HM exist do
           Aggregate the weight
        end if
        Tile.add(pixel)
        tileList.add(Tile);
     end for
   end for
   for Tile in tileList do
       render <pixel,value>Tile in R vector
   end for
```

IOP Conf. Series: Earth and Environmental Science 1004 (2022) 012017 doi:10.1088/1755-1315/1004/1/012017

```
for Tile in tileList do
    render <pixel,R>Tile in Rgba
end for
Return <pixel,rgba> tileList;
```

2.4. *GeoMapViz visualization framework and other APIs*

2.4.1. *The GeoMapViz Framework*

According to the proposed TS_Tile data organization model, we finally formed a set of multiscale massive data visualization frameworks, GeoMapViz, that integrate data aggregation, calculation, storage, service and visualization. The system integrates dynamic visualization rendering of thermal maps, visualization of geographic fences and original flow, and spatiotemporal query association modules.

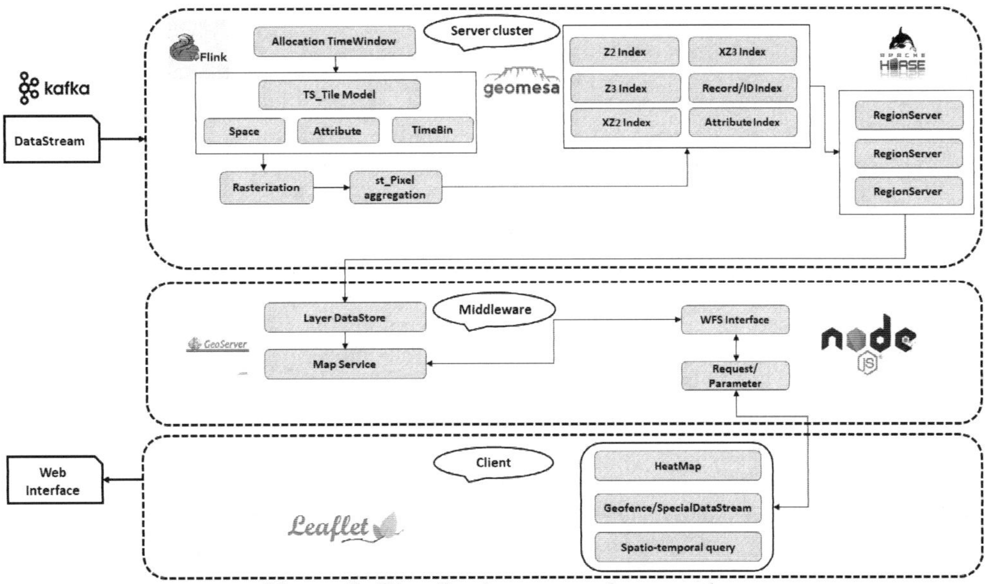

Figure 9. Visualization system framework.

As shown in Figure 9, the GeoMapViz visualization framework includes three parts: background server cluster, middleware and client.

The background server cluster is built in the distributed environment of multiple data nodes, supporting the pretreatment of the original data stream, completing the construction of the TS_Tile model, and providing data storage and index construction. The middleware realizes the management of large-scale thermal diagram tiles, provides a data access interface, and analyses the client's request. The client is the Web port to display the heatmap and the original stream, to visualize the geographic fence and to provide a dynamic query visualization method.

2.4.2. *Thermal Visualization API*

In the provided real-time thermal map service, we transform the data streams based on Flink, a parallel data-processing framework with a time window, and map each data stream by the TS_Tile model. In the heatmap calculation, the tile coordinates map the original individual latitude and longitude points into different pixels to obtain the pixel stream and then forms the heatmap tiles by pixel aggregation, which solves the problem of stacking visualization of large-scale data.

In this paper, the computation is organized in tiles, and the data streams are partitioned and assigned time windows based on Tile_id, enabling the rapid construction of a large-scale parallel-processing process of thermal map tiles on a Flink cluster, forming a multilevel thermal map time-series tile pyramid model without aggregating thermal map tiles level by level and improving computational efficiency.

2.4.3. Geofence Visualization API

Geofence is a new application of location-based services (LBS). By constructing a virtual fence, a virtual geographic boundary is surrounded to monitor the area of interest in real time.

In GeoMapViz, we define some regions of interest in advance and store them in the database permanently in GeoJSON format. After defining the original data stream as SpatialDataStream, it is necessary to register the geographic fence information as a broadcast stream to apply the joint operation of the data stream and the geographic fence to all concurrent instances when the data flow enters. A fence label is added to each data stream after the data flow enters, and then it is stored permanently.

GeoMapViz provides an integrated display of the thermal map and original flow and sets the visualization of spatiotemporal data flow points when the scaling level is greater than LEVEL. The original stream point information within the fence geographic scope is highlighted.

2.4.4. Spatiotemporal Query Association API

GeoMapViz provides a user-defined spatiotemporal query scheme. Users can customize the mapping of geographic fence boundaries and give a time range for spatiotemporal querying of geographic fences. The returned information includes the minimum bounding rectangle range of the geographic fence, the number of points in the geographic fence and the highlight display of the original flow point.

In addition, GeoMapViz also provides the optional function of dynamic query because GeoMapViz is a process of stream reading, processing and visualization of real-time data, which requires real-time display of data calculation results and high real-time data. After defining the geographic fence, when GeoMapViz follows the data flow processing for dynamic visualization, the optional method of dynamic query can be synchronized with the system time to realize the real-time follow-up of the system evolution time and to realize the continuous dynamic query effect.

2.5. External Map Service

Due to the diversification of data types, the data standards of all walks of life are not uniform, resulting in an information island and low efficiency of data integration. Therefore, it is necessary to achieve the unified management and organization of big data through technical means and to realize a standard format for the sharing of data. GeoMapViz uses Geoserver as a platform for data service publishing and management to support OGC map service standards and realize the sharing of spatial geographic information.

Web access to the published map service through the WFS service standard can quickly and accurately obtain the thermal map tile data in the corresponding spatiotemporal unit. The WFS service access is shown in Algorithm 2.

Algorithm 2： WFS service request for heatmap tile data

```
    Input： address
             map
           layerName
           epsg
           filter
    Output： <object >heatmapData
1. Function loadWFS(map， layerName， epsg， cql_filter)
2.  urlString = address；
3.  param = { service: 'WFS',
4.          version: '1.0.0',
```

```
        request: ‘GetFeature’,
        typeName: layerName,
        outputFormat: ‘application/json ’
        cql_filter: filter ,
        srsName: epsg
                };

u = urlString + getParamString(param, urlString)
 $.ajax({
  url:u,
  dataType: 'json',
  success: loadWfsHandler,
      });
function loadWfsHandler(heatmapdata) {
        Return heatmapData;
      }
```

The spatiotemporal range query returns all data in the spatiotemporal range according to the specified spatial range, such as the range of the view window ([X1, Y1], [X2, Y2]), the determined scaling level and a given time point T0 or time range [T1, T2]. Geoserver provides CQL/ECQL filter language that supports spatial queries and attribute queries. In the integrated visualization of the thermal map and original data stream, all tiles in the view are obtained, and a set of ID serial encodings is formed to query the thermal diagram data. In the query of raw data flow, a time range is customized for the data flow through the concept of a time window. In the query, only the data flow information falling in the view space and time window is retrieved.

The spatiotemporal query is shown in Algorithm 3.

Algorithm 3: The spatiotemporal query of heatmap and original flow

```
    Input: latLonBox =(X1,Y1,X2,Y2)
      zoomLevel = L
      queryTime, windowTime
    Output: queryResult
Function spatioTemporalQuery(latLonBox, zoomLevel, queryTime)
 If zoomLevel < L0
   tileBox = getTile(latLonBox);
   timecode = getTime(queryTime);
   foreach tile in tileBox do
     zOrder = getZCoding(tile);
     tileID = stCoding(timecode , zoomLevel , zOrder)
     idBox.add(tileID);
   end for
   While idBox.hasNext() do
      result = getTileData(idBox.next());
     queryResult.add(result);
   end while
 else
   pointBox = getGeo(latLonBox);
   timeRange = getRange(queryTime, windowTime);
   result = getPointData();
   If result.box in pointBox and result.t during timeRange do
     queryResult.add(result);
   end if
```

21. **end if**
22. **Return** queryResult;

3. Experiments and Discussion

3.1. Environment and Dataset Description

The GeoMapViz framework is now deployed in the 4-node open source Tencent Cloud Linux cluster, node configuration and environment, as shown in Table 2.

Table 2. Cluster configuration information.

Node	Configuration	Environment
1	4 Core, 8GB	Geoserver, Node
2	8 Core, 16GB	Flink TaskManager, HBase RegionServer, Kafka
3	8 Core, 16GB	Flink TaskManager, HBase RegionServer, Kafka
4	8 Core, 16GB	Flink JobManager, HBase RegionServer, HMaster, Kafka

In the framework of GeoMapViz, we utilize two experimental datasets: Xiamen and New York. Each dataset includes geospatial, temporal and attribute information. One of the datasets for the Xiamen region is from the Digital China Innovation Contest, DCIC 2020. The geometric feature type is point type, and the field features are listed in Table 3. The New York data were collected by a Taxicab and Livery Passenger Enhancement Programs (TPEP/LPEP) licenced technology provider and provided to the New York Taxi and Limousine Commission (TLC), with fields including pick-up and drop-off date/time, pick-up and drop-off locations, trip distance, line-by-line fare, rate type, payment type, and number of passengers reported by the driver. We preprocessed the dataset to ensure data quality and data validity. Since we focus only on the spatiotemporal representation of the trajectory data streams, we focus on extracting and computing the information in the temporal and spatial dimension fields of the data streams.

Table 3. Xiamen trajectory fields.

Field	Type	Description
RUNNING_STATUS	Int	Cab operation status, such as empty (1), carrying passengers (2)
SPEED	Double	GPS measurement of car driving speed
DIRECTION	Int	Driving direction angle
GPS_DATE	Date	Satellite positioning time
CARNO	String	Car Track ID
PID	String	Car Record ID

Table 4 shows the summary of all datasets, including the number of records, the size of the original data, the time span and the number of aggregation tiles.

Table 4. Dataset Information.

Dataset	Time Span	Record	Data Size	Tile Number
Xiamen	2019.06.07-2019.06.10	70 million	5.9 G	12 million
New York	2009.01.01-2010.01.01	190 million	6.3 G	40 million

From Table 3, it is not difficult to find that the original records from Xiamen and New York are not the same as the final number of aggregated tiles. This difference can be attributed to the different sparse densities of the dataset, and the time window size defined by the two datasets due to different time ranges is also different, resulting in different proportions of aggregated tiles.

3.2. Data-Processing Experiment

GeoMapViz aims to use the streaming processing framework to form a set of real-time distributed data-processing and visualization frameworks. Therefore, it is necessary to coordinate the interaction parameters between the server and client to achieve synchronous access to the front and back ends. In the thermal map, the size of the time window affects the visualization effect of the data. If the time

IOP Conf. Series: Earth and Environmental Science 1004 (2022) 012017 doi:10.1088/1755-1315/1004/1/012017

window is too short, the data may be sparsely distributed due to fewer data. If the time window is too long, it will take a long time to obtain the thermal diagram of the period of interest. It is necessary to select the appropriate time window size according to the range of data. The relationship between parameters is shown in Equation (15).

$$\begin{cases} W = \dfrac{24}{T} \times V \\ S = \dfrac{W \times 60}{V} \end{cases} \tag{15}$$

where V is the display time (s) of a thermal diagram at the front end, T is the total display time (m) of the actual 1-day data volume, W is the size of the time window (m), and S is the ratio of the actual time to the display time.

Figure 10 shows the visualization interface of the GeoMapViz framework, which includes three related views: dynamic visualization rendering of the thermal map, visualization of the geographic fence and original flow, and the spatiotemporal query association module. The system integrates the original flow on the thermal map, forming an integrated display of the thermal map and original flow, which is determined by the scaling level.

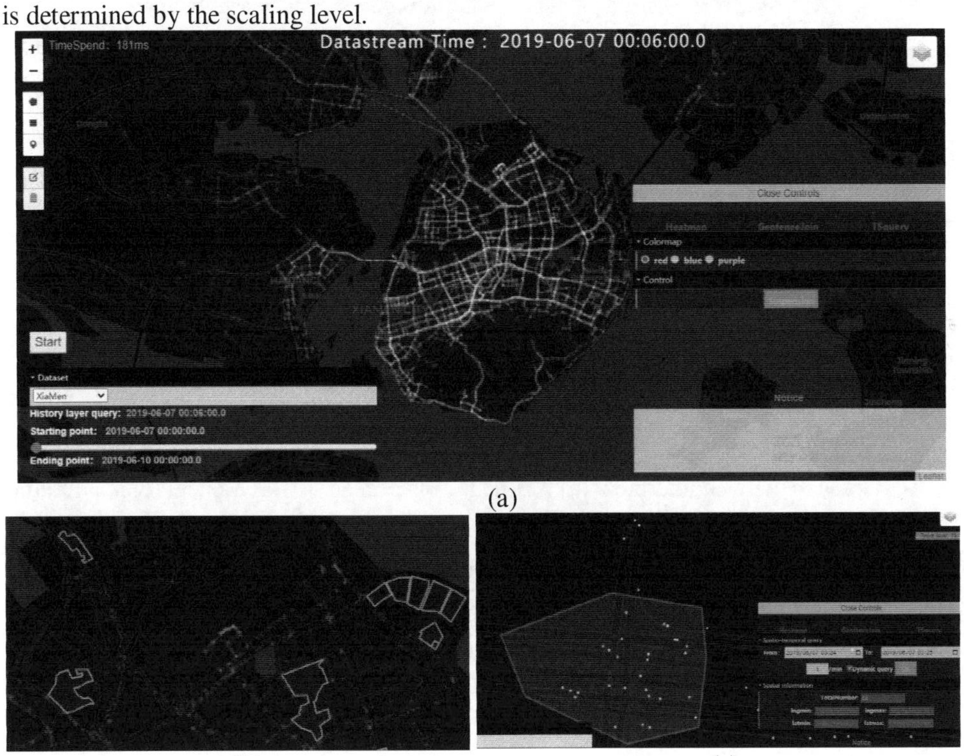

(a)

(b) (c)

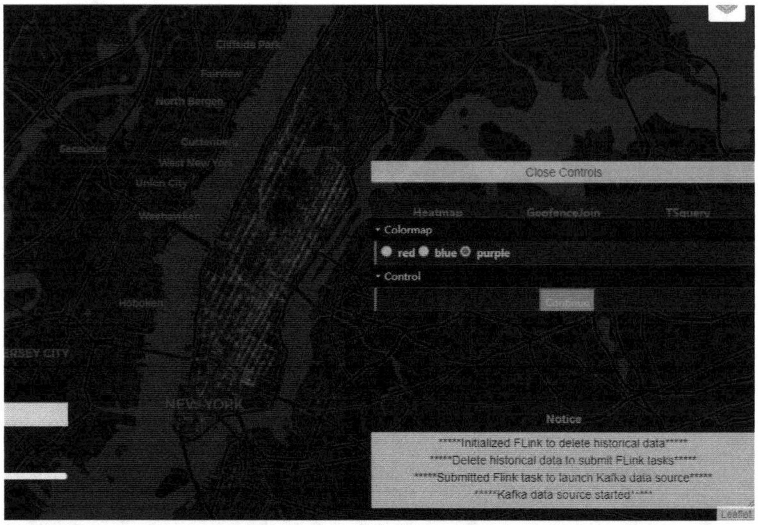

(d)

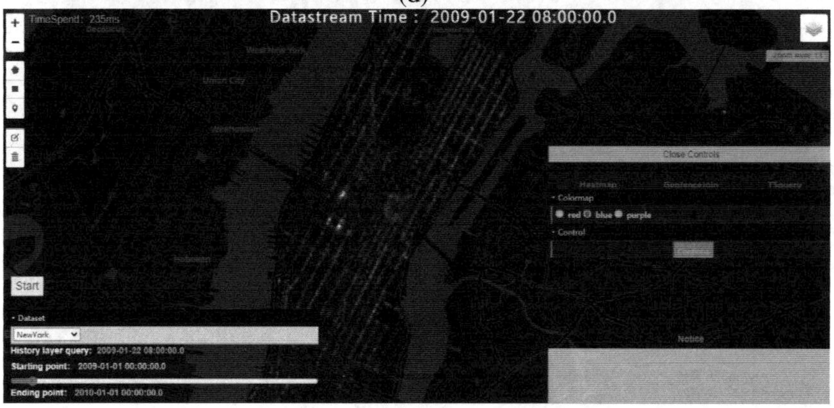

(e)

Figure 10. GeoMapViz web interface. (a) Xiamen thermal map, (b) Geofence, (c) spatiotemporal query, (d) data processing, and (e) New York thermal map.

3.3. Verification of Spatiotemporal Query Efficiency

There are many ways to access thermal tiles. Geomesa can establish an ID index for the main key field of thermal tiles and can also construct an attribute index. Experiments show that compared with the attribute index, using the ID index can improve the efficiency of data retrieval. GeoMapViz's query of thermal map tiles is a spatiotemporal query; that is, the window view is a spatial range and a time window. Since the time window is mapped to the end time of the time window to represent the entire spatiotemporal range, it can be queried by obtaining all data tile IDs in the window to form the ID set.

In this experiment, a variety of spatiotemporal thermal map query cases are defined, and Apache JMeter is used to test the query scheme. The experimental results are shown in Table 5 (total time includes WFS data acquisition, network transmission, thermal map data deserialization, map-rendering visualization, etc.).

Table 5. Thermal map query experiment.

Dataset	Window size	Level	Tile number	Data query(ms)	Total time(ms)
Xiamen	1 minute	12	15	15	280
Xiamen	1 minute	16	24	17	160
New York	1 hour	12	15	16	260

New York	1 hour	16	24	20	130

In the verification of the spatiotemporal query of geofences, this experiment defines several fence regions with different sizes in advance and a variety of time spans (hours/days/months, taking the New York dataset as an example) to test the efficiency of the spatiotemporal query. The results are shown in Table 6 (total time includes WFS data acquisition, network transmission, geographic fence joining, and primitive flow visualization).

Table 6. Spatiotemporal query experiment of the geofence.

Geofence Group	Bounding Box	Time Span	Point Number	Total Time (ms)
1	[-73.9914, 73.9901] [40.7506,40.7512]	1 hour	18	120
		1 day	566	280
		1 month	9671	2500
2	[-73.9955, 73.9950] [40.7495,40.7497]	1 hour	4	110
		1 day	102	220
		1 month	1360	1350
3	[-73.9806, 73.9796] [40.7835,40.7840]	1 hour	19	115
		1 day	267	170
		1 month	2976	1200

It can be seen from the experimental results that in the query experiment of the thermal map, with the continuous change of time and space range, the time for WFS service to obtain data is constant at a level of 10 ms, and the overall visualization time fluctuates slightly at a level of 100 ms according to the number of tiles in the query. This shows that the TS_Tile model can adapt well to different spatiotemporal query sizes and can also ensure a subsecond response. In the spatiotemporal query experiment of geofences, the combined experiment based on different sizes of geofences and time intervals is finally attributed to the statistics of the number of original flow points falling in the spatiotemporal range. With the increase in the number of points, the visualization time of the system increases proportionally. Therefore, GeoMapViz can fit the real-time calculation of spatiotemporal information flow well and can realize the integration of dynamic visualization rendering of thermal maps, visualization of geographic fences and original flows, and spatiotemporal query association modules.

4. Conclusion and Prospects

Spatiotemporal big data have multisource, heterogeneous, high-dimensional and spatiotemporal correlations. Large and dynamic spatiotemporal data to be rendered have always been a popular and difficult topic in the research of spatiotemporal big data. Classical geospatial map visualization requires considerable time and cost to generate large-scale geospatial data maps, which is inefficient. To solve the above problems, based on the concept of the tile pyramid model and spatiotemporal cube, we propose the TS_tile model to achieve scalable storage and efficient retrieval of data flow. On this basis, we designed and implemented GeoMapViz, a distributed management based on massive spatiotemporal data streams and a multiscale geographic spatial visualization framework. GeoMapViz adopts a high-performance Flink stream computing cluster to implement the large-scale parallel construction of hierarchical tile pyramids, implements distributed storage and index construction of data based on HBase and Geomesa, and uses Geoserver to manage the map service to provide a spatiotemporal range query interface. According to the proposed TS_Tile data organization model, we finally form a set of multiscale massive data visualization frameworks, GeoMapViz, that integrate data aggregation, computing, storage, service and visualization. The system integrates dynamic visualization rendering of the thermal map, visualization of the geofence and original flow, and a spatiotemporal query association module.

To a certain extent, this paper solves the problems of management, organization and access of large-scale sequential thermal maps and provides a visualization scheme for thermal maps, which can be optimized from the following aspects in subsequent studies:

(1) GeoMapViz is organized through a time bin in the time dimension, which lacks data aggregation in the time dimension and lacks query support for different time window sizes in the query application scenarios of long-term sequences.

(2) Current visualization schemes, including rasterized pixels of spatiotemporal data streams and pixel aggregation operations, are all aimed at spatiotemporal points. In the future, a multidata-type thermal map tile pyramid structure will be formed to realize the visualization scheme of multithermal map themes.

(3) GeoMapViz focuses on the realization of real-time thermal map organization and visualization. HBase is used as a public data storage framework for historical and current moments, and more methods to improve query efficiency will be considered in the follow-up. Two sets of data storage schemes are formed to support multiple types of query requirements.

References

[1] Li D R, Ma J, Shao Z F 2015 On Spatio-temporal Big Data and Its Application *Satellite Application* **9** 7-11.
[2] Jatana N, Puri S, Ahuja M, et al. 2012 A survey and comparison of relational and non-relational database *International Journal of Engineering Research & Technology* **1** 1-5. [Linkout]
[3] Gyorödi C, Gyorödi R, Sotoc R 2015 A comparative study of relational and non-relational database models in a web-based application *International Journal of Advanced Computer Science and Applications* **6** 78-83. [Linkout]
[4] McKusick M K and Quinlan S 2009 GFS: Evolution on Fast-forward: A discussion between Kirk McKusick and Sean Quinlan about the origin and evolution of the Google File System *Queue* **7** 10-20. [Linkout]
[5] Borthakur D 2008 HDFS architecture guide *Hadoop apache project* **53** 2. [Linkout]
[6] dos Santos M N and Cerqueira R 2006 Gridfs: Targeting data sharing in grid environments vol 2 (Singapore: IEEE) pp 17. [Linkout]
[7] Lei T 2009 Building Low-Cost Electronic Records Management System Based on MogileFS *Modern Computer* **0** 170-1.
[8] Dittrich J and Quiané-Ruiz J A 2012 Efficient big data processing in Hadoop MapReduce *Proceedings of the VLDB Endowment* **5** 2014-15. [Linkout]
[9] Fang J, Li C, Zhao Y, et al. 2018 Research on Distributed Storage Method for Big Data of Geological and Mineral Resources Based on Hadoop (Kunming: IEEE) pp 1-8. [Linkout]
[10] Tian C, Zhou H, He Y, et al. 2009 A dynamic mapreduce scheduler for heterogeneous workloads (Lanzhou: IEEE) pp 218-224. [Linkout]
[11] Liu Y, Chen B, He W, et al. 2013 Massive image data management using HBase and MapReduce (Kaifeng: IEEE) pp 1-5. [Linkout]
[12] Dutton G 1996 Encoding and handling geospatial data with hierarchical triangular meshes (Netherlands: Talor & Francis) pp 43. [Linkout]
[13] Kimerling J A, Sahr K, White D, et al. 1999 Comparing geometrical properties of global grids *Cartography and Geographic Information Science* **26** 271-288. [Linkout]
[14] Liu J, Li H, Gao Y, et al. 2014 A geohash-based index for spatial data management in distributed memory (Kaohsiung: IEEE) pp 1-4. [Linkout]
[15] Xiang L, Wang D, Gong J 2017 Organization and efficient range query of large trajectory data based on geohash *Geomatics Inf. Sci. Wuhan Univ.* **42** 21-27. [Linkout]
[16] Sarwat M and Nandi A 2017 On designing a geoviz-aware database system-challenges and opportunities (Arlington: Springer Cham) pp 384-7. [Linkout]
[17] Yu J and Sarwat M 2020 Turbocharging geospatial visualization dashboards via a materialized sampling cube approach (Dallas: IEEE) pp 1165-76. [Linkout]
[18] Root C and Mostak T 2016 MapD: A GPU-powered big data analytics and visualization platform ACM SIGGRAPH 2016 Talks pp 1-2. [Linkout]

[19] Yu J, Zhang Z, Sarwat M 2018 Geosparkviz: a scalable geospatial data visualization framework in the apache spark ecosystem (the 30th International Conference on Scientific and Statistical Database Management) pp 1-12. [Linkout]

[20] Eldawy A, Mokbel M F, Jonathan C 2016 HadoopViz: A MapReduce framework for extensible visualization of big spatial data (Helsinki: IEEE) pp 601-612. [Linkout]

[21] Eldawy A, Mokbel M F, Alharthi S, et al. 2015 Shahed: A mapreduce-based system for querying and visualizing spatio-temporal satellite data (Seoul: IEEE) pp 1585-96. [Linkout]

[22] Guan X, Xie C, Han L, et al. 2020 MAP-Vis: A Distributed Spatio-Temporal Big Data Visualization Framework Based on a Multi-Dimensional Aggregation Pyramid Model *Applied Sciences* **10** 598. [Linkout]

[23] Gatalsky P, Andrienko N, Andrienko G 2004 Interactive analysis of event data using space-time cube (London: IEEE) pp 145-152. [Linkout]

[24] Gray J, Chaudhuri S, Bosworth A, et al. 1997 Data cube: A relational aggregation operator generalizing group-by, cross-tab, and sub-totals *Data mining and knowledge discovery* **1** 29-53. [Linkout]

[25] Bresenham J 1977 A linear algorithm for incremental digital display of circular arcs *Communications of the ACM* **20** 100-106. [Linkout]

Acknowledgements
Thank you for all author contributions and the support of the project.

Funding
This research was funded by the Hubei Science and Technology Major Project - Key R & D Project, grant number 2020AAA004.

Real-Time Data Infrastructure for Sea Ice Monitoring in Key Arctic Waters

Bo Lin[1,2]

[1] Tsinghua University, Beijing, 100084
[2] National Marine Forecasting Observatory, Beijing, 100081

linbo@nmefc.cn

Abstract. In the theis, the progress of the National Marine Prediction Center in participating in Arctic expeditions and undertaking Arctic expedition forecasting tasks, as well as building an independent real-time data infrastructure for sea ice monitoring in critical Arctic waters are introduced; The creation process, working principles, technical methods, data significance, application directions and service benefits (effects) of the three basic monitoring (observation and detection) information systems in the Arctic sea ice monitoring real-time data collection network—automatic ocean sounding information system, automatic meteorological sounding information system and sea ice long-term observation information system are highlighted; In addition, this paper also gives a general outlook on the future Arctic business and work on the basis of China's Arctic policy and the national marine forecasting station's response and commitment.

1. Introduction

Most of the Arctic waters are covered by ice all year round. Over the past 30 years or so, the rising temperatures in the Arctic have resulted in a continuous decrease in Arctic summertime sea ice, which has not only led to changes in the natural Arctic environment, but may also trigger global problems such as climate warming, sea level rise, increased extreme weather phenomena, and damage to biodiversity. The natural condition of the Arctic and its changes have a direct impact on China's climate system and ecological environment, which in turn relate to China's economic interests in the fields of agriculture, forestry, fisheries, and oceans [1]. Since the late 1990s, the State Oceanographic Observatory (hereinafter referred to as SOA) of China has participated in all Arctic expeditions, including the first Chinese Arctic expedition, and has undertaken and completed all the tasks of marine environmental forecasting for Arctic expeditions. Marine environmental forecasting (modern numerical forecasting) is based on real-time data of marine environmental monitoring (observation and detection) as raw material, which requires a powerful and efficient real-time data monitoring, acquisition, transmission and processing system in multiple dimensions of water, sea surface and space. With the favorable conditions of its direct participation in Arctic expeditions, SOA has established an independent and relatively advanced and perfect marine environment monitoring system in the key Arctic waters. The Arctic critical sea area refers to the important target area of the Arctic expedition, mainly including the three major shipping lanes, glacier area, near the pole. Sea ice is the crystallization of the physical factors of the ocean and the atmosphere, and is one of the most critical elements of the marine environment (disaster) in the Arctic waters. The real-time data infrastructure for sea ice monitoring in key Arctic waters includes: automatic ocean sounding information system, automatic meteorological sounding information system, and long-term sea ice observation information system.

2. Automatic ocean detection information system

In 2012, China conducted the fifth Arctic expedition, and the participating SOA team deployed a set of four Argo buoys in the Norwegian Sea, the marginal sea of the Atlantic sector of the Arctic Ocean (Figure 1). This was the first time that China deployed Argoocean buoys in the Arctic expedition and carried out long-term ocean observation based on ocean buoys in the Norwegian Sea [2]. Argo buoy was customarily called because it was first used in the international Argo program [3], but in fact its full name is "self-sinking floating ocean profiling buoy", which is an automatic ocean exploration system and the most suitable drifting automatic ocean station for remote collection of ocean monitoring information so far. It has a maximum service life of 5 years and a profiling depth of 2000dbar, monitoring elements such as seawater temperature and salinity and ocean surface current velocity. It can not only transmit the monitored ocean profile information and position data to the national ocean station via satellite in real time at any time, but also ensure the long-term continuous operation of the detection information system in the time dimension through regular recovery and re-deployment. With the progress and innovation of new energy battery technology, the life cycle of the automatic ocean station is gradually extended. The Norwegian Sea is at the end and key section of the "Ice Silk Road" (Northeast Arctic Seaway). The Argo buoy array deployed by SOA in the Norwegian Sea and the large ocean-atmosphere coupled observation buoy (1 set) deployed by the Ocean University of China in the Norwegian Sea at the same time, and the sea ice material of Balance buoys (5 sets) and sea ice drift buoys (8 sets) deployed by the China Polar Research Center in the Norwegian Sea [4], the subsequent establishment of the China-Iceland Aurora Observatory (later upgraded to the China-Iceland Arctic Comprehensive Observatory) [5], as well as submarines deployed by the first Institute of Oceanography of the State Oceanic Administration (now the first Institute of Oceanography of the Ministry of natural resources) in the Bering Sea and Chukchi Sea at different times, jointly form the area of "Norwegian Sea-Iceland" in the Arctic Atlantic sector and "Bering Sea-Chukchi Sea" in the Arctic Pacific sector, which realize the integration of underwater, surface and space along the Northeast Passage of the Arctic, and also form fixed stations, drifting stations and long-term stations. It also forms the situation of fixed, drifting and long-term stations. At the same time, it makes a three-dimensional, simultaneous and continuous monitoring network of ocean hydrographic, meteorological and sea ice multi-element parallel measurements (Atlantic sector near Arctic Circle monitoring network / Figure 2). The real-time and quasi-real-time monitoring information obtained has provided a continuous stream of data support for the development of marine environmental security warning and forecasting information products for the Maritime Silk Road - "Silk Road on Ice". The huge amount of historical information accumulated over the years has laid a solid foundation for the research and prediction of Arctic marine climate. Especially, after 2013, with the accelerated melting of Arctic ice in the context of global warming and the development of China's modern shipbuilding industry (marked by the manufacture of mega marine vessels and their advanced navigation technology and ice-breaking capabilities), the commercial shipping of the "Silk Road on Ice" has entered a new historical period. Since 2013, the vessel "Yongsheng" of COSCO SHIPPING SPECILIZED CARRIERS CO., LTD has successfully sailed from China's coastal ports to Europe in the Northeast Arctic Passage under the forecast of SOA. Since the success of the first voyage from China's coastal ports to Europe under the guarantee of SOA, China has had more than 20 merchant vessels on 50 voyages to and from the Northeast Arctic Passage ("Silk Road on Ice"). Compared to the traditional route (South China Sea-South Pacific-Indian Ocean-Red Sea-Mediterranean Sea route), this has greatly reduced the operating costs of ocean-going shipping companies and effectively lowed the fuel consumption and carbon emissions of merchant ships. In terms of scientific research, SOA has made new progress in the research of short-term forecasting technology for Arctic sea ice, evidenced by that in July 2019, a research paper on technological innovation in sea ice ocean data assimilation

IOP Conf. Series: Earth and Environmental Science 1004 (2022) 012018 doi:10.1088/1755-1315/1004/1/012018

with Dr. Liang Xi as the first author and corresponding author was published online in the top international Journal of Geophysical Research oceans [6]. Meanwhile, the research results on the key technology for Chinese merchant ships to open up the Arctic route, which was completed by SOA in cooperation with COSCO SHIPPING SPECILIZED CARRIERS CO., LTD and other units, won the 2018 China Navigation Science and Technology Award Grand Prize [7]. The Arctic Automatic Ocean Exploration Information System (Atlantic sector near the Arctic Circle monitoring network) and its detection data mainly serve the Arctic expedition and the "Silk Road on Ice" navigation safety.

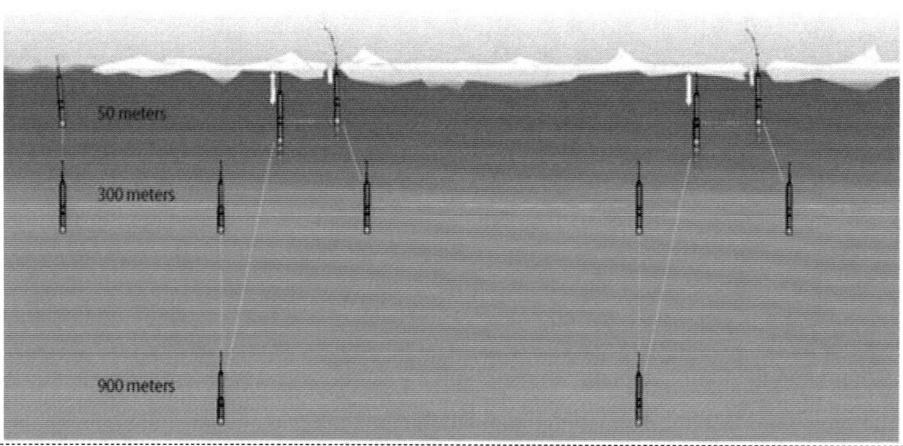

Figure 1. Ice-probing Argo self-sinking profiling buoy measurement process (China Argo real-time information center/photo).

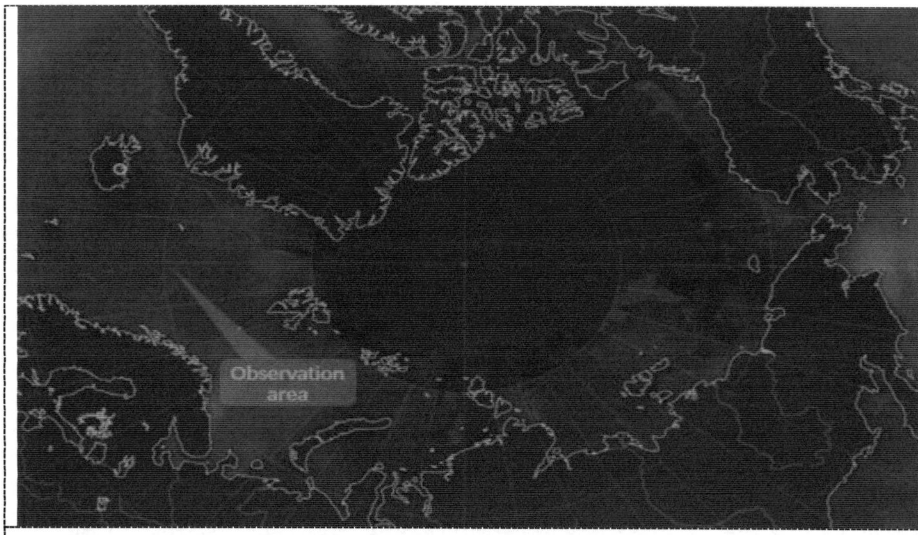

Figure 2. Arctic Atlantic sector near Arctic Circle marine environmental monitoring network.

3. Automatic weather detection information system

In 2004, according to the Svalbard Treaty [8], China established the first comprehensive Arctic scientific research station, Yellow River Station, in Ny-Alesund (78°55′N), Svalbard, Norway, which has the largest space physics and space environment exploration platform in the global polar scientific research. Since 2013, SOA has been involved in the meteorological observation of glaciers in and around the Yellow River Station in conjunction with the Arctic expedition. 2014, taking advantage of the opportunity to participate in the 6th Chinese Arctic expedition, China installed an automatic meteorological station at the East Lovenbreen Glacier (Austre Lovenbreen/A Glacier for short) near the Yellow River Station to carry out the measurement of snow and ice radiation, albedo, atmospheric temperature, humidity, snow and ice temperature, and snow temperature. humidity, snow and ice temperature, and snow thickness, so as to accumulate long-term observations for a comprehensive study of Arctic cryosphere changes [9]. In 2019, China successfully erected another set of automatic weather stations at P Glacier (Pedersenbreen), which is the first time for human to establish an automatic weather detection system for glacier atmosphere monitoring at P Glacier (78°52′N) [10] (Figure 3). The energy control system of this automatic weather station can ensure the automatic weather station to work around the clock, and the satellite communication module can realize the quasi-real-time transmission of observation data. The continuously observed meteorological information obtained on the surface of Arctic glaciers is an important fundamental information for studying global changes [11]. The Chinese Arctic Yellow River Station, A Glacier Automatic Weather Station, P Glacier Automatic Weather Station and other automatic weather detection information systems are networked to form a fixed atmospheric monitoring system of the "near 80°N circle" of the Atlantic fan (Figure 4). The circle (first step) goes to the middle step of the "North Pole". In addition to the fixed network of atmospheric monitoring stations, there are also mobile automatic weather detection information systems, including horizontal and vertical mobile detection systems. In August 2018, SOA installed for the first time an ice-based drifting automatic weather station [12] in the high Arctic latitude (84°47.05′N/off-nadir direction), which is a new type of ice-based low-temperature resistant drifting automatic meteorological detection information system. The equipment is a new type of ice-based low-temperature resistant drifting automatic meteorological detection information system. Equipped with advanced ultra-low temperature batteries and designed with solar panels, this equipment can automatically switch the power supply mode and can realize automatic weather monitoring under long-term unattended conditions. The monitoring elements include temperature, humidity, wind direction, wind speed at 2m height and 4m height, as well as total upward and downward radiation at 2m height, and ice surface pressure and latitude and longitude, etc. The monitoring data are transmitted to the domestic data center in quasi-real time via Argos satellite on a time-by-time basis, as an automatic weather detection system with horizontal motion detection as the main focus. The obtained monitoring data are valuable information for studying the state of Arctic atmospheric motion and optimizing the parameterization scheme of the Arctic sea ice numerical prediction model. During the same expedition, a large (750-gram) sounding balloon was successfully launched by SOA team using the mooring point of the Xuelong research vessel participating in the Arctic expedition as a platform [13], which was used to move vertically to detect the actual Arctic atmospheric motion. This is also the first time that SOA has released a standard 750-gram sounding balloon in the Arctic, which flew at an altitude of more than 26,000 m and reached a maximum altitude of 35,215 m. The Vaisala RS-SG sounding instrument carried by the balloon can collect weather elements such as temperature, humidity, pressure, wind direction and speed from the ground to high altitude on the flight path at any time during the lift-off, and transmit the data to the ground system in real time to carve out the vertical profile information of the local space atmosphere. On this basis, it is possible to study the Arctic atmosphere and boundary layer changes, verify the accuracy of satellite data, check and evaluate the effect of Arctic atmospheric numerical forecasting, and provide more scientific and solid real-time data support for optimizing and improving the Arctic atmospheric numerical model. The Arctic Automated Meteorological Information System (the "Near 80°N Circle" monitoring network in the Atlantic sector) and its data are mainly used for Arctic expeditions, data support for the "Arctic Environmental Multi-Element Spatial Observation and Information Service Research" project of the National Key Research and Development Program--Key Project of Intergovernmental International Science and Technology Innovation Cooperation, navigation security of the Arctic Central Passage, research on "ocean-sea-ice-atmosphere interaction" in the Arctic, and prediction of Arctic ocean climate.

Figure 3. Automatic weather station set up by SOA at the Yellow River Station P Glacier (Polar Room/photo).

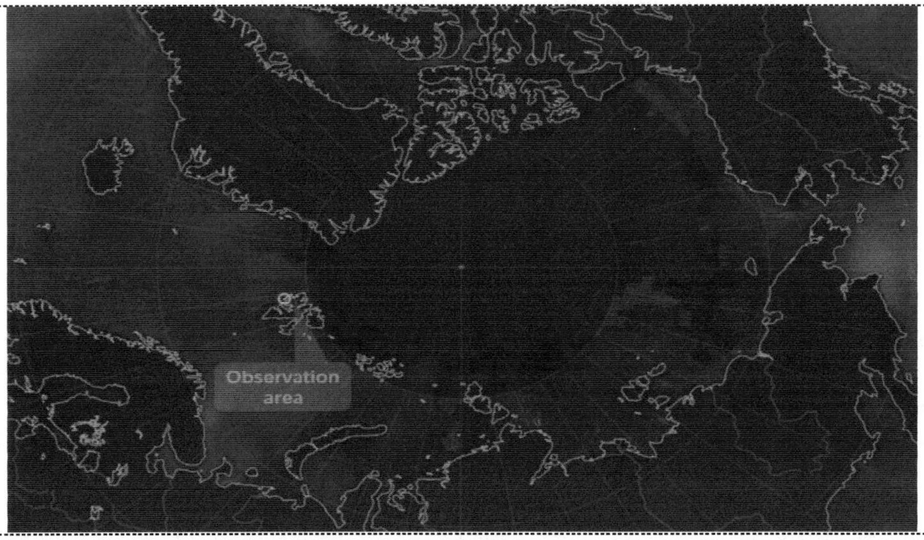

Figure 4. Atmospheric environmental monitoring network of "the near 80°N circle" in the Arctic Atlantic sector

4. Long-term observation information system of sea ice

Sea ice, which is the saltwater ice made by direct freezing of seawater [14], is a product of seawater [15]. However, because it is between seawater and the atmosphere, sea ice is the result of the joint action of the physical mechanisms of the ocean and the atmosphere (sea-air interaction [16]). The Arctic Ocean, where the Arctic is located, is the smallest, shallowest and coldest ocean on Earth, and the huge and deep ice cap in the central waters of the Arctic Ocean has survived for 3 million years [17], and is permanent sea ice. Although sea ice covers only 7% of the ocean, it causes significant changes in heat and momentum and material exchange between the ocean and the atmosphere [18].

The spatial and temporal variability of Arctic sea ice profoundly affects the ocean current system and climate system at high latitudes and even globally, and because sea ice is the most sensitive component of Arctic climate change, it has also become one of the greatest uncertainties in predicting future global changes [19]. Sea ice is a serious marine natural disaster, which can damage marine engineering structures, and then threaten the lives of staff and destroy navigation facilities, resulting in the suspension of marine shipping， damage ship equipment, and accidents; and it can bring incalculable economic losses to the marine fishing and aquaculture industry. Sea ice, as the most important priority option for Arctic expeditions and research, determines the importance of long-term Arctic sea ice observations and the construction of the Arctic Sea Ice Long-Term Observation Information System. The Arctic Point is located in the central sea of the Arctic Ocean, the whole Arctic ice cap is actually a huge ice floe in the Arctic Ocean, and the Arctic sea ice long-term observation information system consists of several long-term ice stations. The so-called long-term ice station is to select a large and relatively fixed ice floe on the Arctic ice cap near the North Pole, and then establish a sea ice monitoring station on this ice floe. As early as 1999, during the first Chinese Arctic expedition, the expedition team established the first Arctic long-term observation ice station in the "74° N circle" of the Pacific sector of the Arctic. The second Arctic long-term observation ice station was established in the "78° N circle" of the Arctic-Pacific sector in 2003 during the second Arctic expedition, the third Arctic long-term observation ice station was established in the "84° N circle" of the Arctic-Pacific sector in 2008 during the third Arctic expedition, the fourth Arctic expedition was established in the "87° N circle" of the Arctic-Pacific sector near the "87° N circle" in 2010, and the fourth Arctic long-term observation ice station was established in the Arctic Pacific sector near the "87° N circle" during the 4th Arctic expedition in 2010. [20] These have advanced the construction of Arctic long-term ice stations to the vicinity of the North Pole (86° 55' N, 178° 53' W/Figure. 5). From the beginning, SOA has been involved in the construction of long-term Arctic ice stations and real-time data collection. Sea ice observation includes fixed ice observation, floating ice observation and iceberg observation, long-term sea ice observation in the Arctic is actually a comprehensive observation that combines the above three. The standard long-term sea ice observation projects include the spatial part above the ice such as temperature, humidity, pressure, wind direction, wind speed, visibility, sunshine (light), various weather phenomena, etc.; the spatial part under the ice such as water color, water temperature, tides, currents (flow direction/velocity), salinity, density, seawater transparency, etc.; the ice part such as ice volume, density, ice type, ice surface characteristics, ice shape, ice size, floating ice drift direction and velocity, ice thickness, ice area edge line (ice boundary), ice accumulation and accumulation height, fixed ice width and thickness, ice radiation and albedo, etc.[21]. With the increase of Arctic sea ice monitoring time and the effective accumulation of data, the research on Arctic sea ice prediction technology has been gradually started. In June 2021, the "High Performance Computing Excellent Youth Innovation Fund Project" [CCF-Huawei HP2021006] , jointly undertaken by Tsinghua University and SOA (hosted by Bo Lin), was officially started. This project, with the monitoring information of Arctic automatic ocean sounding information system, automatic meteorological sounding information system and long-term ice station observation information system as the main data support and combined with China's marine satellite remote sensing information, China Argo real-time information and Global Ocean Observing System (GOOS) shared data, is committed to a comprehensive, extensive and in-depth research on sea ice data science and data application technology in key Arctic waters. The Arctic sea ice long-term observation information system (across the "74° N-87° N" monitoring network/Figure 6) and its detection data are mainly used for Arctic expedition activities and global climate change research.

Figure 5. The construction site of the long-term ice station of the 4th Chinese Arctic scientific expedition (Zhang Jiansong/photo).

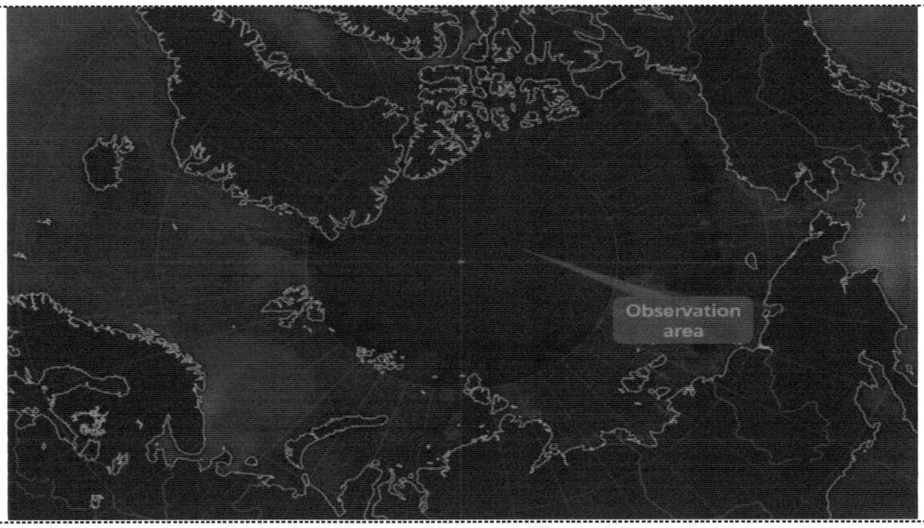

Figure 6. Long-term sea ice monitoring network in the Pacific sector of the Arctic across 74°-87°N.

5. Outlook

Exploration and awareness of the Arctic represents a priority direction and focus area for China's Arctic activities, while technical equipment is a prerequisite for the awareness, utilization, and protection of the Arctic [22]. Even though it is a fact that we have initially achieved the "three-step jump" from the near-Arctic Circle waters in the Atlantic sector to the main ice area at the North Pole, it is only the tip of the iceberg for the Arctic (Arctic Ocean high seas). According to the 14th Five-Year Plan of the National Economic and Social Development of the People's Republic of China and the outline of the 2035 Vision, "deep space, deep earth, deep sea and polar exploration and other frontier areas" are listed as "fundamental core areas of national security and overall development" [23]. The National Marine Prediction Center has made the response and commitment of "global monitoring, global forecasting, and global services" [24], and will carry out in-depth research on Arctic expeditions and sea ice monitoring and forecasting services in a comprehensive and multi-dimensional manner in order to effectively provide marine environmental security services on the "Polar Silk Road". It is committed to contributing to "understanding the Arctic, protecting the Arctic, using the Arctic and participating in its governance, safeguarding the common interests of countries and the international community in the Arctic, and promoting the sustainable development of the Arctic [25]".

References

[1] State Council Information Office official website: white paper on China's Arctic policy (2018.01). Beijing http://www.scio.gov.cn/zfbps/ndhf/37884/Document/1618193/1618193.htm
[2] Chen Hongxia 2016 Report on China's Polar Ocean Long-term Observation Technology and Work (Beijing: Ocean Press) pp 5-6
[3] China Argo real-time information center official website: the latest interpretation of the International Argo Program (2012.12.15). Hangzhou http://www.argo.org.cn/index.php?m=content&c=index&a=lists&catid=36
[4] Chen Hongxia 2016 Report on the Technology and Work of Long-term Polar Ocean Observations in China. (Beijing: Ocean Press) pp 5-6
[5] Chinese Academy of Sciences official website: China will cooperate with Iceland to establish an auroral observatory (2012.08.20). Beijing http://www.cas.cn/xw/kjsm/gndt/201208/t20120820_3629588.shtml
[6] National Marine Environment Forecasting Center official website: Center News (2019.07.15). Beijing http://www.nmefc.cn/nr/cont.aspx?itemid=2&id=6823
[7] National Marine Environment Prediction Center official website: Center News (2019.07.15). Beijing http://www.nmefc.cn/nr/cont.aspx?itemid=2&id=6824
[8] The Polar Research Office of the State Oceanic Administration official website: Svalbard Treaty (2018.12.01). Beijing http://chinare.mnr.gov.cn/catalog/service
[9] National Marine Environment Forecasting Center: 30 Years of Polar Forecasting in China. (Beijing: Ocean Press) p 074
[10] National Marine Environment Prediction Center official website: Center News (2019.07.15). Beijing http://www.nmefc.cn/nr/cont.aspx?itemid=2&id=6724
[11] National Marine Environment Prediction Center official website: Center News (2019.07.15). Beijing http://www.nmefc.cn/nr/cont.aspx?itemid=2&id=6724
[12] National Marine Environment Prediction Cente official website r: Center News (2018.09.05). Beijing http://www.nmefc.cn/nr/cont.aspx?itemid=2&id=6534
[13] National Marine Environment Prediction Center official website: Center News (2018.08.15). Beijing http://www.nmefc.cn/nr/cont.aspx?itemid=2&id=6503
[14] Marine Science and Technology Terminology Review Committee 2016 Marine Science and Technology Terminology. (Beijing: Science Press, vol 02) 2nd edition p 18
[15] (U.S.) Keith A. Sverdrup, (U.S.) E. 2020 Virginia Armbrust: Understanding the Ocean. [M]. Translated by Wei Youyun. (Fuzhou: Fujian Education Publishing House, vol 03) p 139
[16] Song Xiangzhou 2020 Introduction to Sea-Air Interactions. (Beijing: Ocean Press, vol 08) pp 3-5
[17] Hu Jinzhao, Lu Xujie and Xu Gongdi 2018 Introduction to Marine Environmental Science. (Guangzhou: South China University of Technology Press, vol 08) pp 193-194
[18] Hu Jinzhao, Lu Xujie and Xu Gongdi 2018 Introduction to Marine Environmental Science. (Guangzhou: South China University of Technology Press, vol 08) pp 193-194
[19] Zhang Zhanhai 2016 Rapid Arctic sea ice change: observations, mechanisms, and their weather-climate effects. (Beijing: Ocean Press) p 2
[20] Zhang Jiansong 2021 China's 4th Arctic scientific expedition establishes "long-term ice station" (2010.08.16). Beijing http://www.gov.cn/jrzg/2010-08/16/content_1681196.htm
[21] Mao-Chong S, Pei-Liang Li 2018 Marine Survey Methods. (Beijing: Ocean Press, vol 04) pp 154-165

[22] State Council Information Office website: White Paper on China's Arctic Policy (2018.01). Beijing http://www.scio.gov.cn/zfbps/ndhf/37884/Document/1618193/1618193.htm
[23] CSTNet: (Science and Technology News) Outline of the 14th Five-Year Plan and 2035 Visionary Goals for National Economic and Social Development of the People's Republic of China (2021.03.13). [OL/EB]. October 2021, http://www.stdaily.com/
[24] He Liang 2021 Director of the National Marine Environment Forecasting Center talks about marine disaster prevention and mitigation in Fujiang (2020.12.21). Beijing http://www.mnr.gov.cn/dt/hy/202012/t20201221_2595738.html
[25] State Council Information Office website: White Paper on China's Arctic Policy (2018.01). Beijing http://www.scio.gov.cn/zfbps/ndhf/37884/Document/1618193/1618193.htm

Acknowledgments

Thank for the fund projects: (1) National Key Research and Development Program - Key Special Project of Intergovernmental International Science and Technology Innovation Cooperation Research on Multi-Element Spatial Observation and Information Service of Arctic Environment [2017YFE0111700]; (2) China Computer Federation -- Huawei Integrated Research Program "High Performance Computing Excellent Youth Innovation Fund Project", Research on Machine Learning Based Sea Ice Prediction Deviation Revision Method for Key Arctic Seas [CCF-HuaweiHP2021006].

The impacts of trees' canopy occupation on pedestrian thermal environment of township streets

LIU Yamin[1] , LIN Ding[1], ZHU Yongbing[2] and CHEN Chongcheng[1]

[1] Key Laboratory of Spatial Data Mining & Information Sharing of MOE, Fuzhou University, Fuzhou 350116, China
[2] State Key Laboratory for Civilian Nuclear, Biological and Chemical Defence, Beijing 102205, China

Email: 2470589304@qq.com
Email:linding@fzu.edu.cn

Abstract. This paper measures the effect of trees' canopy occupation on township street pedestrian thermal environment by ENVI-met software. The demonstration experiment was conducted in two streets with different features; street A (NE-WS) with 28% buildings coverage and 0.36 AR (street aspect ratios, H/W), street B (N-S) with 63.85% buildings coverage and 1.39 AR. The results show that trees' canopy occupation ratio (TCR, V_{tc}/V_{street}) influences the average temperature at pedestrian height. The average temperature increases in the morning, whereas it decreases at noon and in the afternoon. It seems that TCR needs an offset to work on wider street, such as in street A, where the average temperature drops when TCR is greater than 0.15 (noon case, 12:00) and greater than 0.05 (afternoon case, 16:00). For every 0.1 increase of TCR (V_{tc}/V_{street}), the average temperature drops by 0.06°C(street A, 12:00), 0.118°C(street A, 16:00), 0.124°C(street B, 12:00), 0.1°C(street B), respectively. We can conclude that increasing the space occupation of trees' canopy in the street can improve the pedestrian thermal environment, especially, the street with a small street aspect ratio. At the same time, we suggest that wider and taller trees shall be planted for street greening.

1. Introduction

The street thermal environment is one of the urban microclimate, and the main influencing factors are building distribution, the underlying surface, and vegetation. Climate and environmental characteristics are relatively consistent and easy to improve[1]. The street thermal environment at pedestrian level is closely related to human daily life. Several studies demonstrate trees can help to regulate and improve the thermal comfort of a surrounding environment[2]. Planting trees would reduce the PMV value by 1.75 on average (Ceren Altunkasa et al,2020) [3]. Vaz M et al. found that in London, the cooling distance was most strongly related with tree canopy [4]. Tan Z et al. found that planting trees can lower the temperature by at least 0.3°C in subtropical cities with high-density buildings [5]. In Cairo, 50% canopy cover reduces air temperature in high density built up areas by 0.5 K(Amir A et al, 2020) [6]. And increasing canopy coverage in open spaces by 25% to 50% can reduce PET temperature at pedestrian level by 0.5 °C to 8.7 °C (Ma Xuan et al,2019) [7]. This shows that increasing the number of trees will have a positive impact on the thermal environment.

However, most of the existing thermal environment studies are focused on urban heat island [8, 9], urban thermal comfort[10], and evaluating urban overheating mitigation strategies [11-13]. with only a few studies focusing on townships. With the rapid development of economy today, we pay attention to the greening of township streets. What role do trees play in the pedestrian thermal environment of township streets? In this paper, ENVI-met was used to simulate the air temperature of pedestrian level in two streets in order to explore the value of trees' canopy occupation ratio of street space on pedestrian thermal environment.

2. Methods

2.1 Study area and data

This study focuses on two streets located in Yuexi County, Anhui Province, China(30°39′~31°11′N , 115°55′~116°33′E). The satellite image (Google Earth) was shown in figure 1(a). One is street A (NE-SW) with a street aspect ratio (AR) of 0.36 and trees with low LAD (LAD is the leaf area density). The other one is street B (N-S), which has a street aspect ratio of 1.39, trees planted only at the intersection, no tree planted inside. The detail information of these streets were shown in Table 2, their 3D models were shown in the figure 1 (b) and figure 1 (c). In the 3D models, the red shaded area is the horizontal area of pedestrian level. The blue shaded area is the average building height. For the 3D model of the street, this paper uses buildings of Google Earth image as reference to sketch the base map, field investigation to obtain the building floor data of two streets to estimate the building heights, and estimates the building heights. The 3D model was built in ENVI-met according to the base map and the building heights,

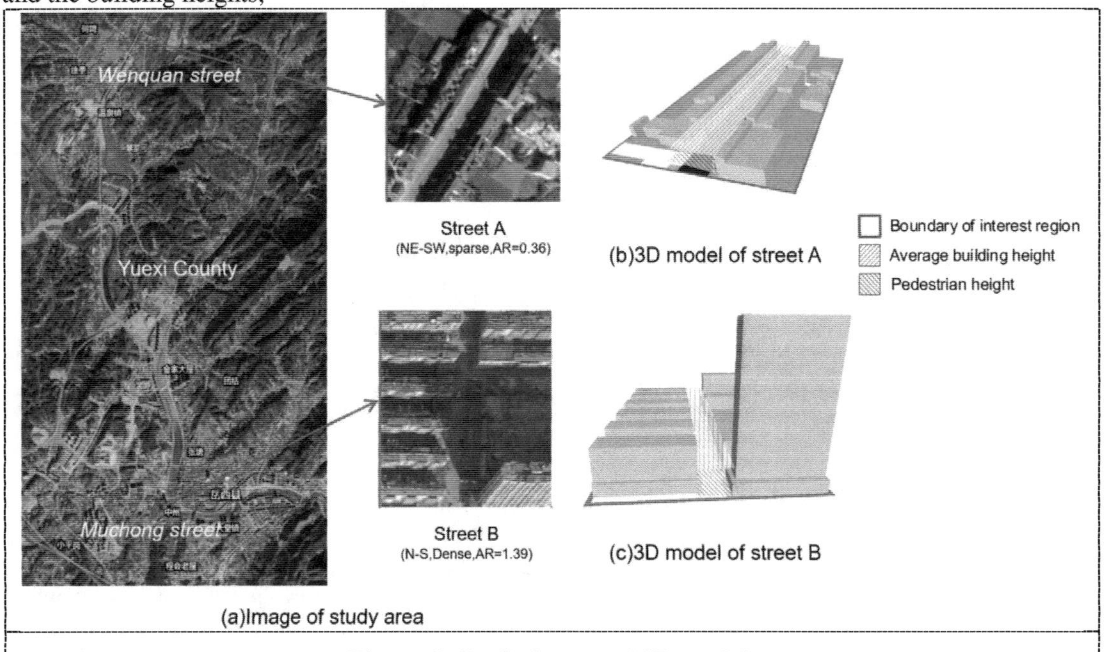

Figure 1. Studied area and 3D model.

Table 1. The information of the two streets.

Street	Street length(L)	Street width(W)	Building density(K)	Average height of buildings on both sides($\overline{H}$)	Street aspect ratio（AR）	Street orientation
A	≈160m	≈28m	≈28%	≈10m	≈0.36	NE-SW

B	≈160m	≈14m	≈63.85%	≈19.57m	≈1.39	N-S

The indexs in the table 1 such as the average building height ($\overline{H}$), building density (K) and street aspect ratio (AR) were calculated according to formulas (1)–(3):

(1) Building density (K)is the ratio of building area ($S_{building}$) to region of interest ($S_{interest}$), showed as formula (1).

$$K = \frac{S_{building}}{S_{interest}} \tag{1}$$

(2) Average height of buildings on both sides ($\overline{H}$), showed as formula (2), Hi is the height of a building, Si is the area of the corresponding building.

$$\overline{H} = \frac{\sum_i^n H_i S_i}{\sum_i^n S_i} \tag{2}$$

(3) Street aspect ratio（AR）is ratio of the average height of buildings on both sides ($\overline{H}$) to street width W, showed as formula (3).

$$AR = \frac{\overline{H}}{W} \tag{3}$$

Collecting the air temperature data at pedestrian level of the two streets from May 17 ~ 23, 2020（from 8:00-9:00 a.m, 11:30-12:30 p.m. and 15:00-16:00 p.m.）2020. The points measured were located in the middle of the streets (point A-1, B-1), under tree (point A-2, B-2) and at the intersection (point A-3, B-3), the location of measuring points is shown in figure 2(b). The measured data will be used to verify the simulation results.

2.2 Simulation case design of ENVI-met

The brief schematic diagram of the computing domain of the software is shown in figure 2(a). As in figure 1, red shaded area represents the horizontal area of pedestrian level, the blue shaded area is the average building level, and its height ($\overline{H}$) is calculated by formula 2. The purple wireframe in figure 2(a) is the region of interest, the range of interest selected for street A and street B is 100 * 160m, with a resolution of 2 * 2 * 3m. The main 3D model is limited to the middle of calculation domain, the height of the calculation domain h > H_{max}, H_{max} is the maximum height of the building. W is the street width of the region of interest. L is the street length of the region of interest.

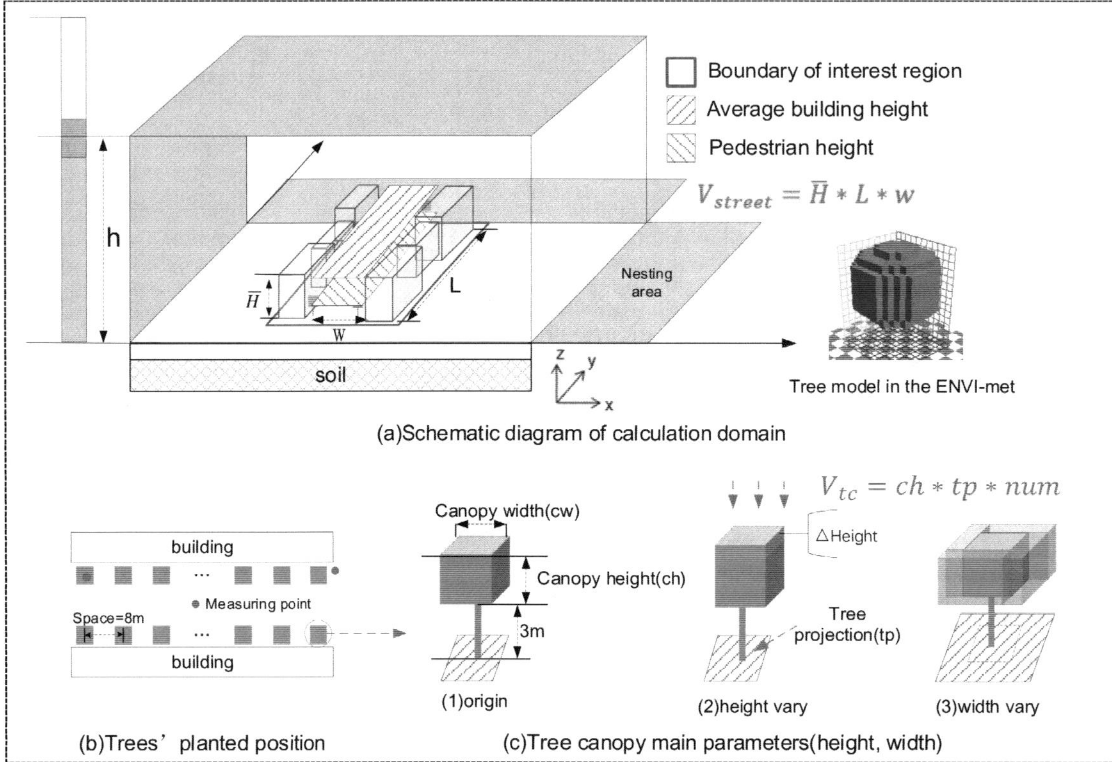

Figure 2. Compute domains and tree's change.

Table 2. Simulation case.

	Case name	Canopy height (ch)	Canopy width (cw)	Trees coverage in street area(S_{tc}/S_{street})	Trees' canopy occupation ratio in street space(TCR,V_{tc}/V_{street})
(1) Reference cases	A-Nt	0m	0m	0	0
	B-Nt	0m	0m	0	0
(2) Real cases	A-current	3m	1m	0.0112	0.0033
	B-current	7m	7m	0	0
(3) Planting cases	Ach2cw1	2m	1m	0.0085	0.0017
	Ach3cw3	3m	3m	0.0763	0.0229
	Ach5cw5	5m	5m	0.1781	0.0891
	Ach5cw7	5m	7m	0.3138	0.1569
	Ach7cw5	7m	5m	0.1781	0.1247
	Ach7cw7	7m	7m	0.3138	0.2197
	Bch2cw1	2m	1m	0.0179	0.0018
	Bch3cw3	3m	3m	0.1607	0.0246
	Bch5cw5	5m	5m	0.3750	0.0958
	Bch5cw7	5m	7m	0.6607	0.1688
	Bch7cw5	7m	5m	0.3750	0.1341
	Bch7cw7	7m	7m	0.6607	0.2363

Table 2 shows the simulation case design in this paper:

(1) There was no tree was planted in the streets of the 'Reference cases'.

(2) The 'Real cases' in table 2 will be constructed according to the current situations of two streets. These cases will simulate the pedestrian thermal environment of street A and street B from May 17, 2020 to May 23, 2020, that simulated data use to compare with measured data by calculating the correlation coefficient R^2 and root mean square deviation RMSD . The closer R^2 is to 1 and RMSD is to 0, the more effective the simulated value is. The RMSD is shown in equation (3), where T_S is the simulated temperature, T_m is the measured temperature recorded in the investigation, and N is the data record items.

$$RMSD = \sqrt{\frac{\sum_1^N (T_s - T_m)^2}{N}} \tag{3}$$

(3) 'Planting cases' in table 2 will changing canopy height and canopy width, as shown in figure 2 (c). In ENVI-met, the entire canopy model is confined to a cube consisting of 1*1*1m cells, with each cell in the model represents a specific LAD value [14], figure 2 (a) shows the tree canopy model in ENVI-met software. According to the study of Li Zhang et al [15], the impact of vegetation on both heat environment and ventilation depended on tree arrangement, LAI (leaf area indexes), crown width and tree height .When changing the canopy height and canopy width, the leaf area density, trunk height (3m) and planting spacing unchanged. And according to CCJJ75-97 code for planning and design of urban road greening [16], trees are usually planted between 4 and 8m and should not block the driving line of sight. In 'Planting cases', the trees distribution is shown in figure 2 (c) with a planting spacing of approximately 8m. In the table 2, the trees coverage in street area (S_{tc}/S_{street}) is shown in equation (4), where W is the width of the street, L is the length of the street, num is the number of trees planted in the street, j is the jth tree, tp_j represents the tree projection area of the jth tree. Trees' canopy occupation ratio in street (TCR, V_{tc}/V_{street}) is the ratio of V_{tc} to V_{street}. The formula as shown in figure 2, $\overline{H}$ is average height of buildings on both sides.

$$S_{tc}/S_{street} = \frac{\sum_j^{num} tp_j}{W \times L} \tag{4}$$

The detail parameters of model are shown in table 3.

Table 3. Model parameter (Simulation time:2020.5.17).

Parameter	Value
Simulation grid settings	210*270*30，1m*1m*3m
Wind speed	1.90m/s
Wind direction	185°
Temperature range	Minimum:16.7°C（6:00），Maximum 31.3°C（14:00）
Average relative humidity	68%

3. Simulation verification

Figure 3 shows the scatterplot graphs of the comparison between the measured and simulated air temperatures [6, 17, 18]. R^2 of each point are > 0.7 and RMSD of each point are < 0.43. According to previous experience [1, 17], the value of RMSD and R^2 in this paper are within the acceptable range. Therefore, it can be considered that the simulation results are relatively reliable.

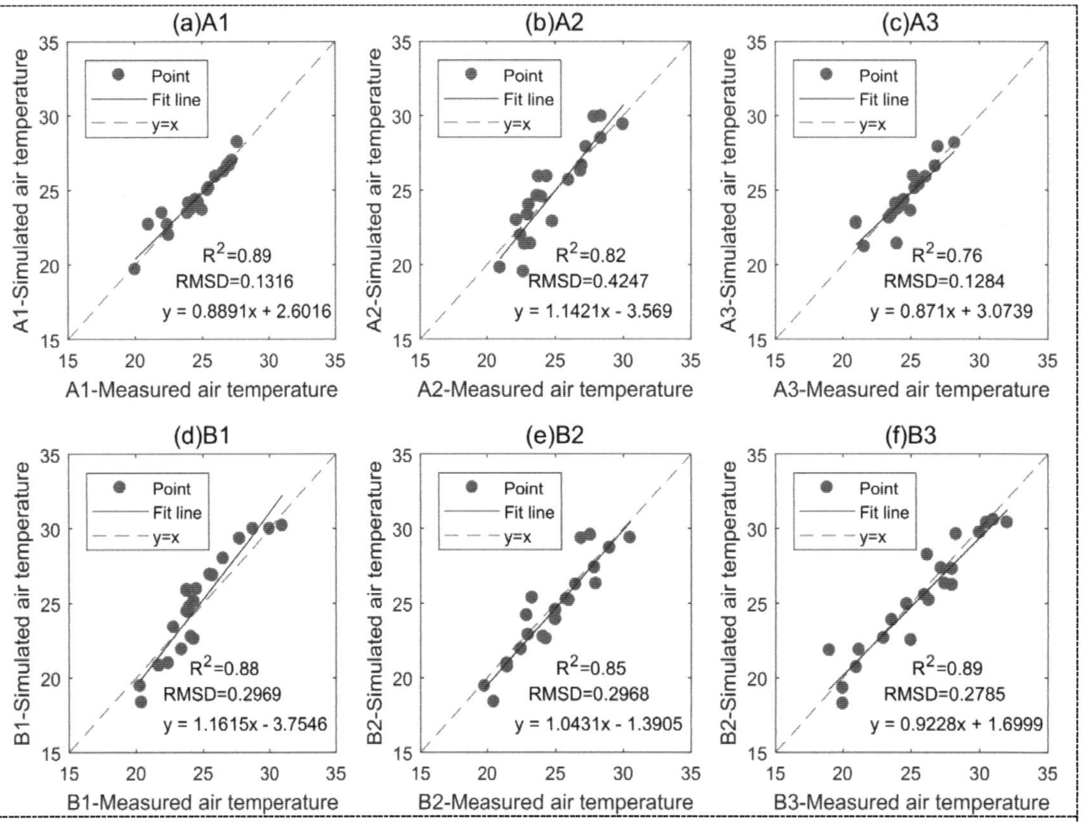

Figure 3. Correlation between measured and simulated data.

4. Simulation results and analysis

4.1 Difference in pedestrian thermal environment between the two streets

Figure 4 and figure 5 respectively show the simulation results of pedestrian temperature at A Street and B Street in the morning (8:00), noon (12:00) and afternoon (16:00). In the morning, the east side of street A is blocked by building shadow due to the influence of sunlight, so the temperature on the west side of the street is higher and the temperature on the east side is lower. The building at the southeast corner of street B is higher, south of the street the street is obscured by the building shadow, so the temperature is lower. At noon, the building shadow area formed by sunlight is smaller. Compared with that in the morning, the east-west temperature distribution of the two streets is relatively uniform, but the temperature on the east side of the street is still slightly lower than that on the west side. The higher temperature area of street valley changed from north of street valley (in the morning) to south of street valley (at noon). In the afternoon, the temperature at the windward junction of two street valley is higher. Compared with street A, the temperature of street B is relatively lower, which could be due to the dense buildings on both sides of street B.

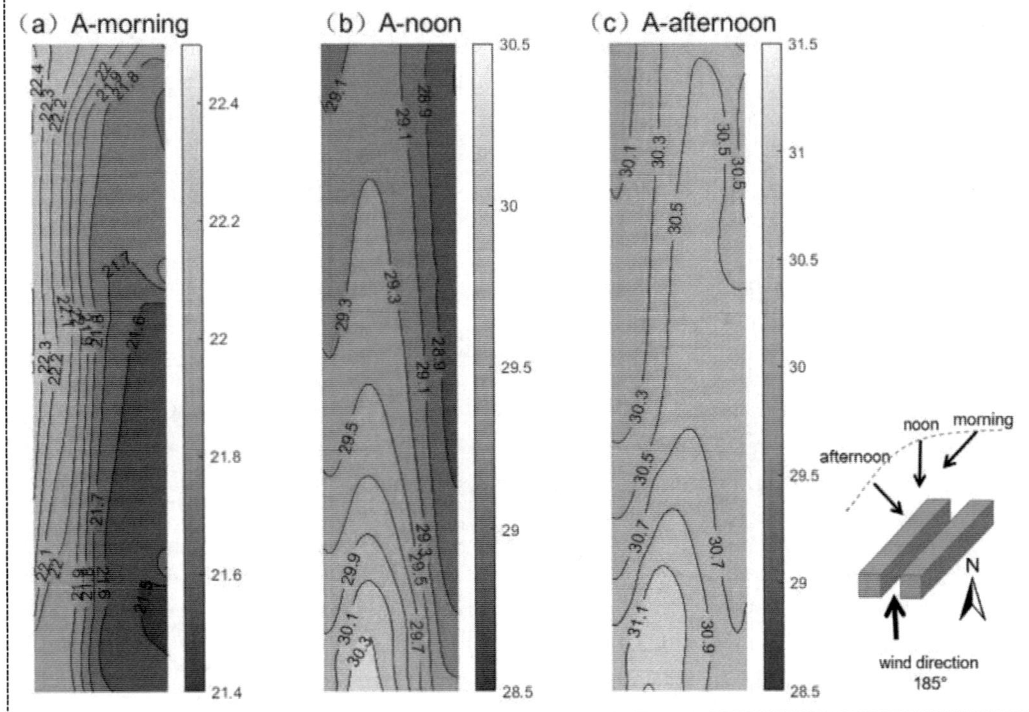

Figure 4. Street A pedestrian thermal environment.

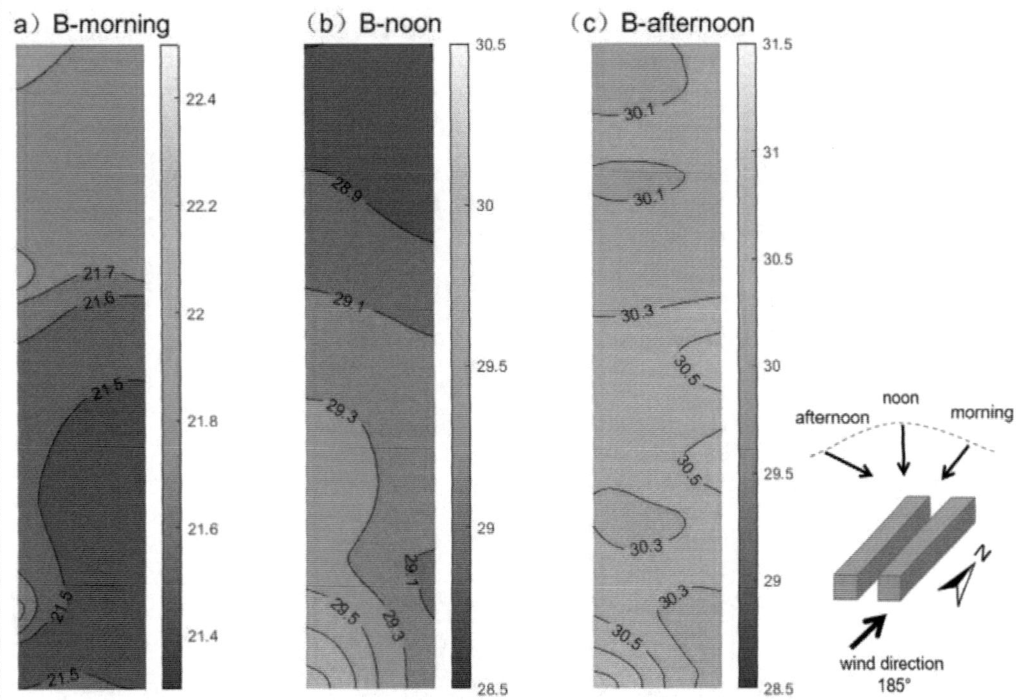

Figure 5. Street B pedestrian thermal environment.

4.2 The impacts of trees' canopy occupation on pedestrian thermal environment of township streets

Taking the difference between 'Planting cases' and 'Reference case' to gain the pedestrian temperature spatial difference C_{ij} of the street which caused by trees. The each grid temperature difference (C_{ij}) and average temperature difference ($\bar{C}$) caused by trees can be obtained from equations (5) and (6):

$$C_{ij} = C_{ij}^{P-x-y} - C_{ij}^{P,Nt}, x \in \{2,3,5,7\}, y \in \{1,3,5,7\} \quad (5)$$

$$\bar{C} = \frac{\sum_0^K C_{ij}}{K}, \quad K = \mathrm{i} * \mathrm{j} \quad (6)$$

Where, i and j represent the abscissa and ordinate of the simulated grid, P represents street A or street B, x is the crown height of the tree, y is the crown width of the tree, P-x-y represents the case in table2, C_{ij}^{P-x-y} is the temperature of each grid point simulated in the scene with trees (Planting cases), C_{ij}^{P-Nt} is the temperature of each grid in the scene without trees (Reference case), C_{ij} is the difference value of each grid, K is the number of grids. The simulation difference of between 'Planting cases' and 'Reference case' was shown in figure 6 and figure 7, respectively. The red squares show where the trees are planted.

It shows that in the two streets, the pedestrian temperature increases with the increase of tree canopy in the morning, but the temperature rises range does not exceed 0.15 °C, and the smaller the tree canopy, the smaller the overall temperature rises range. While an increase in tree canopy causes a local temperature rise, it has a local cooling effect in the tree planting area (blue area in the above figure), and an increase in tree canopy increases the cooling range. Compared to the morning, the cooling range provided by trees at noon and afternoon increases.

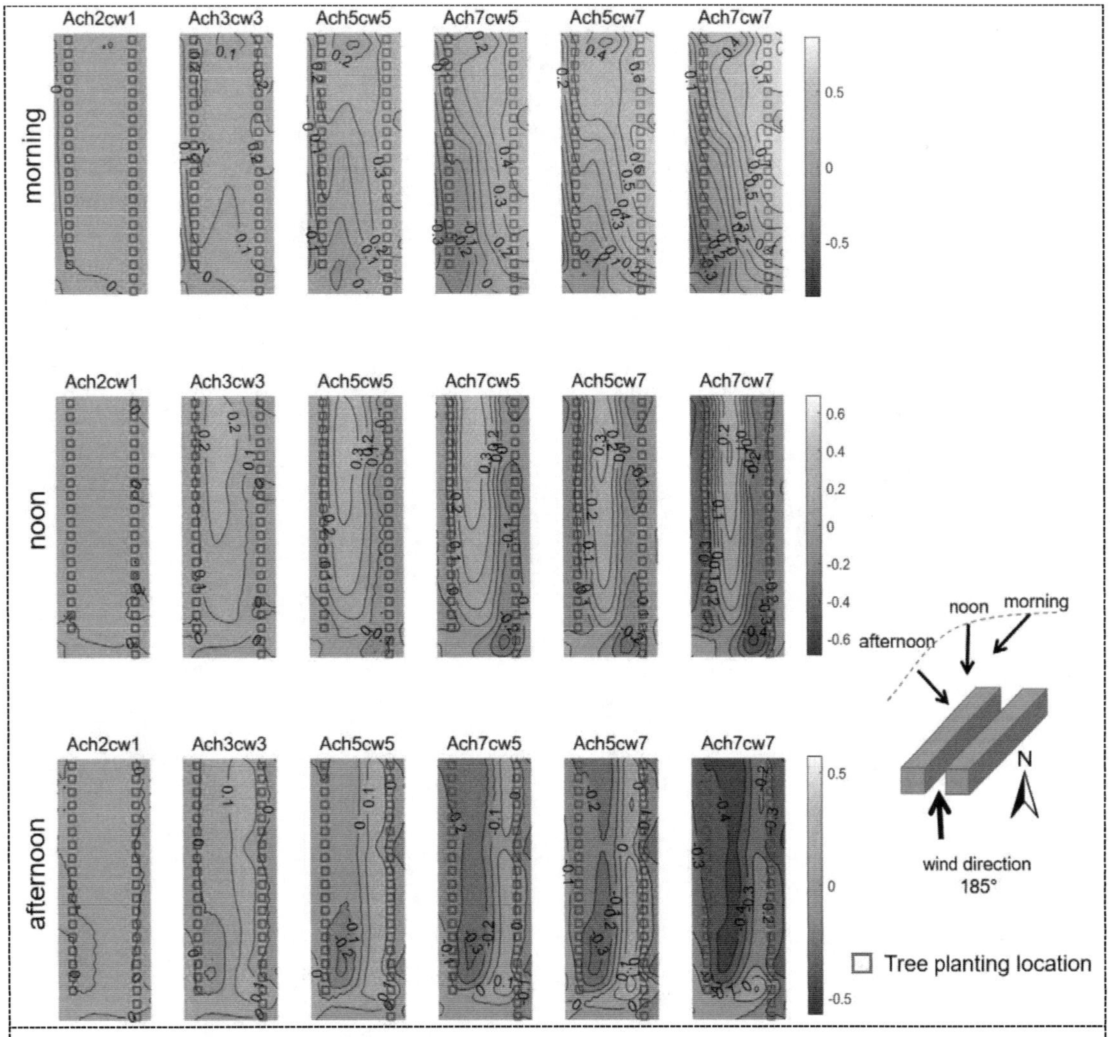

Figure 6. Street A-Effects of trees on pedestrian thermal environment.

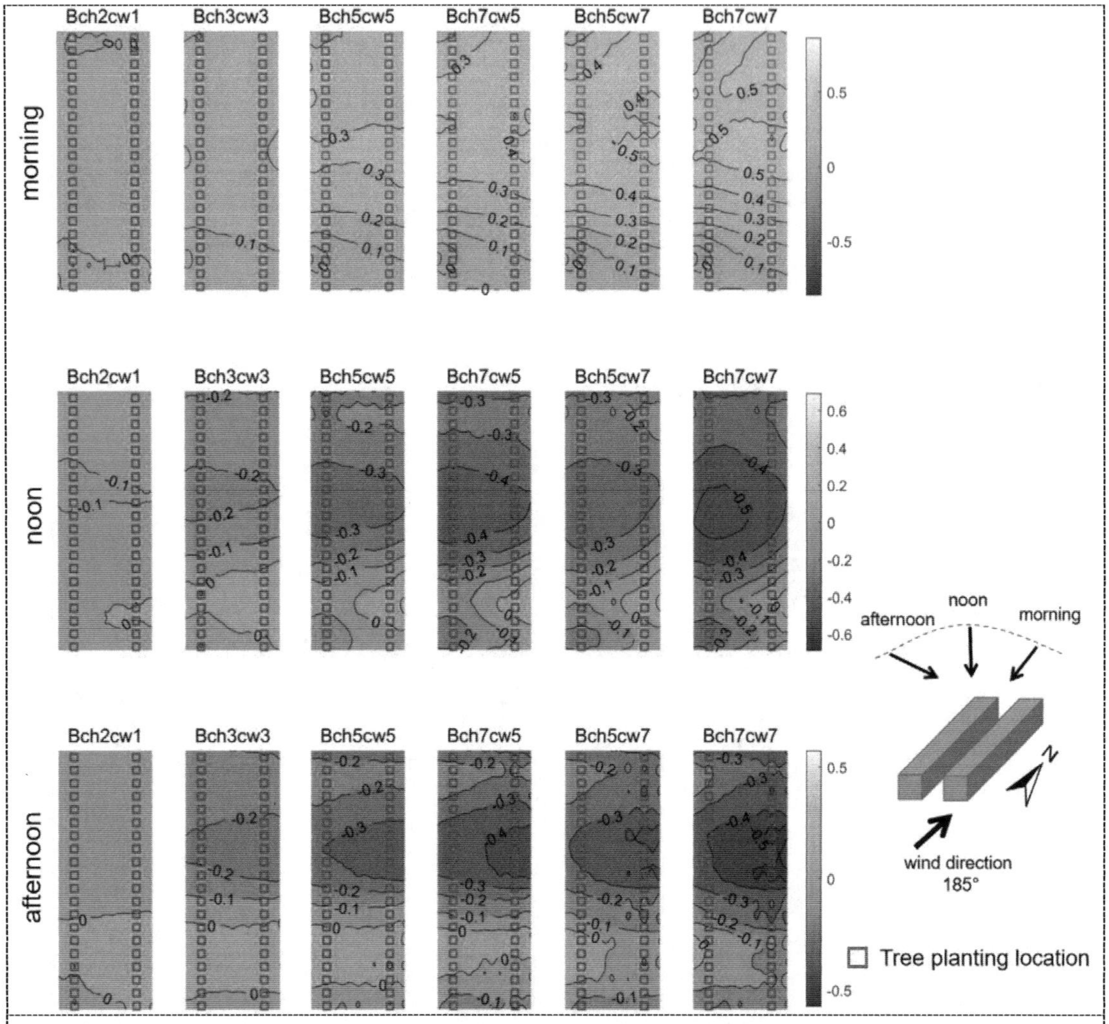

Figure 7. Street B-Effects of trees on pedestrian thermal environment.

In order to explore the relationship between the space occupation of tree canopy in the street and the pedestrian thermal environment, calculate the trees' canopy occupation ratio (TCR, Vtc/Vstreet) of each case and the average temperature difference of pedestrian temperature in the street brought by trees, and create the scatter diagram (figures 8 and 9). This is used to describe the difference of average pedestrian temperature when planting trees or not.

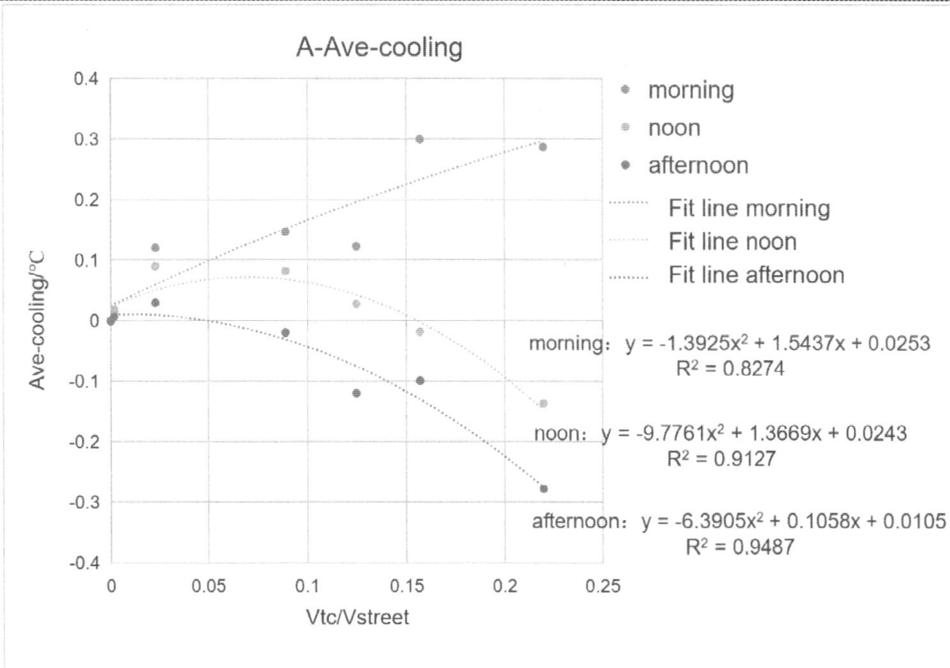

Figure 8. Street A-Effects of crown volume

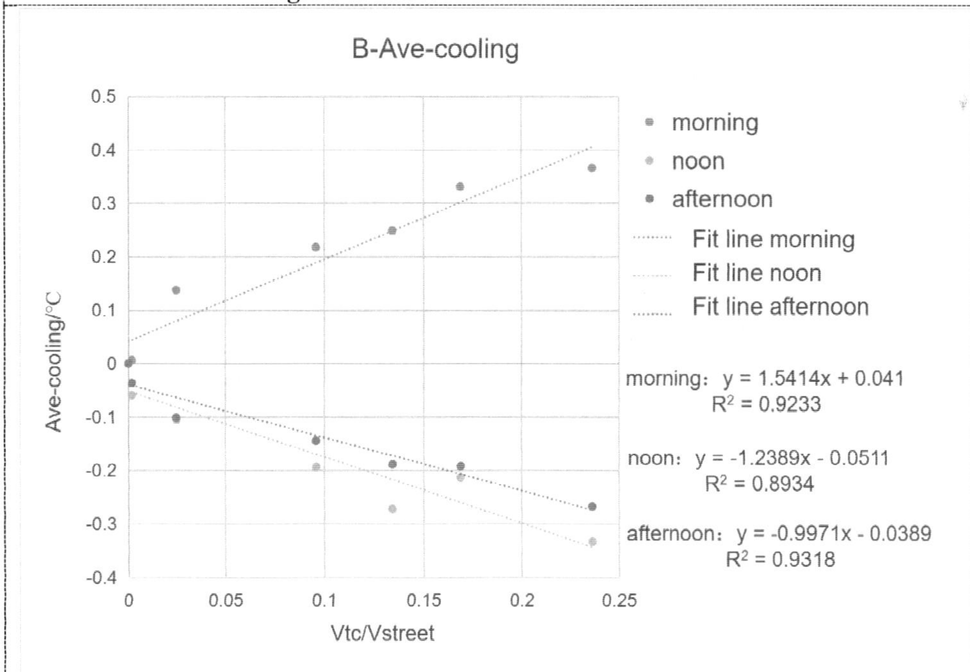

Figure 9. Street B-Effects of crown volume

At 8:00 am, trees increased rather than decreased the average temperature of the pedestrian environment in streets A and B. At 12:00 noon, when the TCR value of street A is greater than 0.15, trees can reduce the average temperature. At noon, the average temperature of street B decreases with the increase of TCR value. Through linear fitting, the average temperature decreases by about 0.124 °C every time the value of TCR increases by 0.1. In the afternoon, when the value of TCR occupied by the canopy space of street A is greater than 0.05, the average temperature of pedestrian level in the street begins to decrease. Through curve fitting, it is found that there is a convex quadratic function relationship between the average cooling value and TCR. When TCR > 0.05, the average temperature of pedestrians in street A decreases by at least 0.05°C every time the value of TCR increases by 0.1. In the afternoon, the average temperature of pedestrian height in street B also decreases with the increase of TCR. Through linear fitting, the average temperature decreases by about 0.1 every time the value of TCR increases by 0.1°C.

The influence of the space occupied by trees on the overall ambient temperature varies between the two streets. The possible reason is that the morphological structure of the two streets is different, resulting in the difference of the thermal environment in the streets. Generally speaking, the shading effect of trees is an important reason why trees can cool the hot environment. Compared with street B, street A has low buildings and wide streets, which will be exposed to more sunlight, resulting in higher temperature. Therefore, when less space is occupied by trees in the street valley, the cooling effect brought by trees is minimal.

4.3 Discussion

The results of this study are in line with previous studies of urban streets [19], which showed that street with smaller AR have higher temperature. Furthermore, the results of this study agree with others urban streets studies， which showed that larger tree canopies can help cool the streets [4, 5, 7]. But the findings of this study differ from those of Amir [20] and Justine et al [21]. There show that trees do not always cool streets in Cairo and no cooling effect from increased canopy occupancy in Manchester. The main reasons for these are geographical location and street shape.

5. Conclusion

This paper designs different tree's crown volume inside the township street to explore the impact of trees' canopy occupation ratio on the pedestrian thermal environment, the following findings are obtained:

The geometry of a street will have an impact on the pedestrian thermal environment. Street B has a good shading effect and low temperature at pedestrian level because of the dense buildings.

Trees can reduce the pedestrians temperature around them at noon and in the afternoon mainly due to their shading effect, thus, taller trees have better cooling effect. Street A is relatively wide and open, its TCR (V_{tc}/V_{street}) is small, the cooling effect of trees is not obvious. It seems that the TCR needs an offset to work on wider street, such as in street A, the average temperature of pedestrian reduces, when its TCR is greater than 0.15 (noon, 12:00) and greater than 0.05(afternoon, 16:00). There is a linear relationship between the TCR and the average cooling value in street B during noon and the afternoon. According to linear fitting, it is found that every time the TCR of street B increases by 0.1, the average temperature decreases by approximately 0.124 °C (noon, 12:00) and by approximately 0.1 °C (afternoon, 16:00).

Based on the above analysis, there is a suggestion for street A; more mature and tall trees should be planted to provide a significant shading effect in the pedestrian area, and the TCR should be greater than 0.15. On the basis of making rational use of street space, it is recommended to plant trees on street B, which lacks tree greening.

Reference

[1] Lao Z M, Ying-Min L I, Deng X J 2017 *J. Environ. Sci.* Numerical simulation of thermal environment in Zhongshan urban streets based on ENVI-met 37 3523-3531

[2] Petri A C, Wilson B, Koeser A 2019 *Land Use Pol.* Planning the urban forest: Adding microclimate simulation to the planner's toolkit. 88 104117

[3] Altunkasa C, Uslu C 2020 *Sust. Cities Soc.* Use of outdoor microclimate simulation maps for a planting design to improve thermal comfort. 57 102137

[4] Vaz Monteiro M, Doick K J, Handley P 2016 *Urban For. Urban Green.* The impact of greenspace size on the extent of local nocturnal air temperature cooling in London. 16 160-169
[5] Tan Z, Lau K K-L, Ng E 2016 *Energy Build.* Urban tree design approaches for mitigating daytime urban heat island effects in a high-density urban environment. 114 265-274
[6] Aboelata A, Sodoudi S 2020 *Build. Environ.* Evaluating the effect of trees on UHI mitigation and reduction of energy usage in different built up areas in Cairo. 168 106490
[7] Ma X, Fukuda H, Zhou D 2019 *Sol. Energy.* The study on outdoor pedestrian thermal comfort in blocks: A case study of the Dao He Old Block in hot-summer and cold-winter area of southern China. 179 210-225
[8] Chatterjee S, Khan A, Dinda A 2019 *Sci. Total Environ.* Simulating micro-scale thermal interactions in different building environments for mitigating urban heat islands. 663 610-631
[9] Aboelata A, Sodoudi S 2019 *Build. Environ.* Evaluating urban vegetation scenarios to mitigate urban heat island and reduce buildings' energy in dense built-up areas in Cairo. 166 106407
[10] Ma X, Fukuda H, Zhou D 2019 *Tourism Manage.* Study on outdoor thermal comfort of the commercial pedestrian block in hot-summer and cold-winter region of southern China-a case study of The Taizhou Old Block. 75 186-205
[11] Battista G, de Lieto Vollaro R, Zinzi M 2019 *Sol. Energy.* Assessment of urban overheating mitigation strategies in a square in Rome, Italy. 180 608-621
[12] Wu Z, Dou P, Chen L 2019 *Sust. Cities Soc.* Comparative and combinative cooling effects of different spatial arrangements of buildings and trees on microclimate. 51 101711
[13] Wang Y, Ni Z, Chen S 2019 *J. Clean Prod.* Microclimate regulation and energy saving potential from different urban green infrastructures in a subtropical city. 226 913-927
[14] Yang Y, Gatto E, Gao Z 2019 *Build. Environ.* The "plant evaluation model" for the assessment of the impact of vegetation on outdoor microclimate in the urban environment. 159 106151
[15] Zhang L, Zhan Q, Lan Y 2018 *Build. Environ.* Effects of the tree distribution and species on outdoor environment conditions in a hot summer and cold winter zone: A case study in Wuhan residential quarters. 130 27-39
[16] Bei-Bei L I, Qin J P, Li-Rong Q I 2018 *J. Environ. Sci.* Emission Characteristics of Wind Erosion Dust from Topsoil of Urban Roadside-Tree Pool. 39 1031-1039
[17] Taleghani M, Marshall A, Fitton R 2019 *Sol. Energy.* Renaturing a microclimate: The impact of greening a neighbourhood on indoor thermal comfort during a heatwave in Manchester, UK. 182 245-255
[18] Liu Z, Cheng W, Jim C Y 2021 *Build. Environ.* Heat mitigation benefits of urban green and blue infrastructures: A systematic review of modeling techniques, validation and scenario simulation in ENVI-met V4. 200 107939
[19] Andreou E 2013 *Renew. Energy.* Thermal comfort in outdoor spaces and urban canyon microclimate. 55 182-188
[20] Aboelata A 2020 *Build. Environ.* Vegetation in different street orientations of aspect ratio (H/W 1:1) to mitigate UHI and reduce buildings' energy in arid climate. 172 106712
[21] Hall J M, Handley J F, Ennos A R 2012 *Landsc. Urban Plan.* The potential of tree planting to climate-proof high density residential areas in Manchester, UK. 104 410-417

Optimal spatial resolution of remote-sensing imagery for monitoring cantaloupe greenhouses

B Cui[1,2,3], W J Huang[1,2], H C Ye[1,2], Q X Chen[2], Z C Li [3], H Y Jiang[3]

[1]Key Laboratory of Earth Observation Hainan Province, Hainan Institute of Aerospace Information Research Institute, Chinese Academy of Sciences, Sanya 572029, China

[2]Aerospace Information Research Institute, Chinese Academy of Sciences, Beijing 100094, China

[3]Sanya Institute of Remote Sensing, Sanya 572029, China

Abstract: Plastic greenhouses are vital agricultural facilities to protect cash crops from disease and insects, especially in the Hainan region of China, which has high temperature and high humidity. Remote-sensing technology is an efficient means to quickly determine the spatial distribution of plastic greenhouses on the regional scale. With the rapid development of remote-sensing technology, and especially the increasing types of high-spatial-resolution remote-sensing imagery, many studies have obtained good results by using remote-sensing technology to monitor plastic greenhouses. However, the best spatial resolution of images for monitoring plastic greenhouses has yet to be studied. To address this issue, we use cantaloupe greenhouses as the research object and GF-2 images with 1m spatial resolution as data source. We then use the re-sampling method to generate images from these data with spatial resolutions of 0.5, 2, 3, and 5 m. The details of the spatial distribution (texture features and shape features) and the spectral features of the plastic greenhouses were then extracted from images of varying spatial resolution, and a remote-sensing monitoring method for cantaloupe greenhouses was constructed based on the object-oriented random forest algorithm, which combines spectral, texture and shape features, and the monitoring results are compared. The results show that the use of 2 m spatial resolution provides the highest monitoring accuracy of cantaloupe greenhouses (overall accuracy = 94.85% and KIA = 0.92). This study thus provides a theoretical basis for remote-sensing monitoring of greenhouse cantaloupes that satisfies the current demands of production accuracy.

Keywords: plastic greenhouse; cantaloupe; object-oriented method; random forest algorithm

1. Introduction

Plastic greenhouses are important agricultural facilities that fulfill the functions of pest control, rain control, heat preservation, and humidity control. Greenhouses reduce the number of seedlings killed by heavy rain and reduce the direct harm caused by disease and pests, thereby reducing the use of pesticides, especially in high-temperature and -humidity areas. Timely and accurate knowledge of the spatial distribution and dynamic changes of plastic greenhouses is not only necessary to ensure the steady implementation of the "vegetable basket" project but is also the basis for relevant government agencies to formulate agricultural subsidy policies. The traditional method is to collect and report statistics by administrative unit level through field investigations. However, this approach implies a long working cycle and high cost, requires tedious investigation work and is inoperable over large regions. With the development of remote-sensing technology and especially its wide application in agriculture, it is now suitable for monitoring agricultural facilities.

Remote-sensing technology has already been used to monitor plastic greenhouses in numerous studies. Lu et al. [1]constructed a threshold discrimination model based on the normalized difference vegetation index (NDVI) and cumulative days by using long-term MODIS image data and exploiting spectral changes caused by variations in crop growth in plastic greenhouses. Ou et al. [2] used Landsat images from 1990 to 2018 and used spectral reflection and vegetation index to carry out spatial-temporal dynamic monitoring based on the Google Earth Engine (GEE) platform of plastic greenhouses in protected farmland. Zhu et al. [3]introduced the texture features of agricultural greenhouses from multi-temporal Landsat images and extracted the spatial distribution of agricultural greenhouses in the Shandong province in the past 30 years by using the random forest algorithm. The average overall accuracy reached 91.63%, which improved the extraction accuracy of large-scale agricultural greenhouses by remote sensing. Wu et al. [4] used texture features and spectral features of Landsat images to construct an object-oriented plastic greenhouse extraction method based on support vector machine, and the recognition accuracy is significantly higher than that based on spectral features only.

With the development of high-resolution remote-sensing technology, more and more studies now use high-resolution remote-sensing images to extract the spatial distribution of plastic greenhouses and obtain higher recognition accuracy. Commonly used remote-sensing data sources with high spatial resolution include aerial image data [5], low-altitude unmanned aerial vehicle (UAV) image data [6], and satellite image data GF-2 [7-10] (Gao et al., 2018; Gonzalez et al., 2018; Zhao et al., 2019; Wu et al., 2019), spot-7 [11] and worldview-2 [12]. Sun et al. [6] used UAV remote-sensing images as data source to monitor agricultural facilities based on pixel and object-oriented methods. The results show that the recognition accuracy of the latter is significantly better than that of the former. Wu et al. [10] reported that, when using a single-texture algorithm to identify plastic greenhouses, the recognition accuracy of a local binary model exceeds that of the gray-level co-incidence matrix and the pixel shape index algorithm. Shi et al. [13] proposed a three-step stratified model to extract plastic greenhouses based on GF-2 remote-sensing data; that is, they first distinguished plastic greenhouses, vegetation, and other feature types by using the double-coefficient vegetation screening index. Next, they used the high-density vegetation suppression index to screen out high-density vegetation features. Finally, the NDVI was used to distinguish low-density vegetation from plastic greenhouses.

The above studies all involve remote-sensing extraction of spatial distribution information on plastic agricultural greenhouses. With the increase of remote sensing images with high spatial resolution, many studies found that the identification accuracy is not always high when using images with higher spatial resolution, which may be because excessive spatial resolution increases the unimportant information from ground objects. Based on this, the present study selected plastic cantaloupe greenhouse as research objects and resampled GF-2 images to obtain images of varying spatial resolution, then built an object-oriented random forest algorithm with comprehensive use of the spectrum, texture, and shape features to monitor cantaloupe greenhouses. Finally, we report the best spatial resolution for monitoring cantaloupe greenhouses.

IOP Conf. Series: Earth and Environmental Science 1004 (2022) 012020 doi:10.1088/1755-1315/1004/1/012020

2. Materials and Method

2.1 Study region

The study area is in Ledong Li Autonomous County of Hainan Province, which is known as the "Hometown of Cantaloupe" in China. It has a tropical monsoon climate, abundant light, and warm temperatures. The annual average temperature of the coastal plain is 25.2 ℃, and the annual average precipitation is 1075 mm (Tan, 1981). To improve the yield and quality of the cantaloupe and reduce the damage caused by disease and insects, the cantaloupe is planted in a steel-frame plastic greenhouse. The shoulder height of the shed is 1.7 m, the top height is 3 m, and the span is 4 m. The distance between adjacent sheds is 40–60 cm. Insect-proof nets span between the sheds, and the top of each shed is covered with plastic film. The consistency of planting patterns of cantaloupe in the study area facilitates remote sensing identification. In this study, the study area is Ledong Li Autonomous County in the coastal area, which hosts densely planted cantaloupe (see Figure 1).

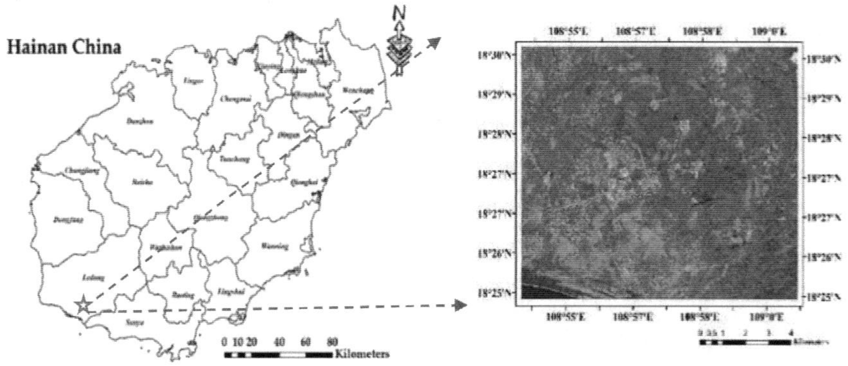

(a) Location of research region. (b) GF-2 image with B4(R)B3(G)B2(B) combination.

Figure 1. Location of study region and GF-2 image.

2.2. Data acquisition and processing

This study uses a GF-2 remote-sensing image acquired on January 16, 2020 as data source. GF-2 is the first civil optical remote sensing satellite independently developed by China with submeter spatial resolution. It is equipped with a 1 m panchromatic camera and a 4 m multispectral camera. The panchromatic band spectrum has a range of 450–900 nm and the multispectral bands include a blue band (450–520 nm), green band (520–590 nm), red band (630–690 nm), and a near-infrared band (770–890 nm). To eliminate atmospheric interference and improve the spatial resolution of the image, the GF-2 image was preprocessed by radiometric calibration, atmospheric correction, image registration, fusion, and other techniques to obtain a spatial resolution of 1 m. Images with spatial resolutions of 0.5, 2, 3, and 5 m were obtained by resampling and provided the data source for the follow-up study of optimal spatial resolution.

The field survey of cantaloupe greenhouses in Ledong Li Autonomous County was undertaken on January 14, 2020 and yielded a total of 30 cantaloupe greenhouse samples. To satisfy the classification requirements, we also comprehensively considered the Google historical high-spatial-resolution images of November 2019 and February 2020, and the visual interpretation of the study area was carried out to obtain a sample set containing the "true ground values" to verify classification and accuracy. Finally, six object categories containing a total of 1277 samples were selected including early cantaloupe greenhouses (period of low vegetation coverage), middle-late cantaloupe greenhouses (period of high vegetation coverage), vegetation, water, soil, and other (land for construction, road, beaches, and other) (Table 1). In this study, about two-thirds of the total sample was selected as the training samples to build the classification model, and the remaining one-third of the total samples was used as the validation samples to validate the accuracy of classification model.

Table 1 Samples used in this study

Type	Number of training samples	Number of validated samples	Total
Early cantaloupe greenhouses	66	33	99
Middle-late cantaloupe greenhouses	61	30	91
vegetation	255	128	383
water	60	30	90
Bare soil	249	124	373
Other	161	80	241
Total	852	425	1277

2.3. Segmentation algorithm

Image segmentation is a key step in the object-oriented method. The accurate segmentation of different ground objects directly affects the recognition accuracy. The multiresolution segmentation algorithm proposed by [15] is a bottom-up segmentation method for merging two regions. This method takes a single pixel as the starting point and iteratively merges pixels into larger units. It is an optimized process that minimizes the average heterogeneity of a given number of image objects to obtain maximum homogeneity. In the process of segmentation, the spectral, spatial and neighbourhood information of remote-sensing images are considered comprehensively, which has been widely used in remote-sensing image segmentation. Segmentation scale, shape index and compactness index are the main parameters that affect the segmentation accuracy of the multiresolution segmentation algorithm. The specific segmentation process requires many attempts to set different factor parameters before obtaining the best segmentation results.

2.4. Feature selection

The spectral reflection from plastic greenhouses is influenced both by the plastic film and by the growth state of the crops within the greenhouses. In the early growth stage of the crops, the spectral reflection from plastic greenhouses is greatly affected by soil and is like that of bare soil. With increasing crop coverage, the spectral reflectance of plastic greenhouses increases gradually under the influence of vegetation, and its reflection spectrum becomes like that of vegetation. Therefore, it is difficult to carry out remote-sensing recognition of plastic greenhouses based only on spectral characteristics. Because the spatial distribution of plastic greenhouses is regular and their size and shape are consistent, the introduction of texture and shape features in the remote-sensing recognition procedure helps to improve the recognition accuracy.

This study uses spectral features, including the brightness, the normalized difference vegetation index (NDVI) [16] and the normalized plastic greenhouse index (NDPG) [17]. We used eight texture features extracted by using the gray-level co-occurrence matrix [18], including homogeneity, contrast, dissimilarity, entropy, angular second moment, mean, variance and correlation. Shape features include asymmetry, border index, compactness, density, elliptic fit, radius of largest enclosing ellipse, radius of smallest enclosing ellipse, rectangular fit, roundness and shape index.

$$NDVI = (NIR - RED)/(NIR + RED) \quad (1)$$

$$NDPG = (RED - GREEN)/(RED + GREEN) \quad (2)$$

Table 2 The spectral features used in this study

Type	Feature	Type	Feature
Spectral features	Brightness	Shape features	Asymmetry
	NDVI		Border index
	NDPG		Compactness
Texture features	Homogeneity		Density
	Contrast		Elliptic fit
	Dissimilarity		Radius of largest enclosing ellipse
	Entropy		Radius of smallest enclosing ellipse
	Angular second moment		Rectangular fit
	Mean		Roundness
	Variance		Shape index
	Correlation		

2.5. Building the classification model

This study uses the random forest (RF) algorithm to construct a classification model for plastic greenhouses. RF was proposed by [19] and is based on the decision tree algorithm, and its integrated classification was done by combining multiple classification and regression decision trees. The RF algorithm is robust against noise and does not easily over-fit. It is also good at dealing with outliers and collinear variables and has been widely used to classify ground objects. In terms of accuracy of remote-sensing extraction of plastic greenhouses, the RF algorithm is better than other machine learning algorithms (e.g., maximum likelihood method, support vector machine, K-proximity method, decision tree) [2] [7] [11]. For these reasons, the RF algorithm is used herein to monitor cantaloupe greenhouses.

The RF algorithm requires few parameters; the key parameters are *Ntree* and *mtry*. *Ntree* is the number of decision trees constituting the RF, and *mtry* is the number of features randomly selected during the training of each decision tree. Based on results of the relevant experiments, *Ntree* is taken as 50 herein, and *mtry* is the square root of the total number of features.

2.6. Accuracy evaluation

The confusion matrix method is commonly used to verify the classification accuracy of remote-sensing images and is used herein to evaluate the classification accuracy. The confusion matrix is a comparison matrix that represents the number of samples in a certain category and the number of real samples in the category. The evaluation metrics used herein include overall accuracy (OA), producer's accuracy (PA), user's accuracy (UA) and the kappa coefficient (KIA).

3. Results

3.1. Results of image segmentation

The results of segmenting (be it over-segmentation or under-segmentation) remote-sensing images directly affect the classification accuracy. This study uses the multiresolution segmentation algorithm. After several tests to ensure to the extent possible the purity of greenhouse objects, the optimal segmentation parameters were finally determined for remote-sensing images with different spatial resolutions (see Table 3). Higher spatial resolution corresponds to a larger optimal segmentation scale, shape index and compactness index, mainly

IOP Conf. Series: Earth and Environmental Science 1004 (2022) 012020 doi:10.1088/1755-1315/1004/1/012020

because the image with high spatial resolution contains more detailed information. To avoid disturbance from unimportant information detail, a higher segmentation parameter should be used in this study to prevent over-segmentation.

Table 3. Optimal segmentation parameters for images of different spatial resolution.

Spatial resolution	Segmentation scale	Shape	Compactness
0.5 m	130	0.9	0.9
1 m	90	0.9	0.9
2 m	80	0.8	0.9
3 m	70	0.6	0.8
5 m	70	0.6	0.8

3.2. Classification results

This study uses the object-oriented RF classification algorithm to identify plastic cantaloupe greenhouses from images of spatial resolution of 0.5, 1, 2, 3, and 5 m. The results appear in Figures 2-6. Given the reduction in spatial resolution and the influence of mixed pixels, some small objects in the study area are not identified, such as the road between the plastic greenhouses. The boundary of identified objects is also rough. However, images with higher spatial resolution correspond to more fragmented ground object, especially for areas with bare soil and/or other ground objects.

(a) Classification results for entire region. (b) Classification results for key region.

Figure 2. Classification results for remote-sensing image with 0.5 m spatial resolution.

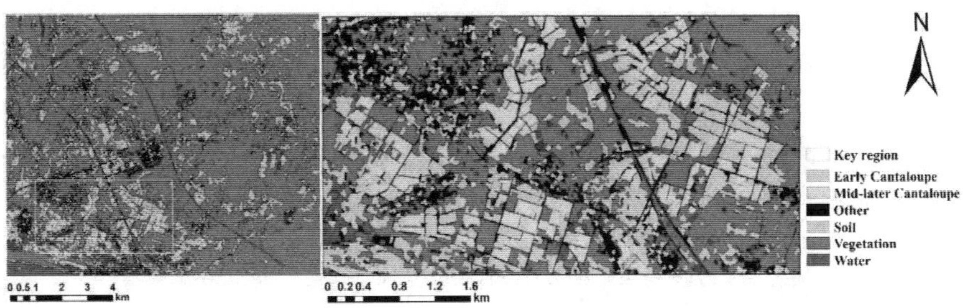

(a) Classification results for entire region. (b) Classification results for key region.

Figure 3. Classification results for remote-sensing image with 1 m spatial resolution.

(a) Classification results for entire region. (b) Classification results for key region.

Figure 4. Classification results for remote-sensing image with 2 m spatial resolution.

(a) Classification results for entire region. (b) Classification results for key region.

Figure 5. Classification results for remote-sensing image with 3 m spatial resolution.

(a) Classification results for entire region. (b) Classification results for key region.

Figure 6. Classification results for remote-sensing image with 5 m spatial resolution.

The validation sample data were used to evaluate the recognition accuracy of the remote-sensing images with different spatial resolutions (see results in Table 4). With decreasing spatial resolution of the images, the classification accuracy first increases and then decreases. The classification accuracy is highest (overall accuracy = 94.85%, KIA = 0.92) for a spatial resolution of 2 m, followed by a spatial resolution of 1 m (overall accuracy = 93.88%, KIA = 0.92). The images with 5 m spatial resolution produce the worst accuracy (overall accuracy = 87.57%, KIA = 0.81). When the spatial resolution of an image is too high, the classification accuracy cannot be improved, possibly because images with high spatial resolution contain too much detailed information on ground objects, which interferes with ground-object recognition. For early cantaloupe greenhouses and middle-late cantaloupe greenhouses, the recognition accuracy for images with 1 m spatial resolution slightly exceeds that for images with 2 m spatial resolution, especially for early-cantaloupe greenhouses. During the early growth stage of cantaloupe, the low coverage means that the spectral characteristics of greenhouses are strongly affected by soil, which is like bare soil, increasing the difficulty of identifying plastic greenhouses. However, images with 1 m spatial resolution have more texture and shape features than images with 2 m spatial resolution, which improves the recognition accuracy.

Table 4. Classification results based on images of varying spatial resolution.

Spatial resolution of images	Type	Producer's accuracy	User's accuracy	Overall accuracy	KIA
0.5 m	Early cantaloupe	96.27%	90.99%	92.49%	0.90
	Mid-later cantaloupe	70.41%	100.00%		
1 m	Early cantaloupe	100.00%	92.06%	93.88%	0.92
	Mid-later cantaloupe	82.61%	97.05%		
2 m	Early cantaloupe	97.83%	85.91%	94.85%	0.92
	Mid-later cantaloupe	84.25%	97.00%		
3 m	Early cantaloupe	97.19%	87.04%	92.96%	0.90
	Mid-later cantaloupe	73.32%	100.00%		
5 m	Early cantaloupe	86.02%	78.10%	87.57%	0.81
	Mid-later cantaloupe	63.81%	71.02%		

4. Discussion

Remote-sensing images with high spatial resolution have gradually become the mainstream data source for mapping greenhouses. Significant research has been done to monitor greenhouses by using high-spatial-resolution images, such as GF-2 with 1 m spatial resolution, QuickBird with 2.5 m resolution, IKONOS with 4 m resolution and GeoEye-1with 0.5 m resolution. All of these have obtained a good identification accuracy with an overall accuracy of 84.20%–97.34% [13] [20-23]. Note that remote-sensing images of higher resolution cost more than their low-resolution counterparts, and images with higher spatial resolution may include redundant detail that could hinder the effort to map greenhouses. The present work studies this question by considering images with different spatial resolutions. The results indicate that higher spatial resolution is not always for mapping greenhouses. For mapping cantaloupe greenhouses, the most accurate classification was obtained with a spatial resolution of 2 m (overall accuracy = 94.85%, KIA = 0.92). This result can provide a theoretical basis for the selection of remote sensing image to estimate greenhouse. However, images with different spatial resolutions acquired by resampling method are different from real images, more tests should be to carried out in future.

5. Conclusion

This study is based on GF-2 images constructed by using an object-oriented random forest classification method with the spectral, texture and shape features of remote-sensing images with varying spatial resolution. The images are used to monitor cantaloupe greenhouses to determine the best spatial resolution suitable for such a task. The results show that, with spatial resolution decreasing from 0.5 to 5 m, the classification accuracy first increases and then decreases. The best classification accuracy is obtained for 2 m spatial resolution (overall accuracy = 94.85%, KIA = 0.92). However, for greenhouses with crops in the early growth stage (i.e., low crop coverage), a better classification accuracy of plastic greenhouses is obtained with 1 m spatial resolution.

References

[1] Lu L, Huang D, Di L 2015 Threshold model for detecting transparent plastic-mulched landcover using moderate-resolution imaging spectroradiometer time series data: a case study in southern Xinjiang, China *J APPL REMOTE SENS.* **9** p 097094
[2] Ou C, Yang J, Du Z, et al. 2020 Long-Term Mapping of a Greenhouse in a Typical

Protected Agricultural Region Using Landsat Imagery and the Google Earth Engine *Remote Sensing* **12** p 55

[3] Zhu D, Liu Y, Feng Q, et al. 2020 Spatial-temporal Dynamic Changes of Agricultural Greenhouses in Shandong Province in Recent 30 Years Based on Google Earth Engine *Transactions of the Chinese Society for Agricultural Machinery.* **051** p 168-175.

[4] Wu C F, Deng J S, Wang K, et al. 2016 Object-based classification approach for greenhouse mapping using Landsat-8 imagery *International Journal of Agricultural & Biological Engineering.* **9** p 79-88.

[5] Tarantino E and Figorito B. 2012 Mapping Rural Areas with Widespread Plastic Covered Vineyards Using True Color Aerial Data *Remote Sensing* **4** p 1913-1928.

[6] Sun Y, Han J, Chen Z, et al.2018 Monitoring Method for UAV Image of Greenhouse and Plastic-mulched Landcover Based on Deep Learning *Transactions of the Chinese Society for Agricultural Machinery.***49** p 133-140.

[7] Gao M, Jiang Q, Zhao Y, et al. 2018 Comparison of plastic greenhouse extraction method based on GF-2 remote-sensing imagery *Journal of China Agricultural University.* (in Chinses) **23** p 125-134.

[8] González-Yebra óscar, Aguilar M, Abderrahim N, et al. 2018 Methodological proposal to assess plastic greenhouses land cover change from the combination of archival aerial orthoimages and Landsat data *Biosystems Engineering* **175** p 36-51.

[9] Zhao L, Ren H, Yang L. 2019 Retrieval of agriculture greenhouse based on GF-2 remote sensing images *Remote sensing technology and application* **34** p 677-684.

[10] Wu J, Liu X, Bo Y, et al. 2019 Plastic greenhouse recognition based on GF-2 data and multi-texture features *Transactions of the Chinese Society of Agricultural Engineering* **35** p 173-183.

[11] Balcik F, Senel G, and Goksel C.2020 Object-Based Classification of Greenhouses Using Sentinel-2 MSI and SPOT-7 Images: A Case Study from Anamur (Mersin), Turkey *IEEE J-STARS.***13** p 2769-2777.

[12] Aguilar M, Nemmaoui A, Novelli A, et al. 2016 Object-Based Greenhouse Mapping Using Very High Resolution Satellite Data and Landsat 8 Time Series *Remote Sensing* **8** p 513.

[13] Shi L, Huang X, Zhong T, et al.2020 Mapping Plastic. Greenhouses Using Spectral Metrics Derived From GaoFen-2 Satellite Data *IEEE J-STARS.***13** p 49-59

[14] Tan W. 1981 Climate and agriculture in Ledong County, Hainan Island *Tropical geography* **1** p 33-38.

[15] Baätz M, Schäpe A, Strobl J, et al. 2000 Multiresolution segmentation: An optimization approach for high quality multi-scale image segmentation *Angewandte Geographische* **Informations-Verarbeitung XII** p 12-23.

[16] Rouse J, Haas R, Schell J, et al. 1974 Monitoring the Vernal Advancement of Retrogradation of Natural Vegetation NASA/GSFC Type III, Final Report. Greenbelt, MD, USA p 1-371.

[17] Cui B, Huang W J, Ye H C, Li Z C and Chen Q X. 2021 Object-oriented greenhouse cantaloupe identification by remote sensing technology 9th International Conference on Agro-Geoinformatics p 1-6.

[18] Haraclick R. 1973 Texture features for image classification *Studies in Media and Communication* **3** p 610-621.

[19] Breiman L . 2001 Random Forests *MACH LEARN.* **45** p 5-32.

[20] Agüera F, Aguilar F J, and Aguilar M A. 2008 Using texture analysis to improve per-pixel classification of very high resolution images for mapping plastic greenhouses," *ISPRS J. Photogrammetry Remote Sens.* **63(6)** p 635-646.

[21] Mura M D, Benediktsson J A, Waske B, and Bruzzone L. 2010 Morphological attribute profiles for the analysis of very high resolution images *IEEE Trans. Geosci. Remote Sens.* **48(10)** p 3747-3762.

[22] Agüera F. and Liu J G. 2009 Automatic greenhouse delineation from QuickBird and Ikonos satellite images *Computers and Electronics in Agriculture* **66(2)** 191-200.

[23] Aguilar M A, Bianconi F, Aguilar F J, Fernández I. 2014 Object-based greenhouse classification from GeoEye-1 and WorldView-2 stereo imagery *Remote Sens.* **6** 3554-3582.

Acknowledgments

This work was supported by Hainan Provincial Key R&D Program of China (ZDYF2019084), Hainan Provincial Major Science and Technology Program of China (ZDKJ2019006), National special support program for high-level personnel recruitment (Wenjiang Huang) and Ten-Thousand Talents Program (Wenjiang Huang).

Remote-sensing monitoring of rubber plantations using object-oriented characteristics from the vigorous period

B Cui,[1,2,3] W J Huang,[1,2] H C Ye,[1,2] Q X Chen,[1] Z C Li[3] and H Y Jiang[3]

[1]Aerospace Information Research Institute, Chinese Academy of Sciences, Beijing 100094, China
[2]Key Laboratory of Earth Observation Hainan Province, Hainan Institute of Aerospace Information Research Institute, Chinese Academy of Sciences, Sanya 572029, China
[3]Sanya Institute of Remote Sensing, Sanya 572029, China

Abstract. Natural rubber is one of the four major industrial raw materials in China, and the demand for it is increasing rapidly in China. However, due to geographic and climatic limitations, the rate of natural rubber production in China is significantly less than required to satisfy this demand. Therefore, to ensure the healthy development of China's rubber industry, it is urgent to develop a method to rapidly and accurately monitor the planting and distribution of rubber forests in China. Existing studies have exploited the unique phenological characteristics of rubber plantations manifested by the spectral characteristics of the vigorous growth period and the deciduous period to identify rubber plantations through remote sensing. Unfortunately, the cloudy and rainy climate of rubber-growing regions makes it is difficult to obtain remote-sensing data in the optimal period. Therefore, the present study uses Planet images as the basic data and extracts typical spectral and textural features of rubber forests during the single vigorous growth period to construct a remote-sensing method to monitor rubber forests based on the object-oriented random forest algorithm, which uses all the spectral and texture features. The results show that the proposed random forest classification model achieves a high classification accuracy: the total classification accuracy reaches 89%, and the rubber producer's accuracy and user's accuracy both exceed 92%. This study can effectively solve the problem of lack image data with cloudless from deciduous period and provides a good theoretical basis for remote-sensing monitoring of China's rubber forests, thereby facilitating the development of China's rubber industry.
Keywords: rubber; vigorous period; texture features; object-oriented method; random forest algorithm

1. Introduction

The rubber tree (*Hevea brasiliensis*) is a tall perennial tree that secretes abundant latex, which is an important industrial raw material used to produce various consumer goods. Monitoring the spatial distribution of rubber plantations in a region is thus vital to ensure the steady development of the rubber industry and a stable market supply of latex.

Remote-sensing technology offers a rapid and timely way to monitor ground objects over large areas. In previous work in this area, Li and Fox [1] used the Mahalanobis topicalities method with MODIS time-series normalized difference vegetation index (NDVI) to monitor rubber plantations across Southeast Asia. Liu et al. [2] confirmed that the time period from early January to late March is optimal for monitoring rubber plantations and established decision tree (DT) classification models for young rubber (<10 years old) and mature rubber forest (≥10 years old) by using the original spectral characteristics, normalized indexes, K-T transform indexes, and texture features obtained from Landsat images. Fan et al. [3] and Li et al. [4] extracted rubber forests based on the difference in the vegetation index of Landsat ETM+/OLI images obtained during the defoliation and leaf-extraction periods of rubber forests. Although data products with good accuracy are obtained under cloud-free conditions, it remains difficult to achieve the same results over an arbitrary specified time range and for a large area. Chen et al. [5] used cloud-free phased-array-type L-band synthetic aperture radar (PALSAR) data together with multi-temporal Landsat images to map rubber plantations in Hainan Island, China. They first identified and mapped forests based on the structure information obtained from PALSAR and the yearly maximum NDVI obtained from Landsat data and then mapped the rubber plantations from the deciduous features in the forest map.

Because different objects may produce the same spectral characteristics and the same objects may produce different spectral characteristics [6], using only spectral information to classify land cover cannot provide perfect results. However, the increasing spatial resolution of remote-sensing images now allows rubber tree plantations to be monitored based not only on spectral features but also on textural features. For example, Yang and Tong [7] extracted the rubber tree distribution by using the pixel textural feature and spectral features of high-resolution IKONOS images acquired on May 4, 2010. The arithmetic C5.0 DT proposed by the authors allowed for accurate monitoring, obtaining a producer's accuracy, user's accuracy and total accuracy of 81.00%, 82.65% and 83.5%, respectively. This accuracy is mainly due to the introduction of textural features and the influence of evergreen trees. Furthermore, Zhang et al. [8] pointed out that the producer's accuracy of rubber plantations based on both textural features and spectral features of GF-1, Sentinel-2, and Landsat8 images improved by 8.07%–9.44% over that obtained using spectral features alone.

Most existing studies are based on multitemporal or long time series remote-sensing images with medium or low spatial resolution and exploit the changes in phenology of rubber trees to monitor rubber plantations. However, given that rubber trees are mostly planted in cloudy and rainy zones, it is difficult to obtain good remote-sensing data during the critical phenological period, which hinders rubber monitoring in real applications. Otherwise, the classification method based on pixel information is the most common method for monitoring rubber plantations by remote sensing, although, being mainly based on spectral information, this method is prone to the "salt and pepper" phenomenon. In contrast, the object-oriented classification method uses multiple adjacent pixels as the processing unit (i.e., the object). The classification process then uses both the spectral information and the spatial relationship between the objects, so the semantic information of the object itself is used, which overcomes the limitations of the traditional classification method that uses only the pixel as the basic processing unit [9-10]. Thus, to monitor rubber plantations, this study applies an object-oriented remote-sensing method that uses the characteristics of the rubber-growing period to provide a practical solution for monitoring rubber plantations on a regional scale.

2. Materials and Method

2.1. Study area

Danzhou city is in the northwest of Hainan Province and spans from 19°11'–19°52' N to 108°56'–109°46' E. It has the largest land area (3398 km^2) and the longest coastline (307 kilometres) in Hainan province. Danzhou city has a tropical monsoon marine climate, with an average annual temperature of 23.1 °C and an average annual rainfall of 1823 mm. It is one of the areas least affected by typhoons in Hainan. The forest area of the city is 2 810 100 mu, and the forest coverage is 54.7%. The terrain is mainly hilly, accounting for 76% of the city's area, with plain covering 23.6% of the terrain, and mountainous areas covering the remaining 0.4%

of the terrain. The terrain gradually descends from southeast to northwest, with the southeast region covered by small hills.

Danzhou is one of the major rubber-plantation areas in China. According to statistics, the rubber-plantation area in Danzhou covers 87 084 hectares, and the total yield was 54 579 tons in 2019. Throughout the history of rubber production in China for over 100 years, Danzhou has been among the best in terms of area, quantity and yield. The Rubber Research Institute of the Chinese Academy of Tropical Agricultural Sciences is located in Danzhou and is the only national research institute in China that uses natural rubber as the main research object. In addition, it leads the development of the technical system for China's natural rubber industry. Adhering to inheritance and innovation, the institute has built a scientific and technological innovation system covering the entire industrial chain for natural rubber, including resources, breeding, physiology, biochemistry, cultivation ecology, soil fertilizer, plant protection, biology, material engineering, gum picking, machinery and equipment, processing, comprehensive wood use and industrial development.

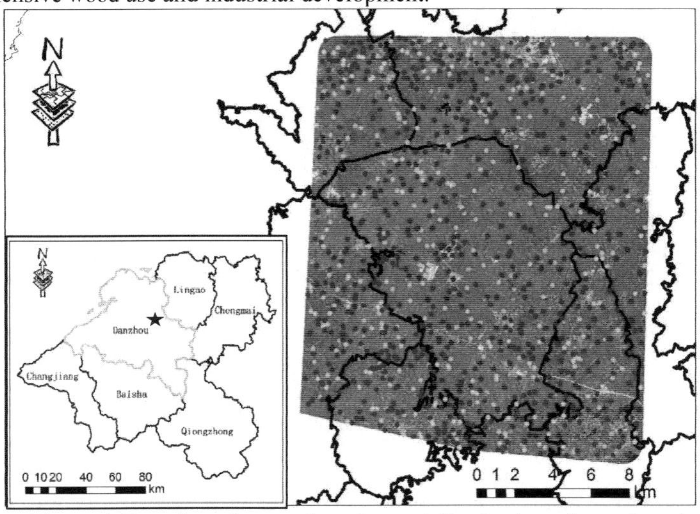

Figure 1. Location of research region and distribution of sample points
Note: Green dots are the validated samples, blue dots are the training samples.

2.2. Data acquisition and processing

2.2.1. Planet satellite image acquisition

Planet images were acquired by Planet Inc., which is the world's largest satellite constellation with more than 170 satellites in orbit, allowing it to monitor the whole world once a day. Planet satellites and its imaging products are the world's only remote-sensing satellites with global high-resolution, high-frequency, full coverage capabilities. Planet satellite image data from June 8, 2019 were selected for this study. The image is clear, cloud-free and of good quality, with a spatial resolution of 3 m. It includes four spectral bands: blue, green, red and near-infrared. The planet image selected for this study is an ortho-grade (3B) data product, which has been pre-processed to obtain a true and reliable surface reflectance image. Table 1 lists the main parameters of the Planet satellites.

Table 1. Main parameters of Planet satellites.

Parameter	Value	Parameter	Value
Orbit	International Space Station's Sun-synchronous orbit	Orbit altitude	400 km 475 km
Orbital inclination	52° 98°	Sensor type	Bayer filter CCD camera
Spatial resolution	3–4 m	Breadth	24.6 km × 16.4 km
Spectral band	Band 1: blue (455–515 nm) Band 2: green (500–590 nm) Band 3: red (590–670 nm) Band 4: near-infrared (780–860 nm)		

2.2.2. Collection of ground sample data

The data from three field investigations from January to April 2019, combined with the visual interpretation of high-resolution Google historical images, indicate that sufficient samples of different types were obtained. Figure 2 shows photographs of rubber plantations taken during the vigorous period. The final selected sample categories include forest, rubber, farmland, water and building land, for a total of five categories of 1014 sample sites. About 2/3 of these samples served as training samples to build the classification model, and the remaining 1/3 served as verification samples to validate the classification model. Table 2 details the samples used in this study.

Table 2. Samples used in this study.

Type	Number of training samples	Number of validated samples	Total
Forest	130	64	194
Rubber	215	100	315
Farmland	175	87	262
Building	92	46	138
Water	70	35	105
Total	682	332	1014

Figure 2. Photographs of rubber plantations in research region during vigorous period.

2.3. Segmentation algorithm

This study uses the multiresolution segmentation algorithm proposed by [11]. Three important parameters affect the segmentation results of the multiresolution segmentation algorithm:

segmentation scale, shape index and compactness index. To determine the optimal segmentation scale, a basic principle to follow is that a specific feature category should be expressed by one or more objects so that objects that are significantly broken or mixed should not appear. Homogeneity is represented by two factors: colour (spectrum) and shape, with the shape factor being represented by smoothness and compactness. The weights for the colour factor and the shape factor sum to unity, as do the weights for smoothness and compactness. The colour factor is the most important factor to generate the object, whereas the shape factor controls the degree of fragmentation of the image object. Generally, during the parameter setting, the larger colour factor value is setted as far as possible, and the shape factor value can be set bigger for objects with clear spatially distributed characteristics. Several iterations of the specific segmentation process are necessary to set the different parameters and obtain the best segmentation results.

2.4. Selection of feature variables

2.4.1. Spectral feature

This study uses the spectral reflection features and vegetation indexes of Planet images, including near-infrared band reflectance, brightness, the normalized difference vegetation index (NDVI), the enhanced vegetation index (EVI) and the ratio vegetation index (RVI). Note that, in the object-oriented method, an object is a computational unit rather than a pixel, so we use the average reflectance of all pixels contained in the object to calculate the spectral feature or the vegetation index.

Table 3. Vegetation indices used in this study.

Vegetation index	Formula	Reference
Normalized difference vegetation index	$(R_{NIR} - R_R)/(R_{NIR} + R_R)$	[12]
Enhanced vegetation index	$2.5(R_{NIR} - R_R)/(R_{NIR} + 6R_R - 7.5R_B + 1)$	[13]
Ratio vegetation Index	R_{NIR}/R_R	[14]

Note: R_R is the red band; R_G is the green band; R_B is the blue band; R_{NIR} is the near-infrared band.

2.4.2. Texture features

A texture feature is a type of spatial feature and is identified by the frequency shift of a spectral feature in a small area. For remote-sensing images with high spatial resolution, different objects can be distinguished according to the difference of texture features. Generally speaking, each type of object has a unique texture feature, which solves the problem of spectral features whereby the same object can have different spectral features or different objects can have the same spectral feature. Therefore, texture information is helpful to distinguish different objects and improve the classification accuracy. This study uses the algorithm based on the grey-level co-occurrence matrix [15] to extract the texture features of different ground objects. Eight common texture indexes were obtained from the Planet image: homogeneity, contrast, dissimilarity, entropy, angular second moment, mean, standard deviation and correlation.

2.5. Building the classification model

(1) Random forest algorithm

The random forest (RF) algorithm was proposed by [16] and is based on the decision tree (DT) algorithm. The RF algorithm integrates multiple trees through ensemble learning with a DT as the basic unit. The RF algorithm offers numerous advantages, such as good noise tolerance, resistance to over-fitting, a capacity to deal with high-dimensional data, strong adaptability to data sets and no strict demand for the form of prediction variables, which can be either quantitative variables or qualitative descriptions. Therefore, the RF algorithm has been widely used for ground feature classification.

Constructing a stochastic model involves two very important custom parameters: the number K of classification trees and the number m of random variables of segmentation nodes. These parameters must be optimized to minimize the occurrence of routine errors during data processing.

(2) Decision tree algorithm

The DT algorithm is typically a process of recursively selecting the optimal feature and segmenting the training data according to the feature to optimally classify each sub-data set. It is essentially a process of categorizing data through a set of rules. The best DT usually has less contradiction with the training data and good generalizability. Moreover, the selected conditional probability model should not only fit the training data well but also predict the unknown data well. The DT-generation process includes three processes: feature selection, DT generation and DT pruning. Feature selection refers to selecting a feature from many features in the training data as the split standard of the current node. Numerous different quantitative evaluation criteria exist for selecting features, which produces different DT algorithms. Decision-tree generation refers to the generation of child nodes recursively from top to bottom according to the selected feature-evaluation criteria until the dataset is no longer separable and the DT stops growing. Because DTs are easy to overfit, pruning is generally needed to reduce the tree size and alleviate overfitting. Two types of pruning techniques are available: pre-pruning and post-pruning.

2.6. Evaluating the classification accuracy

To evaluate the classification accuracy, we use the confusion matrix, which is a comparison matrix representing the number of samples in a certain category and the number of real samples in the same category. The confusion matrix has n rows and n columns. Usually, the total of each row represents the true number of samples in the class, and the total of each column represents the predicted number of samples in the class. The overall accuracy (OA), producer's accuracy (PA), user's accuracy (UA) and Kappa are the most used evaluation indexes for remote-sensing image classification and are calculated as follows:

$$OA = \frac{\sum_{i=1}^{k} N_{ii}}{N}, \tag{1}$$

$$PA = \frac{N_{kk}}{N_{+k}}, \tag{2}$$

$$UA = \frac{N_{kk}}{N_{k+}}, \tag{3}$$

$$\text{Kappa} = \frac{N\sum_{i=1}^{k} N_{ii} - \sum_{i=1}^{k}(N_{i+}\times N_{+i})}{N^2 - \sum_{i=1}^{k}(N_{i+}\times N_{+i})}, \tag{4}$$

where k is the total number of categories, N is the total number of samples, the diagonal elements N_{kk} give the number of samples correctly assigned to real samples in the category, N_{k+} is the total number of samples of i class, N_{+k} is the total number of samples assigned to class j, and N_i and N_j is the value in element (i, j) of the matrix.

IOP Conf. Series: Earth and Environmental Science 1004 (2022) 012021 doi:10.1088/1755-1315/1004/1/012021

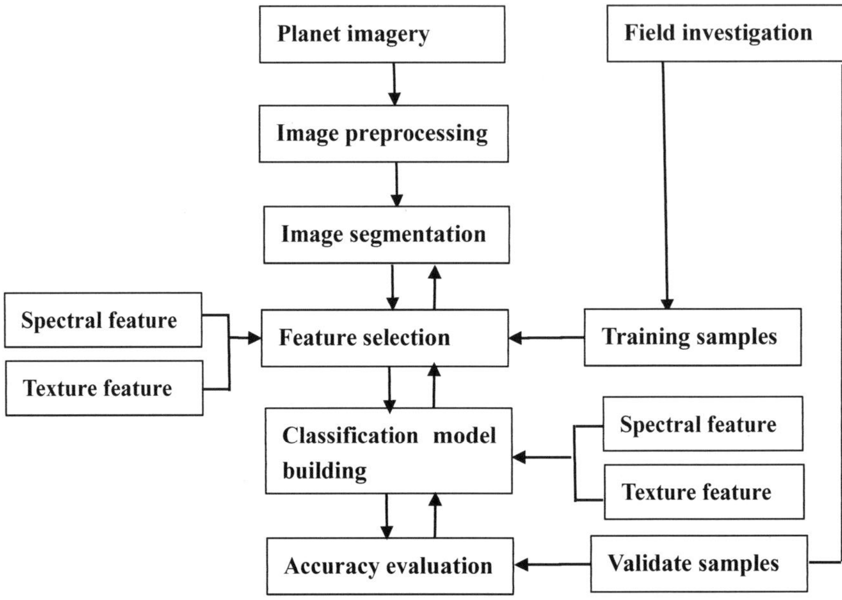

Figure 3 The flowchart of classification

3. Results

3.1. Segmentation results

In the object-oriented method, the segmentation of remote-sensing images directly affects the subsequent classification accuracy. A multiresolution segmentation algorithm was used in this study, with weights of 1, 1, 1, 2 assigned to the four bands (blue, green, red, and near-infrared, respectively) in the Planet image. Because the study area contains complex object types with different shapes, sizes and hues, it is difficult to optimally segment all objects simultaneously. Thus, to ensure the purity of rubber objects, the optimal segmentation parameters are determined after many tests using the following parameters: segmentation scale 100, shape index 0.8 and tightness index 0.8. Figure 4 shows the segmentation results of two partial regions in the study area containing a concentrated rubber plantation.

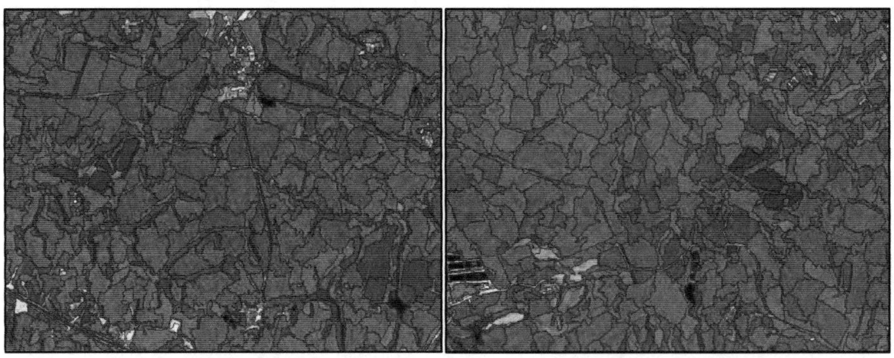

Figure 4. Segmentation results of two partial research regions.

3.2. Classification results

This study extracts rubber plantations from the research area by using the constructed RF model. In addition, to evaluate the classification of the RF model, we compared its accuracy with that of a DT classification model constructed based on the same training samples.

Figures 5(A) and 3(B) show the classification results based on the RF model and on the DT model, respectively.

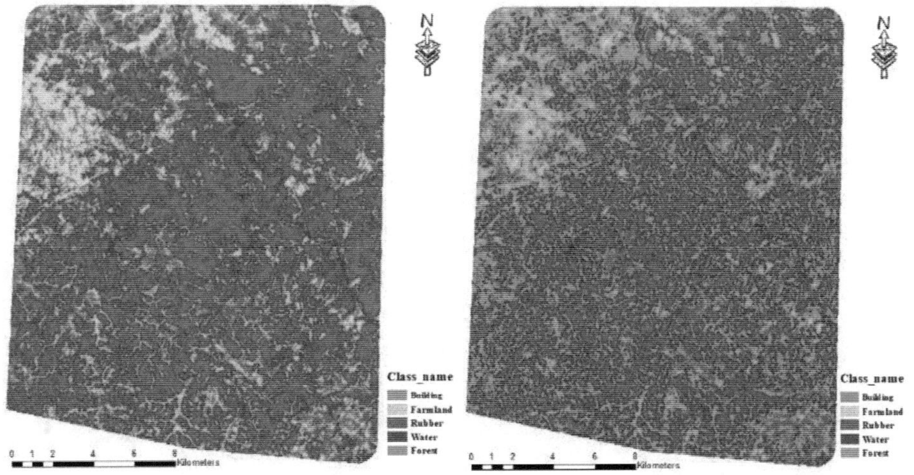

(A) Classification result based on RF model. (B) Classification result based on DT model.

Figure 5. Classification results.

We established a confusion matrix based on the validation sample set and selected four indexes (producer's accuracy, user's accuracy, Kappa coefficient and total accuracy) to evaluate the classification results (Table 4). The results show that the RF classification results are significantly more accurate than the DT classification results. The overall accuracy of the former is 3.81% greater than that of the latter, and the Kappa coefficient increases by 5.50%. For the classification results of rubber objects, the user's accuracy for the RF classification model is slightly lower than that for the DT classification model, whereas the producer's accuracy of RF classification model is significantly greater than that for the DT classification method. Thus, forest remains the main object type that is confused with rubber objects.

Table 4. Comparison of classification accuracy between RF and DT.

	Random forest classification		Decision tree classification	
Type	Producer's accuracy	User's accuracy	Producer's accuracy	User's accuracy
Water	96.09%	95.99%	96.09%	100.00%
Building	86.11%	91.61%	100.00%	88.11%
Rubber	97.02%	92.02%	89.58%	93.30%
Farmland	93.17%	84.03%	85.41%	87.57%
Forest	63.36%	85.23%	68.70%	62.89%
Overall Accuracy	89.32%		86.04%	
Kappa	0.84		0.80	

4. Discussion

The conventional supervised classifiers are the most common methods to map rubber plantations via remote-sensing technology [17]. The classification results presented herein show that one can estimate rubber plantations only by using the object-based characteristics of Planet images acquired during the vigorous rubber period, which gives a producer's accuracy and user's accuracy for rubber of 97.02% and 92.02%, respectively. However, the overall accuracy and Kappa of the RF and DT classification methods are both less than some existing research results by supervised classifiers, which provide overall accuracies of

90.0%–97.2% [18-20]. This result is attributed primarily to the relatively low identification accuracy of forest (only 63.36% for producer's accuracy for the RF classifier and 68.70% for producer's accuracy for the DT classifier).

Recently, several studies used object-oriented classification to map rubber plantations. However, all existing research based on object-oriented methods yielded a better identification accuracy for mapping rubber trees. For example, Dai et al. [18] used Landsat images and the *K*-nearest neighbours' method to map rubber plantations during defoliation and obtained an overall accuracy of 90%. Using SPOT-7 images, Dibs et al. [21] found that object-based classifiers are better than pixel-oriented classifiers for estimating rubber plantations and produce a high overall classification accuracy of 97.48%. Although the overall accuracy of our classification method is not better than that of existing studies, the producer's accuracy and user's accuracy for rubber does offer a significant advantage.

Estimation rubber plantation using the feature of rubber foliation period by remote sensing technology, on the one hand, it avoids the lack of high-quality and free-cloud remote sensing images at rubber defoliation period; on the other hand, it can reduce the impact of undergrowth vegetation in rubber plantation at a certain extent than using defoliation stage characteristics because of the sense understory vegetation (see Figure 2).Therefore, the method proposed herein is feasible and advantageous for estimating rubber plantations.

5. Conclusion

Natural rubber is an important raw material for Chinese industry and is of strategic importance. This study used a single Planet image acquired during the growth period of rubber forests as the data source. By extracting the typical spectral and textural features of rubber forests, we construct an object-oriented random forest (RF) classification model for automated remote-sensing recognition of the spatial distribution of rubber forests. The results show that the proposed RF classification model produces a total classification accuracy of 89.32% and a producer's accuracy and user's accuracy that both exceed 92%. This method overcomes the lack of suitable remote-sensing data when using the deciduous feature of rubber forests for remote-sensing recognition.

References

[1] Li, Z. and J. M. Fox. 2012 Mapping rubber tree growth in mainland Southeast Asia using time-series MODIS 250 m NDVI and statistical data *Applied Geography* **32** p 420-432.

[2] Liu X, Feng Z, Jiang L, et al. 2013 Application of decision tree classification to rubber plantations extraction with remote sensing *Transactions of the Chinese Society of Agricultural Engineering* **29** p 163-172.

[3] Fan, H, et al. 2015 Phenology-Based Vegetation Index Differencing for Mapping of Rubber Plantations Using Landsat OLI Data *Remote Sensing* **7** p 6041-6058.

[4] Li, P, Zhang, J, Feng, Z. 2015 Mapping rubber tree plantations using a Landsat-based phenological algorithm in Xishuangbanna, southwest China *Remote Sensing Letters* **6** p 49-58.

[5] Chen, B, et al. 2016 Mapping tropical forests and deciduous rubber plantations in Hainan Island, China by integrating PALSAR 25-m and multi-temporal Landsat images *INT J APPL EARTH OBS.***50** p 117-130.

[6] Zhou C, Luo J. 2009 GeoComputation of High-resolution satellite remote sensing image *Beijing: Science Press.* p 14-15.

[7] Yang H and Tong X. 2014 Distribution information extraction of rubber woods using remote sensing images with high resolution *Geomatics and information science of Wuhan University* **39** p 411-416.

[8] Zhang, C, et al. 2020 Effect of Textural Features in Remote Sensed Data on Rubber Plantation Extraction at Different Levels of Spatial Resolution *Forests* **11** p 399.

[9] Cao M. 2015 Based on the feature elements of intelligent recognition and extraction of remote sensing image. *Xian: Chang'an university.*

[10] Wu C, Deng J, Wang K, et al. 2016 Object-based classification approach for greenhouse mapping using Landsat-8 imagery *International Journal of Agricultural & Biological Engineering* **9** p 79-88.

[11] Baätz M, Schäpe A, Strobl J, et al. 2000 Multiresolution segmentation: An optimization approach for high quality multi-scale image segmentation *Angewandte Geographische Informations-Verarbeitung* **XII** p 12-23.

[12] Rouse J, Haas R, Schell J, et al. 1974 Monitoring the vernal advancement of retrogradation of natural vegetation *NASA/GSFC.* **Type III Final Rep**.

[13] Verstraete M, Pinty B, Myneni R. 1996 Potential and imitations of information extraction on the terrestrial biosphere from satellite remote sensing *Remote Sensing of Environment* **58** p 201-214.

[14] Pearson R, Miller L. 1972 Remote mapping of standing crop biomass for estimation of the productivity of the shortgrass prairie *Proceedings of the English International Symposiumon Remote Sensing of Enviroment.* p 1375-1381.

[15] Haraclick R. 1973 Texture features for image classification *Studies in Media and Communication* **3** p 610-621.

[16] Breiman L. 2001 Random Forests *MACH LEARN.***45** p 5-32.

[17] Azizan F A, et al. 2021. Application of Optical Remote Sensing in Rubber Plantations: A Systematic Review *Remote Sensing* **13(3)** p 429.

[18] Dai S P, Luo H X, Fang J H, Cao J H, Li H L, Li M F, Wang L L, Luo W. 2014 Object-oriented classification of rubber plantations from Landsat satellite imagery. *In Proceedings of the 2014 The 3rd International Conference on Agro-Geoinformatics,* Beijing, China, pp 11-14.

[19] Suratman M N.2003 Applicability of Landsat TM Data for Inventorying and Monitoring Rubber (Hevea Brasiliensis) Plantations in Selangor, Malaysia: Linkages to Policies. Ph.D. Thesis, The University of British Columbia, Vancouver, BC, Canada, 2003.

[20] Meti S, Rao D V K N, Nair N U U, Jacob J. 2008 Distribution of natural rubber cultivation in relation to soil and landscape attributes in India. *In Proceedings of the 29th Asian Conference on Remote Sensing 2008*, Colombo, Sri Lanka, 10-14 November 2008; **1** pp 7-16

[21] Dibs H, Idrees M O M O, Alsalhin G B A. 2017 Hierarchical classification approach for mapping rubber tree growth using per-pixel and object-oriented classifiers with SPOT-5 imagery. *Egypt. J. Remote Sens. Sp. Sci.* **20** pp 21-30.

Acknowledgements

This work was supported by National Natural Science Foundation of China (41801352), Hainan Provincial Major Science and Technology Program of China (ZDKJ2019006), National special support program for high-level personnel recruitment (Wenjiang Huang) and Ten-thousand Talents Program (Wenjiang Huang).

IOP Conf. Series: Earth and Environmental Science 1004 (2022) 012022 doi:10.1088/1755-1315/1004/1/012022

Exploring spatial entropy in the analysis of landscape pattern and ecological sustainability in watershed ecosystem

X T Liu[1,2] **and H R Zhao**[1,2,3]

[1] Institute of Transportation Engineering and Geomatics, Department of Civil Engineering, Tsinghua University, Beijing 100084, China
[2] 3S Center, Tsinghua University, Beijing 100084, China

[3] E-mail: zhr@tsinghua.edu.cn

Abstract. Ecological sustainability research from the perspective of entropy has received extensive attention in recent years. Spatial entropy ($\boldsymbol{H_S}$) is a new form of spatial heterogeneity analysis tool that applies entropy to distinguish and characterize different landscape patterns. In order to further explore the heterogeneity expression and ecological implication of $\boldsymbol{H_S}$ in the analysis of watershed ecosystem, this research takes Yanhe watershed as the study area. Based on the land use data of years 2000, 2010, and 2020, $\boldsymbol{H_S}$ in combination with 10 landscape metrics was applied for the study of landscape pattern and ecological sustainability. The results showed that: 1) $\boldsymbol{H_S}$ is correlated with landscape richness, aggregation and shape complexity, and it reflects the comprehensive heterogeneity information with high sensitivity to landscape variations, and also describes ecosystem stability with negative correlation; 2) The value of $\boldsymbol{H_S}$ showed a gradual increase from upstream to downstream in Yanhe watershed, and it dropped as a whole during the 20 years; 3) The stability of this basin from 2010 to 2020 had a greater variation range compared with the previous ten years, and the regions with large variation were mainly distributed in the midstream and downstream areas.

1. Introduction

It is known that watershed has an important impact on the water cycle, biodiversity and ecological environment, and it has been recognized as an appropriate and representative scale for ecological sustainability research [1-2]. Landscape change of watershed regarding the natural environment and anthropogenic influence is a highly integrative indicator for sustainability, which affects the quality and evolution of ecosystem and is of great significance to maintaining and improving regional ecological balance [3]. Therefore, characterizing and distinguishing different landscape patterns accurately have long been the main concern of landscape ecology [4-5].

The characteristic of landscape pattern also refers to landscape heterogeneity. In recent decades, a great variety of metrics have emerged for the measurement, analysis and interpretation of spatial heterogeneity [6,7], but there are three critical issues in the understanding and prediction of ecological processes: conceptual flaws in landscape pattern analysis, inherent limitations of landscape indices, and improper use of pattern indices [8]. Therefore, in addition to focusing on the theoretical basis behind the landscape metrics, it is also important to explore the mechanism of ecological process.

Information theory and entropy-related indicators provide insights for complex system analysis, which have high relevance in the analysis of landscape ecology [9]. Shannon entropy is a quantitative measure of the diversity of categorical data, which constitutes the core of information theory [10]. 'Life depends on negative entropy for survival', due to the internal connection with living system and the ability of measuring the order and complexity of ecosystems, entropy was rapidly introduced into the field of landscape ecology [11-13], and regarded as an important method with in-depth ecological process mechanism.

Since the research object of information entropy is one-dimensional, while the reality is three dimensional or even multidimensional, so it is difficult to reflect the characteristics of actual spatial pattern [14]. To solve this problem, combined with the First Law of Geography [15], Wang and Zhao (2018) proposed the spatial entropy ($\boldsymbol{H_S}$) considering spatial proximity, which is a quite useful attempt to apply entropy and information theory to landscape ecology research [16]. It has been proved that $\boldsymbol{H_S}$ can distinguish and quantify different landscape patterns and measure the order degree of spatial pattern [16], while its ecological implication and practical applicability still need to be further studied.

Yanhe watershed, which is a typical loess hilly and gully region and the key area of the Grain for Green Project, was selected as the study area. $\boldsymbol{H_S}$ in combination with classical landscape metrics was applied for the analysis of landscape pattern and ecological sustainability in the protection and development of the watershed ecosystem.

2. Study area and data

Yanhe watershed is located in the hinterland of the Loess Plateau in China, covering an area of 7,687 km^2. The topography in this area is undulating, showing the characteristics of high northwest and low southeast. Loess landform in this region is crisscrossed with ravines, loose soil, poor anti-erosion ability and prone to be dispersed and transported. The annual average rainfall is about 520 mm with uneven seasonal distribution. Vegetation presents a north-south gradient distribution, which can be divided into three zones: forest zone, forest-grassland transition zone and grassland zone. In the past, due to the influence of unreasonable human activities and natural factors, the ecosystem in this watershed was significantly degraded with a sharp decrease of natural vegetation and severe water and soil erosion [17]. Since 1999, a series of ecological construction policies and the Grain for Green Project has been carried out on a large scale in this region [18].

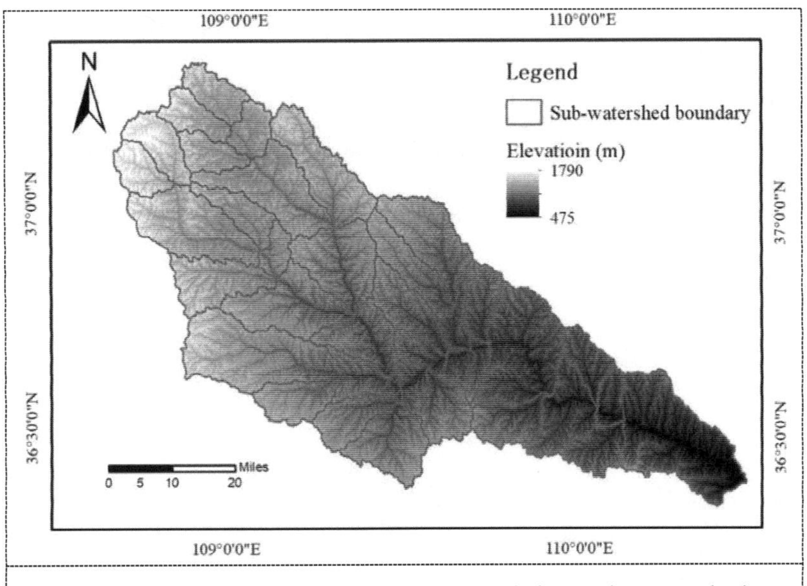

Figure 1. Elevation and the sub-watersheds in Yanhe watershed.

The basic data selected for the study included the 10-m resolution digital elevation model (DEM), and the years 2000, 2010 and 2020 land cover maps with a spatial resolution of 30 m, which was obtained from the GlobeLand30 database (http://globeland30.org) [19]. Yanhe watershed was divided into 39 sub-watersheds using the Hydrology Tool of ArcGIS 10.6, as shown in figure 1.

3. Methodology

3.1. Spatial entropy

Due to its mathematical theoretical characteristics, Shannon entropy has been widely used in the ecological analysis [11-13]. However, it can only reflect landscape composition, not spatial configuration. Based on Shannon entropy and the First Law of Geography [15], $\boldsymbol{H_S}$ takes proximity, which contains the total edge length between different classes and the distance between different class centers, as the key factor to quantitatively describe the influence of spatial attributes. It is applicable to discrete patterns with specific classification [16]. The calculation formula is as follows:

$$H_S = -\sum_{i=1}^{n} \frac{L_i}{d_i} p_i \log_2 p_i \quad (1)$$

$$L_i = \sum_{\substack{k=1 \\ i \neq k}}^{n} L_{ik} \quad (2)$$

$$d_i = \sum_{\substack{k=1 \\ i \neq k}}^{n} d_{ik} \quad (3)$$

where n is the number of different classes and p_i is the proportionality of class i in a landscape. L_i/d_i is defined as 'proximity', where L_i is the total amount of edges between different landscape classes and d_i is the average distance between centroids of different classes. $\boldsymbol{H_S}$ is semi-bounded by the interval $[0, +\infty)$, and will change along with the 'proximity' for a given spatial pattern of the constant number and proportion of categories.

3.2. Landscape metrics

Landscape metrics can provide objective and quantitative descriptions of structure composition and spatial configuration of ecological systems and are useful in landscape monitoring and planning [6-7, 20]. Su et al. [21] summed up three key criteria for selecting landscape metrics: 1) metrics can represent different aspects of landscape characteristics; 2) metrics should not be highly redundant; 3) metrics should be documented in related studies. Since $\boldsymbol{H_S}$ is an indicator at the landscape level, 10 representative landscape metrics were selected from four aspects according to the above criteria and previous studies [6-7, 20-22]: area and density (including Largest Patch Index (LPI), Patch Density (PD) and Edge Density (ED)), shape (including Landscape Shape Index (LSI) and Perimeter-Area Fractal Dimension (PAFRAC)), aggregation (including Contagion Index (CONTAG), Splitting Index (SPLIT) and Aggregation Index (AI)), and diversity (including Shannon's Diversity Index (SHDI) and Shannon's Evenness Index (SHEI)).

LPI reflects the dominant patch effect and PD describes the degree of spatial heterogeneity and fragmentation of landscape. ED extracts complex boundary information, which is the most suitable index for quantifying the spatial configuration of heterogeneity [23], similar to L_i in the calculation formula of $\boldsymbol{H_S}$. LSI and PAFRAC reflect the complexity of landscape patch shape. CONTAG and AI reflect the degree of aggregation and extension of landscape, and SPLIT describes the separation between landscape patches. SHDI and SHEI, based on information theory, were selected as the compositional metrics without spatial configuration information, which are both popular measures of diversity in ecology [20, 23]. The selected metrics were calculated by FRAGSTAT 4.2 at the landscape level.

IOP Conf. Series: Earth and Environmental Science 1004 (2022) 012022 doi:10.1088/1755-1315/1004/1/012022

4. Results

4.1. Landscape heterogeneity analysis

Taking sub-watershed as the geographical unit, H_S and 10 landscape metrics of each unit regarding the years 2000, 2010, and 2020 were calculated. Pearson correlation coefficient method [20] was used for analyzing the correlations between H_S and each metric. Two significance levels (0.05 and 0.01) were set and a two-tailed test was performed. Table 1 shows the significances were all under 99% confidence level, indicating that H_S was closely related to both composition and configuration of the landscape. The correlation coefficient between H_S and LSI was the highest, reaching 0.635, and another shape index, PAFRAC, was also highly correlated with H_S with the coefficient of 0.519. CONTAG and AI, representing aggregation, had a negative correlation with H_S, and the absolute values of correlation coefficients were both greater than 0.5. ED, the measure of boundary abundance, was the index most correlated with H_S in the area and density category with a correlation of 0.518. The coefficients of SHDI and SHEI with H_S were both less than 0.5, implying a relatively weak relationship.

Table 1. Analysis results of correlation between H_S and landscape pattern.

Index type	Landscape metric	Correlation coefficient
Area and density	LPI	-0.417[a]
	PD	0.291[a]
	ED	0.518[a]
Shape	LSI	0.635[a]
	PAFRAC	0.519[a]
Aggregation	CONTAG	-0.544[a]
	SPLIT	0.383[a]
	AI	-0.517[a]
Diversity	SHDI	0.385[a]
	SHEI	0.492[a]

[a] Means significance under 99% confidence level.

Therefore, in most cases, the larger the H_S, the lower the dominant patch effect and the degree of aggregation and extension, the higher the dispersion and shape complexity of patch landscape and the landscape diversity. As for the landscape system, the richer the patch type and the higher the degree of fragmentation, the more uncertainty contained in the landscape, so the weaker the stability of ecosystem [6]. Thus, it can be concluded that the H_S value is inversely proportional to the stability of watershed ecosystem.

In order to compare the sensitivity of H_S and other indexes in characterizing spatial heterogeneity, the coefficients of variation of H_S and 10 landscape metrics of each sub-watershed of years 2000, 2010, and 2020 were calculated respectively, and then the mean coefficients of variation of 39 sub-watersheds were obtained as shown in figure 2. It shows that the mean coefficient of variation of H_S is the highest and is much higher than most landscape metrics, indicating that H_S is quite sensitive to the variation of spatial pattern.

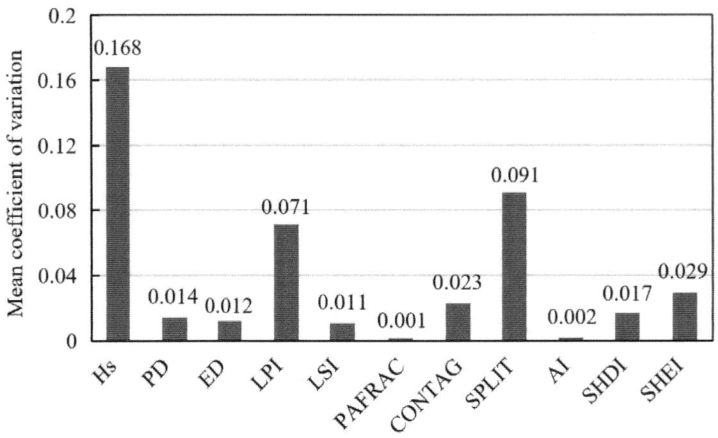

Figure 2. Mean coefficients of variation of $\boldsymbol{H_S}$ and 10 landscape metrics.

4.2. Temporal-spatial variation of $\boldsymbol{H_S}$ *in the study region*

The spatial distribution of $\boldsymbol{H_S}$ in each sub-watershed of the Yanhe watershed was shown in figure 3. It can be seen that there were obvious spatial variations of $\boldsymbol{H_S}$ in the study region, showing a gradually increasing trend from upstream to downstream. From 2000 to 2020, the maximum entropy value of sub-watersheds was greatly dropped from 548.44 to 340.33, showing that the overall stability of ecosystem had been improved.

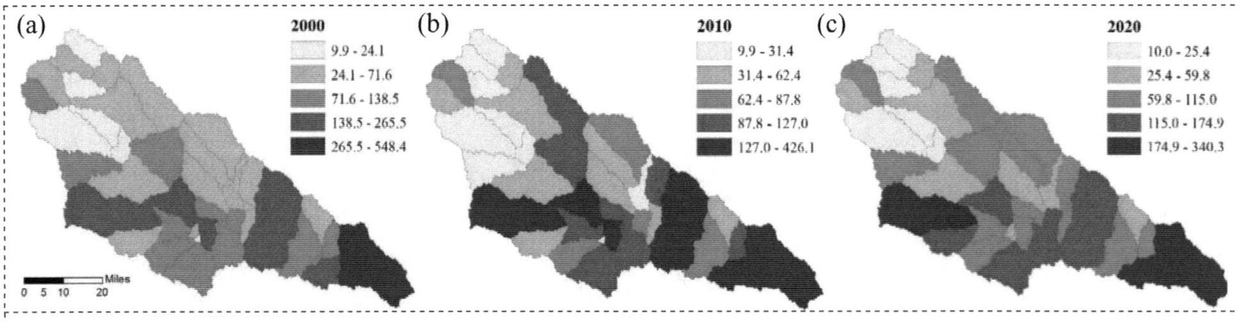

Figure 3. Spatial distributions of $\boldsymbol{H_S}$ of sub-watersheds in Yanhe watershed. (a) 2000; (b) 2010; (c) 2020.

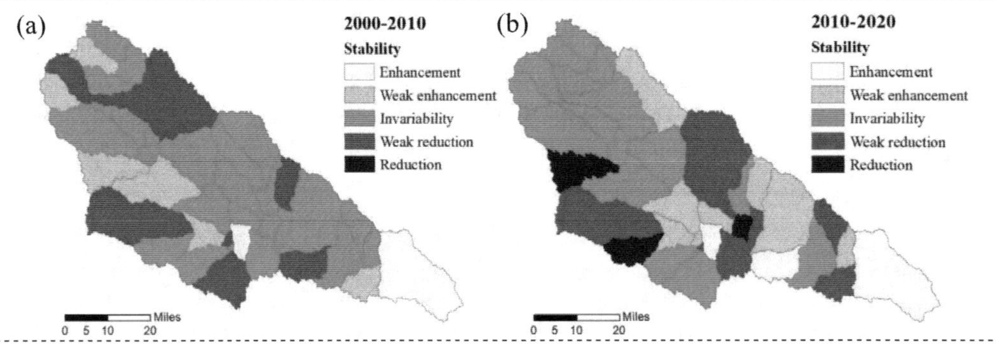

Figure 4. Spatial distributions of stability variation of sub-watersheds in Yanhe watershed. (a) 2000-2010; (b) 2010-2020.

In order to further analyze the impact of environmental governance in recent years, the difference of entropy values of sub-watershed landscapes in different years was further calculated. According to the entropy variations and the conclusion of section 4.1 that H_S is inversely proportional to the stability, the watershed was further divided into stability enhancement area (<-50), stability weak enhancement area (-50~-10), stability invariability area (-10~10), stability weak reduction area (10~50) and stability reduction area (>50) (figure 4). It shows that from 2000 to 2010, the ecological stability of most regions remained stable, and the areas with large changes were mainly located in the upstream and downstream. From 2010 to 2020, the change of ecosystem stability was more obvious, and the scope of the stability enhancement area and stability reduction area were both larger than the previous decade. Moreover, the regions of great variation were transferred to the midstream and downstream areas.

5. Discussion and conclusion

The spatial entropy based on information theory is a new entropy-related index and has been preliminarily used in ecological sustainability research. In order to further explore its spatial heterogeneity expression and ecological implication for the watershed ecosystem, this research took Yanhe watershed as the study area and sub-watershed as the research scale. Based on the land use data of the years 2000, 2010, and 2020, H_S was applied to the analysis of landscape pattern and ecological sustainability.

The results show that H_S can characterize and distinguish different landscape patterns with high sensitivity to the land-use change, and simultaneously reflect the composition and configuration information. The value of H_S is correlated with landscape richness, aggregation and shape complexity. The larger the H_S, the higher the shape complexity and landscape diversity, and the lower the dominant patch effect and landscape aggregation degree, so that H_S is a comprehensive index that can reflect the spatial characteristics of landscape pattern and describe ecosystem stability with negative correlation.

In the Yanhe watershed, the H_S increased gradually from upstream to downstream, which is mainly since the downstream is the gully region of the residual plateau with flat terrain and fertile soil, and has frequent agricultural activities and lack of enough vegetation protection. In recent years, agricultural activities have been rationally carried out to improve land production efficiency, which is consistent with the significant reduction of H_S in the downstream. Since the implementation of the Grain for Green Project, the H_S of Yanhe watershed was dropped generally. From 2000 to 2010, the region with a large range of stability variation was mainly located in the upstream and downstream, but from 2010 to 2020, it was transferred to the midstream and downstream, which is mainly because the early management of the project mainly paid attention to the farmland with the slope greater than 25°, of which more than 70 % was located in the upstream [24], while in the later stage, it mainly focused on the ecological governance and urban construction in the midstream and the rational agricultural activities in the downstream.

In particular, it should be noted that though the change of H_S can explain the change variations of ecosystem stability in some degree, the analysis of ecological quality and sustainability still need to be further analyzed in combination with the actual land use situation and the state of ecosystem (such as stable period, growing period or degradation period).

References

[1] Shen J, *et al.* 2021 Prospect theory in an evolutionary game: Construction of watershed ecological compensation system in Taihu Lake Basin *J. Clean Prod.* **291** 125929

[2] Kumar P, Liu W, Chu X, Zhang Y and Li Z 2019 Integrated water resources management for an inland river basin in China *Watershed Ecology and the Environment* **1** 33–8

[3] Renetzeder C, Schindler S, Peterseil J, Prinz M A, Mücher S and Wrbka T 2010 Can we measure ecological sustainability? Landscape pattern as an indicator for naturalness and land use intensity at regional, national and European level *Ecol. Indic.* **10** 39–48
[4] Cushman S A 2016 Calculating the configurational entropy of a landscape mosaic *Landsc. Ecol.* **31** 481–9
[5] Wu J 2006 Landscape Ecology, cross-disciplinarity, and sustainability science *Landsc. Ecol.* **21** 1–4
[6] Zhao Q, Wen Z, Chen S, Ding S and Zhang M 2019 Quantifying land use/land cover and landscape pattern changes and impacts on ecosystem services *Int. J. Environ. Res. Public Health* **17** 126
[7] Uuemaa E, Antrop M, Roosaare J, Marja R and Mander Ü 2009 Landscape metrics and indices: an overview of their use in landscape research *Living Rev. Landsc. Res.* **3** 1–28
[8] Li H and Wu J 2004 Use and misuse of landscape indices *Landsc. Ecol.* **19** 389–99.
[9] Rodríguez R A, Herrera A M, Quirós Á, Fernández-Rodríguez M J, Delgado J D, Jiménez-Rodríguez A, Fernández-Palacios J M, Otto R, Escudero C G and Luhrs T C 2016 Exploring the spontaneous contribution of Claude E. Shannon to eco-evolutionary theory *Ecol. Model.* **327** 57–64
[10] Claramunt C 2012 Towards a spatio-temporal form of entropy *Lecture Notes in Computer Science* Castano S and Vassiliadis P *et al* (Berlin, Heidelberg: Springer) pp 221–30
[11] MacArthur R 1955 Fluctuations of animal populations and a measure of community stability *Ecology* **36** 533–6
[12] Huang F, Ochoa C G, Chen X, Cheng Q and Zhang D 2020 An entropy-based investigation into the impact of ecological water diversion on land cover complexity of restored oasis in arid inland river basins *Ecol. Eng.* **151** 105865
[13] Li B, Zhang W, Wang Z, Xie H, Yuan X, Pei E and Wang T 2020 Effects of landscape heterogeneity and breeding habitat diversity on rice frog abundance and body condition in agricultural landscapes of Yangtze River Delta, China *Curr. Zool.* **66** 615–23
[14] Parvinnezhad D, Delavar M R, Claramunt C and Pijanowski B C 2019 A modified spatial entropy for urban sprawl assessment *Geocarto Int.* **36** 1804–19
[15] Tobler W R 1970 A computer movie simulating urban growth in the detroit region *Econ. Geogr.* **46** 234–40
[16] Wang C and Zhao H 2018 Spatial heterogeneity analysis: Introducing a new form of spatial entropy *Entropy* **20** 398
[17] Feng Q, Zhao W, Hu X, Liu Y, Daryanto S and Cherubini F 2020 Trading-off ecosystem services for better ecological restoration: A case study in the Loess Plateau of China *J. Clean Prod.* **257** 120469
[18] Bai J, Zhou Z, Zou Y, Pulatov B and Siddique K H.M 2021 Watershed drought and ecosystem services: spatiotemporal characteristics and gray relational analysis *ISPRS Int. Geo-Inf.* **10** 43
[19] Jun C, Ban Y and Li S 2014 Open access to Earth land-cover map *Nature* **514** 434
[20] Ma L, Bo J, Li X, Fang F and Cheng W 2019 Identifying key landscape pattern indices influencing the ecological security of inland river basin: The middle and lower reaches of Shule River Basin as an example *Sci. Total Environ.* **674** 424–38
[21] Su S, Wang Y, Luo F, Mai G and Pu J 2014 Peri-urban vegetated landscape pattern changes in relation to socioeconomic development *Ecol. Indic.* **46** 477–86
[22] Zhou Z X and Li J 2015 The correlation analysis on the landscape pattern index and hydrological processes in the Yanhe watershed, China *J. Hydrol.* **524** 417–26
[23] Yu J, Peng S, Zhang W and Kang S 2020 Index for the Consistent Measurement of Spatial Heterogeneity for Large-Scale Land Cover Datasets *ISPRS Int. Geo-Inf.* **9** 483

[24] Zhu Q, Zhou Z X, Liu T and Bai J Z 2021 Vegetation restoration and ecosystem soil conservation service value increment in Yanhe Watershed, loess Plateau *Acta Ecol. Sin.* **41** 2557–70

Acknowledgments
This work was supported by the Fund of National Natural Science Foundation of China [Grant number 41971379].

Urban forest canopy height modeling using spaceborne laser ICESAT-2 LIDAR

Wang Shufan, Liu Chun*

College of Surveying and Geo-informatics, Tongji University, Shanghai, PR China

E-mail Address. liuchun@tongji.edu.cn

Abstract. Urban forests are an important part of urban ecosystems. Carbon sequestration in urban forests helps reduce the concentrations of greenhouse gases in the region where they are present. Forest height is an important structural parameter for calculating the forest carbon sequestration capacity. Based on this, our study proposes a space-borne laser fusion multi-source remote sensing inversion model of urban forest tree height based on urban space environmental characteristics. This paper mainly consists of three parts: (1) First, a variety of highly correlated tree feature factors were extracted from ICESat2 satellite-borne laser data, LandSat8 multi-spectral data, and spatial environment auxiliary data, and a feature database was constructed. (2) The importance of the feature factors in the feature base was analyzed, and a large-scale forest height inversion model of Shanghai was constructed using a support vector machine (SVM), random forest (RF), and backward propagation neural network (BP-ANN). (3) The accuracy of the urban forest height inversion model was improved by introducing urban spatial environmental features such as texture features. Ablation experiments show that the texture features considered in this study can improve the accuracy of each model to varying degrees, and the accuracy of the BP neural network can reach $\boldsymbol{R}^2$=0.61, RMSE=3.6589. The accuracy of the urban tree height inversion model was $\boldsymbol{R}^2$ =0.6433, RMSE=1.0967, which proves the effectiveness of the space-borne laser fusion multi-source remote sensing urban forest height inversion model considering the characteristics of the space environment.

Keywords: Sustainability, Carbon neutrality, Spaceborne lasers, ICESat-2, Urban Forests, Canopy Height, Machine Learning

1. Introduction

The Paris Agreement, signed in 2016, calls on all countries to reduce carbon emissions to slow down the process of climate warming. To this end ,China has put forward the goal of achieving carbon neutrality by 2060.To achieve carbon neutrality, we need to increase carbon sequestration while reducing carbon emissions. Forests are one of the main carbon sinks on Earth. The carbon sequestration capacity of forests is a measure of the amount of carbon dioxide absorbed, and the height of trees is directly related to the carbon sequestration capacity of forests . Therefore, building a model that can quickly and accurately retrieve the height of trees has important research significance.

In the past few decades, a lot of research has been conducted on the calculation methods of vegetation height. Traditional measurements of tree characteristics rely on field measurements [2], which are limited to an area, have high technical requirements, and have high costs. The emergence of optical remote sensing makes it possible to rapidly measure tree height on a large scale [3], but because of the saturation effect of optical remote sensing, the inversion accuracy is limited [4]. With the development of active remote sensing—Light Detection and Ranging (Lidar), tree height can be measured by a laser on an aircraft for acquiring the three-dimensional structure of a regional forest [4-6]. However, the cost of airborne laser flight missions is high, and the data range obtained in a single mission is still small, which is the main challenge to its large-scale application. The Ice, Cloud, and Land Elevation Satellite-2 (ICESat-2) mission has provided a solution for estimating forest tree heights using a space-borne laser. Lefsky et al. [7] have also demonstrated the feasibility of tree height inversion using spaceborne lidar.

However, there is currently a lack of tree height models for large-scale cities. Urban forests are different from natural forests, and their growth is affected by more spatial elements. Therefore, the spatial distribution characteristics of urban elements and vegetation spatial environmental information should be considered when constructing urban tree height models. As a common measurement feature of the spatial distribution of urban elements, texture features are effective in inverting various tree parameters in optical remote sensing [8, 9].

In addition, the current commonly used vegetation height calculation model mostly adopts a linear function [10, 11], but the linear function is too simple, and the accuracy of the calculation results cannot be guaranteed. Therefore, this study also aimed to address this problem by analyzing the feature factors that contribute more to the modeling of vegetation height, and then inputting SVM\RF\BP and other machine learning methods for modeling, comparing and analyzing the modeling results, and finally establishing a model more suitable for inversion of urban vegetation height.

This paper is structured as follows: Section 2 introduces the process of data preparation and processing. Section 3 briefly describes the modeling methods and processes. Section 4 presents the experimental results, and at the end of this paper, we discuss the significance and implications of the conclusions.

IOP Conf. Series: Earth and Environmental Science 1004 (2022) 012023 doi:10.1088/1755-1315/1004/1/012023

2. Materials

2.1 Study site

The study area, Shanghai, is located in eastern China (120°52E,30°40N,31°53′N,122°12E), with a total area of over 6,340 square kilometers. Shanghai has a large population and is one of the most prosperous cities in China. The study area can be classified as a subtropical monsoon climatic region, with an average annual temperature of 16.6℃, and annual precipitation of approximately 1200 mm [12]. The forest area of Shanghai is 89,000 ha, and the forest coverage is 14.04%. The main vegetation is subtropical evergreen broad-leaved forest and deciduous broad-leaved mixed forest, which originated from artificial forests, and the main tree species include camphor wood, metasequoia, and poplar. Figure 1 shows the location of the study area and the selected ICESat-2 track.

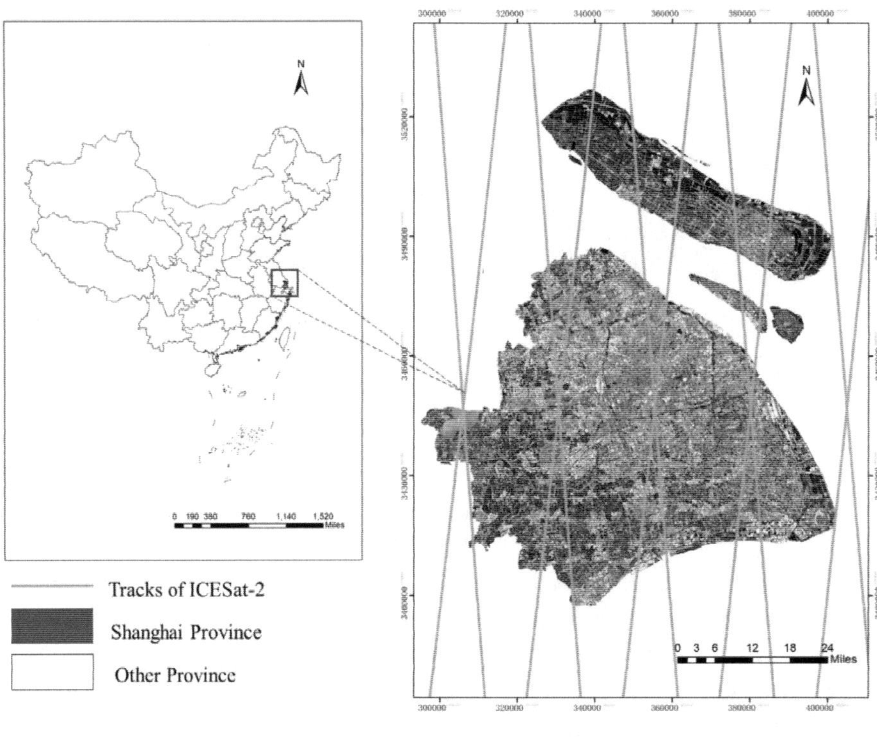

Figure 1. The location of the study area and ICESat-2 tracks

2.2 Data and processing

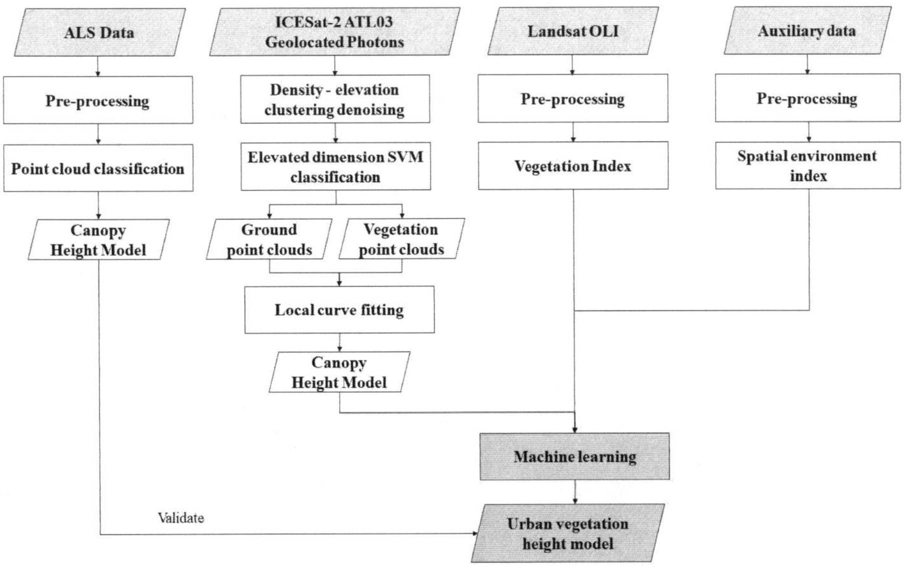

Figure 2. The flowchart of generating the urban forest height map

2.2.1 ICESat-2 data and processing

ICESat-2 is the world's first laser altimeter based on a photon-counting system. The single-photon sampling method increases the density of on-orbit sampling, and the surface elevation accuracy can reach 10 cm [13]. The Level-2 product of global geolocated photons (ATL03) measures the latitude, longitude, time, and absolute height above the World Geodetic System 1984 (WGS84) for each photon. In this study, ICESat-2 ATL03 data were obtained from January 2019 to December 2020, and the track of ICESat-2 in the study site is shown in Figure 1.

After denoising, classification, and height extraction by our algorithm, we obtained the height data of the surface vegetation. First, three-dimensional units are divided in the spatial range, and the index relationship between the disordered point clouds was established. At the same time, the density in the photon neighborhood is calculated to achieve the expansion of the photon characteristics. The data were projected into a high-dimensional space with expanded features, and the background noise was removed by clustering. The support vector machine (SVM) algorithm was used to solve the maximum hypersegmentation plane of the geometric interval between the training sets to realize the classification of vegetation point clouds and ground point clouds. The local curve least square fitting method was used to fit the highest position of the classified ground features, and the height of the vegetation canopy was extracted. Figure 3 shows a comparison between the vegetation height extraction results of the proposed algorithm and the ATL08 results.

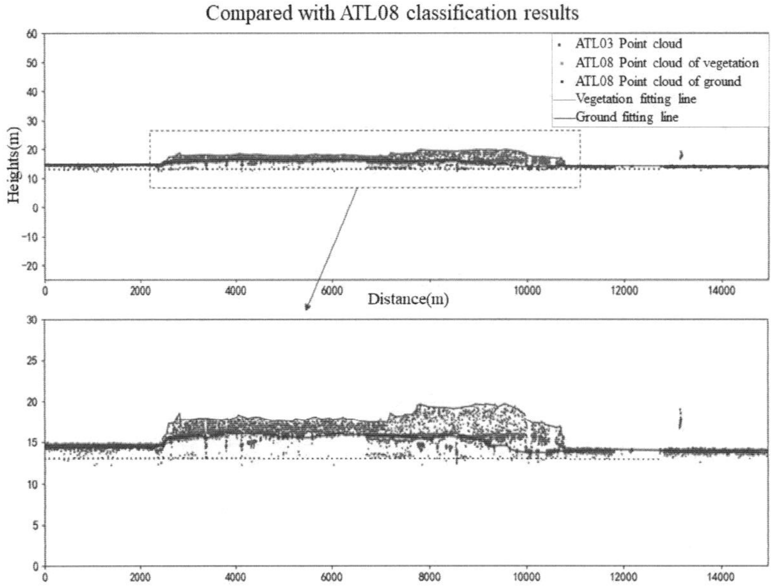

Figure 3. Compared extraction effect of vegetation canopy height with ATL08

2.2.2 Landsat 8 OLI and processing

Although the vertical height data of the ground features can be retrieved from the ICESat-2 point cloud, the orbit distribution of ICESat-2 is discrete (Figure 1), and the data cannot cover the entire region. Therefore, to build a model that can retrieve the vegetation height in the entire region, space-borne laser data and optical remote sensing images were combined [13]. In this study, Landsat8 was used as the source of the optical remote sensing data. The Landsat8 satellite was launched on February 11, 2013, with a total of nine bands, with a spatial resolution is 30m, including a 15 m panchromatic band. Landsat8 images from January 2018 to January 2019 were downloaded using the cloud-based computation platform Google Earth Engine (GEE, https://code.earthengine. google.com/).

After radiometric correction and atmospheric correction of the Landsat8 images, vegetation index features were calculated to obtain a spectral index database. The characteristics of the spectral index database are presented in Table 4.

Table 1. List of vegetation indices (VI) used in this study

Vegetation Index	Description
Normalized difference vegetation index (NDVI) [14]	(NIR-RED)/(NIR+RED)
Atmospherically resistant vegetation index (ARVI) [14]	NIR-(2×RED-BLUE)/NIR+(2×RED-BLUE)
Enhanced vegetation index(EVI) [14]	2.5×(NIR-RED)/(NIR+6RED-7.5×BLUE+1)
Ratio vegetation index (RVI) [15]	NIR/RED
Green normalized difference vegetation index (GNDVI) [16]	(NIR-GREEN)/(NIR+GREEN)
Transformed Normalized Difference Vegetation Index (TNDVI) [17]	Sqrt((SWIR1-NIR)/(SWIR1+NIR)+0.5)

2.2.3 Auxiliary data and processing

The auxiliary data used in this research mainly reflect the differences in the spatial environment information of different urban plots. The urban functional area distribution data were obtained from the Shanghai Master Plan (2017-2035). The air quality index (AQI) data from 2010-2018 were obtained from the China Meteorological Data Service Center, and Kriging interpolation was used to obtain the summer AQI distribution for the whole city. The latitude and longitude data were calculated to indicate spatial position differences.

2.2.4 ALS data and processing

Airborne laser scanning (ALS) data of some regions within the study site were collected to verify the accuracy of the vegetation height retrieved in this study. The ALS data collection region is located in the Qingpu District of Shanghai, which is located southwest of the research site. It is one of the areas with the most abundant vegetation in Shanghai. The tree species within the area are mainly evergreen broad-leaved forests and deciduous coniferous forests, which represent the dominant tree species in Shanghai. Airborne flights were conducted between June and July 2021. Repetitive flights were conducted with a DJI M300 RTK mounted on an L1 airborne laser scan system, producing an average point density of 150 pts/m2 of a triple-echo point cloud. The elevation is the WGS-84 ellipsoid, and the projection coordinate system is the UTM Zone 51 N coordinate system as a geographical reference.

The point cloud model is generated by airband splicing through the overlapping area between the airbands. After point cloud denoising, clipping, and other preprocessing for the original point cloud, point clouds are classified into ground point clouds and vegetation point clouds. According to the categories of the point clouds, a digital elevation model (DEM) and digital surface model (DSM) were generated by interpolation according to a spatial resolution of 2 m, and the relative canopy height model (CHM) with a spatial resolution of 2m was obtained by the difference in the overlapping area between the DSM and DEM [18], which was used as the verification data source for vegetation height inversion.

3. Methodology

3.1 Texture feature

Texture features are commonly used in visual image analysis. By quantifying a certain rule of gray or color change within a region, the feature gap within a class can be narrowed to the maximum possible extent, and the feature gap between classes can be enlarged. The gray-level co-occurrence matrix (GLCM) is a commonly used method to describe texture features. GLCM counts the high-order statistical characteristics of the gray level by counting the gray-level properties of the pixels and the neighborhoods in the moving window. Statistics based on GLCM are often used as texture classification features. In this study, eight feature quantities, such as mean, variance, contrast, homogeneity, anisotropy, entropy, angular second moment, and correlation were calculated based on the Landsat 8 OIL extraction of texture features at the study site. The eight-term characteristic quantity formula is presented in Table 3.

Table 2. List of Texture measures used in this study

	Expression of characteristic quantity
Mean	Mean $\frac{1}{MN}\sum_{i=0}^{M-1}\sum_{j=0}^{N-1} f(i,j,d,\theta)$
Variance	Var $\sum_{i=0}^{M-1}\sum_{j=0}^{N-1}$
Contrast	Con $\sum_{i=0}^{M-1}\sum_{j=0}^{N-1}$
Homogeneity	Hom $\sum_{i=0}^{M-1}\sum_{j=0}^{N-1}\frac{f(i,j,d,\theta)}{1+}$
Anisotropy	Dis $\sum_{i=0}^{M-1}\sum_{j=0}^{N-1} \vee i-j \vee f(i,j,d,\theta)$
Entropy	Ent $= -\sum_{i=0}^{M-1}\sum_{j=0}^{N-1} f(i,j) lg\, f(i,j,d,\theta)$
Angular second moment	ASM$\sum_{i=0}^{M-1}\sum_{j=0}^{N-1} f$
Correlation	AutoCor $\sum_{i=0}^{M-1}\sum_{j=0}^{N-1}(ij) f(i,j,d,\theta)$

3.2 Tree height model

The traditional calculation method of vegetation height requires a lot of field measurements, and usually, only a simple fitting method is used to express the vegetation height. However, the intrinsic relationship among the parameters of vegetation is often nonlinear and difficult to express using a direct formula. Therefore, this study uses a variety of machine learning models to construct a more accurate vegetation height model.

Support vector machines (SVMs) use a nonlinear kernel function to project input data onto a high-dimensional feature space, where complex nonlinear patterns can be simply represented. In this study, the Gaussian radial basis kernel function and E1071 software package of the R software package were used to construct the SVM model.

Random forest (RF) is an integrated learning technique proposed by Breiman [19]. Random forests can measure the importance of variables, and the measurement results can reflect the extent to which each variable plays a role in the prediction. In this study, the random forest model was constructed, the Ntree value was set to 500, and the randomForest package in the R software package was used to achieve it.

A back-propagation (BP) neural network is a model that uses the error backpropagation algorithm to train the network. In the training process, the weight matrix is adjusted iteratively by adjusting the error between the predicted and actual values of the sample output to make the output results closer to the sample value. In this study, the MATLAB R2018b neural network toolbox was used to obtain the network with the best computational accuracy when the number of neurons in the hidden layer was (35 27 39) through cyclic iteration. The inter-neuron transfer function was tansig( ), the training time was 1000, the learning rate was 0.001, and the minimum error of the training target was 0.0001.

4. Results

Pearson correlation calculations (Table 4) and order of importance of random forest (Table 5) were used to calculate the correlation between each characteristic parameter and vegetation height. Tables 4 and 5 list the top 15 features of the correlation calculated by the two methods. It can be seen from the results that the correlation between vegetation height and texture features, geographical location features, and spatial environment features considered in the tree height inversion model in this paper is generally higher than that of vegetation parameters. Longitude and AQI contributed the most to the vegetation height model, and the Pearson correlation and random forest importance were 0.580 and 0.502, respectively, and 21.698 and 16.434, respectively. Figure 4 shows the order of importance of the random forest.

Table 3. Pearson correlation calculation

Index	Correlation
Longitude	0.580**
AQI	0.502**
B1_Mean	-0.375**
B3_Mean	-0.366**
B2_Mean	-0.357**
Latitude	-0.334**
B4_Mean	-0.310**
GNDVI	0.210**
B6_Mean	-0.196**
B2_ Angular second moment	0.192**
B7_Mean	-0.191**
B4_ Angular second moment	0.190**
B4_ Entropy	-0.184**
B2_ Entropy	-0.177**
NDVI	0.171**

Table 4. Random forest order of importance

Index	%IncMSE	IncNodePurity
Longitude	21.697599	4520.2487
AQI	16.434411	2917.7648
Latitude	13.151821	2115.28
B2_Mean	11.7307792	1655.3
B3_Mean	9.495894	1030.1747
B6_Mean	8.546149	501.0962
B1_Mean	8.48029	769.4016
B5_Mean	7.837627	619.5983
B3_ Variance	7.560645	336.6816
B3_ Contrast	7.276603	340.1824
B4_Mean	6.953395	362.0563
B7_Mean	6.778966	465.1347
DVI	6.691078	286.8608
Urban Function Area	6.665913	202.7566
B4_ Contrast	6.648684	226.8497

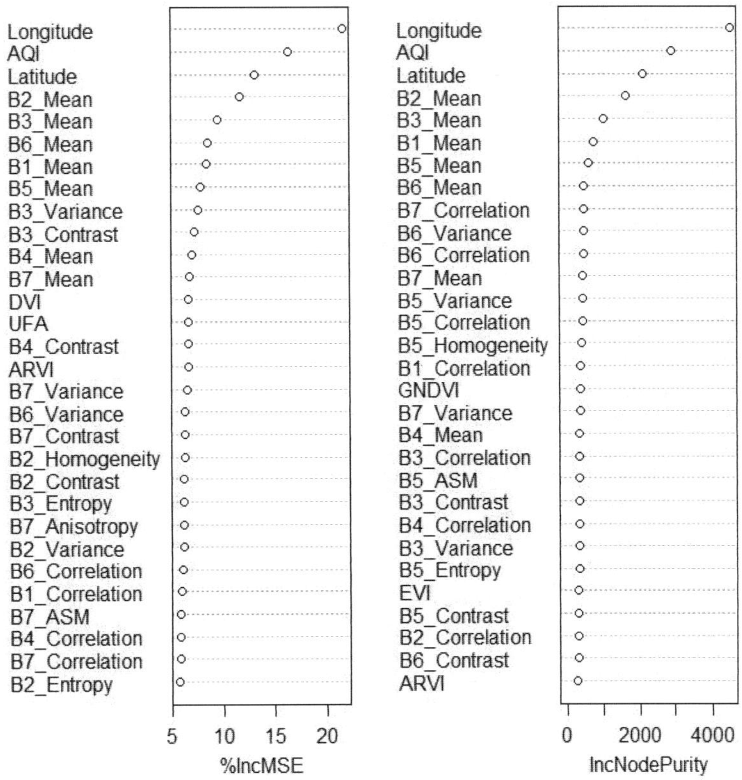

Figure 4. Random forest order of importance

In this study, an ablation experiment was designed to demonstrate the importance of spatial distribution characteristics in the inversion model of urban vegetation height. When only the vegetation index was used to construct the vegetation height model without considering the spatial distribution characteristics (SVM Model: R^2=0.402 RMSE=4.696, RF model: R^2=0.426, RMSE=4.553, BP Model: R^2=0.454, RMSE=4.869), after adding texture features, geographical location features, and spatial environment features to the construction of the urban vegetation height model, the accuracy of the model constructed by the machine learning method was significantly improved. The precision order of the three algorithms for model construction is SVM\RF\BP (SVM Model: R^2 =0.457, RMSE=3.363, RF model: R^2=0.527, RMSE=3.409, BP Model: R^2=0.61, RMSE=1.66), and the accuracy of the urban vegetation height model constructed using the BP neural network was the highest. The regression results of the fitting data of each model and the true value of the verification set are shown in Figure 5.

IOP Conf. Series: Earth and Environmental Science 1004 (2022) 012023 doi:10.1088/1755-1315/1004/1/012023

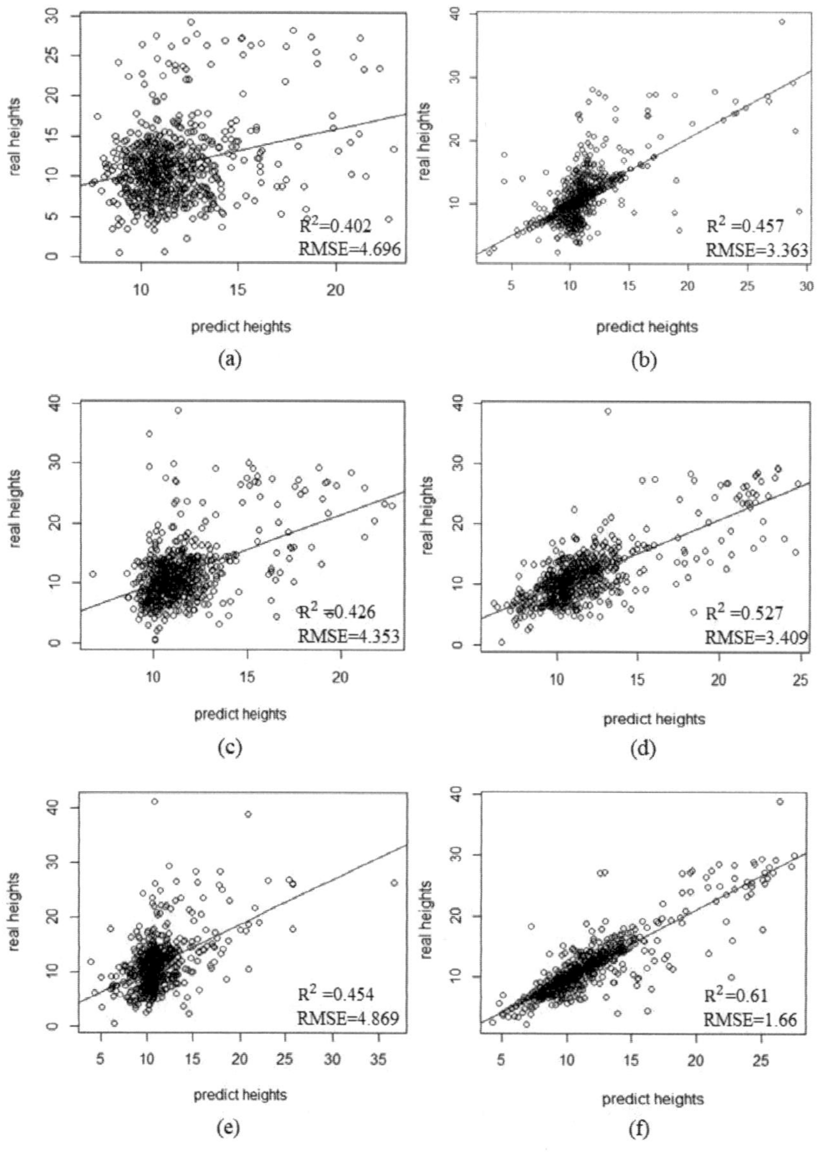

Figure 5. Comparison between the reference canopy heights of the validation datasets and the canopy height estimates using our model: (a)(c)(e) SVM\RF\BP model without spatial distribution characteristics; (b)(d)(f) SVM\RF\BP model with spatial distribution characteristics

The vegetation height calculated by the urban vegetation height model was compared with the canopy height generated by airborne laser scanning in the same region. The regression results are shown in Figure 6. The accuracy (R^2=0.643, RMSE=1.097) between the calculated results of the model and the canopy height model scanned in the field proved that the model had a good inversion effect on the height of urban vegetation.

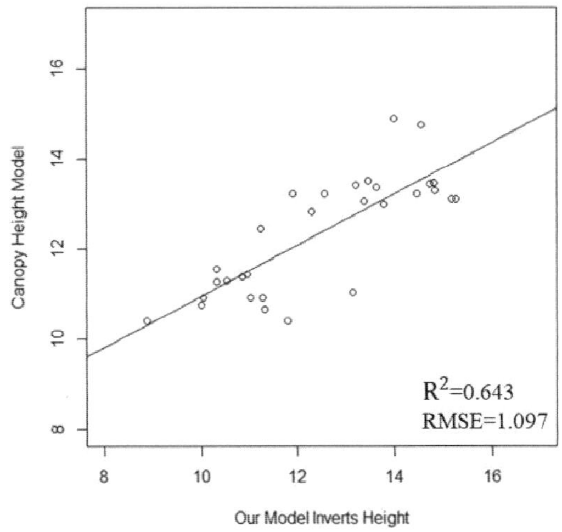

Figure 6. The evaluation results of our forest height models with validation datasets

5. Conclusion

We aimed to construct an accurate inversion model of urban vegetation height. This study discussed the potential of ICESat-2 space-borne laser data fusion with multi-source remote sensing data in the inversion of urban vegetation height and proved the importance of urban spatial environment characteristics and spatial distribution characteristics in the model of urban vegetation height. Finally, the inversion model of vegetation height suitable for a city was constructed using the BP neural network algorithm.

In this study, in the process of constructing an urban vegetation height inversion model, we considered the urban forest vegetation distribution was discrete and included many human factors and characteristics of space environmental factors (including texture features, AQI, geographical location data, etc.) We analyzed the characteristics of importance by experimental and ablation experiments, and it was proved that the spatial characteristic factors considered in this study have a greater impact on the vegetation height data than the traditional vegetation index characteristics.

In this study, a variety of machine learning models were also used to build a more accurate nonlinear correlation relationship between characteristic factors and urban vegetation height. We used SVM, RF, and BP neural networks to build a vegetation height inversion model, which proves that the BP neural network is the optimal model construction method. The regression analysis of the urban vegetation height model and canopy model data obtained by airborne laser inversion was carried out, R^2=0.6433, RMSE=1.0967, which proves that the model constructed in this study is efficient, reliable, and effective.

Acknowledgements

This work was supported by the Major Program of the National Natural Science Foundation of China (Grant No. 42130106).

References

[1] Buckley D S, Isebrands J G and Sharik T L 1999 Practical field methods of estimating canopy cover, PAR, and LAI in Michigan Oak and pine stands. Northern Journal of Applied Forestry 16 25-32

[2] Hall R J, Skakun R S, Arsenault E J and Case B S 2006 Modeling forest stand structure attributes using Landsat ETM+ data: Application to mapping of aboveground biomass and stand volume. Forest Ecology and Management 225 378-390

[3] Donoghue D N M and Watt P J 2006 Using LiDAR to compare forest height estimates from IKONOS and Landsat ETM+ data in Sitka spruce plantation forests. International Journal of Remote Sensing 27 2161-2175

[4] Næsset E 1997 Determination of mean tree height of forest stands using airborne laser scanner data. ISPRS Journal of Photogrammetry and Remote Sensing 52 49-56

[5] Næsset E and Økland T 2002 Estimating tree height and tree crown properties using airborne scanning laser in a boreal nature reserve. Remote Sensing of Environment 79 105-115

[6] Lefsky M A, Harding D J, Keller M, Cohen W B, Carabajal C C, Del Bom Espirito-Santo F, Hunter M O and de Oliveira Jr R John Wiley & Sons, Ltd. 2005 Estimates of forest canopy height and aboveground biomass using ICESat. Geophysical Research Letters 32

[7] Singh M, Evans D, Friess D, Tan B and Nin C 2015 Mapping above-ground biomass in a tropical forest in Cambodia using canopy textures derived from Google earth. Remote Sensing 7 5057-5076

[8] Boutsoukis C, Manakos I, Heurich M and Delopoulos A 2019 Canopy height estimation from single multispectral 2D airborne imagery using texture analysis and machine learning in structurally rich temperate forests. Remote Sensing 11 2853

[9] Inoue A and Yoshida S 2004 Allometric model of the height—Diameter curve for even-aged pure stands of Japanese cedar (Cryptomeria japonica). Journal of Forest Research Taylor & Francis 9 325-331

[10] Khan M N I, Islam M R, Rahman A, Azad M S, Mollick A S, Kamruzzaman M, Sadath M N, Feroz S M, Rakkibu M G and Knohl A 2020 Allometric relationships of stand level carbon stocks to basal area, tree height and wood density of nine tree species in Bangladesh. Global Ecology and Conservation 22 e01025

[11] Liu M, Zhang D, Pietzarka U and Roloff A 2021 Assessing the adaptability of urban tree species to climate change impacts: A case study in Shanghai. Urban Forestry and Urban Greening 62 127186

[12] Li W, Niu Z, Shang R, Qin Y, Wang L and Chen H 2020 High-resolution mapping of forest canopy height using machine learning by coupling ICESat-2 LiDAR with Sentinel-1, Sentinel-2 and Landsat-8 data. International Journal of Applied Earth Observation and Geoinformation 92 102163

[13] Zhou J, Zhou Z, Zhao Q, Han Z, Wang P, Xu J and Dian Y 2020 Evaluation of Different Algorithms for Estimating the Growing Stock Volume of Pinus massoniana Plantations Using Spectral and Spatial Information from a SPOT6 Image. Forests 11 540

[14] Matinfar H R, Maghsodi Z, Mousavi S R and Rahmani A 2021 Evaluation and Prediction of Topsoil organic carbon using Machine learning and hybrid models at a Field-scale. CATENA 202 105258

[15] Elmetwalli A H, Tyler A N, Moghanm F S, Alamri S A M, Eid E M and Elsayed S 2021

Integration of radiometric ground-based data and high-resolution QuickBird imagery with multivariate modeling to estimate maize traits in the Nile Delta of Egypt. Sensors 21 3915

[16] Akhtar A M, Qazi W A, Ahmad S R, Gilani H, Mahmood S A and Rasool A 2020 Integration of high-resolution optical and SAR satellite remote sensing datasets for aboveground biomass estimation in subtropical pine forest, Pakistan. Environmental Monitoring and Assessment 192 584

[17] Wang P, Fan W Y, Li M Z, Liu F and Zhang Q 2011 Extraction of individual tree height using a combination of large-scale aerial photo and LiDAR. Advanced Materials Research 268-270 1157-1162 1157-1162

[18] Breiman L 2001 Random forests. Machine Learning 45 5-32

IOP Conf. Series: Earth and Environmental Science 1004 (2022) 012024 doi:10.1088/1755-1315/1004/1/012024

Analysis of Ecological Environment Changes and Driving Forces in Boao town of Hainan Province

J Y Yang[2,3],Q JWang[1, 2, 3*], Z CLi[1], J J Xie[2,3] and D K Chang[2,3]

1. Key Laboratory of Earth Observation of Hainan Province, Hainan Research Institute, Aerospace Information Research Institute, Chinese Academy of Sciences, Hainan,572029, China
2. Key Laboratory of Digital Earth Science, Aerospace Information Research Institute, Chinese Academy of Sciences, Beijing, 100094, China
3. University of Chinese Academy of Sciences, Yanqi Lake Campus of Chinese Academy of Sciences University, Beijing, 101408, China

*Corresponding author: wangqj@radi.ac.cn

Abstract. Based on the three phases of remote sensing imageries in 2000, 2010 and 2020, taking Boao town as the study area, this paper analyzed the changes in the ecological environment and the driving forces of ecological environment. The results showed that: (1) The main land use type of Boao town was forest, followed by cultivated land. The main change of land use types in Boao town from 2000 to 2020 is from forest to other land types. (2) The average vegetation coverage of Boao town was 70.68% in 2000, 62.49% in 2010 and 71.65% in 2020, which indicated a trend of first decreasing and then increasing. (3) The changes of landscape pattern index of type level from 2000 to 2020 were as follows: the landscape shape index and landscape fragmentation of cultivated land decreased gradually, while the landscape shape index and landscape fragmentation of construction land and forest land increased. The trend of landscape pattern index of landscape level from 2000 to 2020 was as follows: landscape fragmentation continued to increase, landscape spread continued to decline, landscape separation continued to rise, and landscape diversity and evenness continued to increase. (4) In the driving force analysis of ecological environment changes in Boao Town, the results of principal component analysis showed that human factors were the leading factors of ecological environment changes in Boao Town, and economic factors such as GDP, output value of the third industry and per capita disposable income of farmers were the main driving forces of landscape pattern changes in Boao town.

1. Introduction

Ecological environment is the basis of human existence and development [1], and the long-term development of human society must be guaranteed by maintaining the stability and balance of the environment. Studying the changing trend of ecological environment is conducive to make more reasonable environmental protection policies and adopt more appropriate environmental protection measures, so as to protect environment to make healthy and sustainable development, and to lay a good foundation for the further development of human society.

With all kinds of gradual development and mature of the theory and technology, such as environmental science, computer science, remote sensing technology, in view of the ecological environment change research can be mainly divided into climate changes[2], carbon emissions[3], environmental pollutions, land use/cover changes (LUCC)[4-5], the landscape pattern changes[6-8], such as a driving forces of the environmental changes[9-12]. In recent years, the focus of research on ecological environment change tends to be the quantitative analysis and introduction model of all kinds of information extracted from remote sensing data [13], so as to provide scientific decision-making basis for ecological environment protection and regional sustainable development. Remote Sensing (RS) technology has become a common method to analyze and evaluate the level of ecological environment, and has been widely used in various ecological fields.

This paper aims to use multi-temporal remote sensing data to analyze the spatial-temporal changes of land use and landscape pattern in Boao Town, and to explore the driving forces of ecological environment changes, so as to provide reference for the ecological environment protection in this area.

2. Materials and methods

2.1. Materials

2.1.1. Study area

As shown in the Figure 1, the study area of this article is Boao Town, located in the eastern coastal zone of Hainan Province, where Wanquan River flows into the South China Sea. Its central latitude and longitude is 110°35'34"E, 19°9'55"N, with an area of 86 km^2. According to the information released by the Qionghai Municipal People's Government, the total number of households in the town in 2019 was 8,873, with a total population of 31,884. Boao Town is adjacent to the South China Sea to the east, Longgun Town of Wanning City to the south, Jiaji Town and Zhongyuan Town of Qionghai City to the west, and Tanmen Town of Qionghai City to the north. It is about 110 kilometers away from Haikou City, 160 kilometers away from Sanya City, and 19 kilometers away from Jiaji Town. The Hainan East Ring High-speed Railway passes through Boao Town, so it has strong traffic connectivity. At the same time, it is also the location of the "Boao Forum for Asia" and has relatively complete tourist reception facilities.

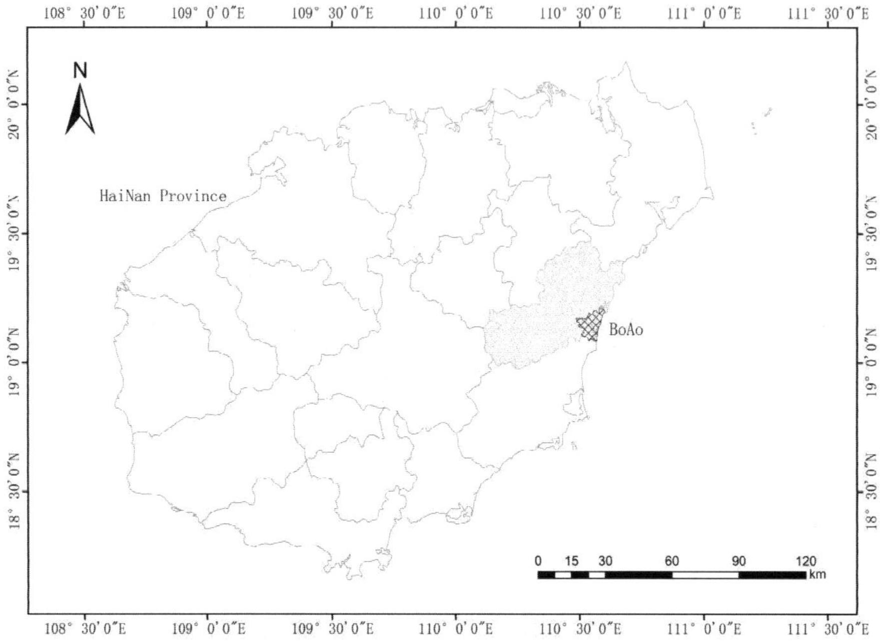

Figure 1. Location of study area.

Boao Town has a significant tropical monsoon climate, and the climate change throughout the year is not obvious. The sunshine is sufficient, and the average annual total sunshine are 1972.6 hours. The average annual temperature of the town is 24.6℃. Days with temperature more than 37℃ are mostly in June, July and August, and extreme low temperatures are between 7-11 ℃, which are mostly in December and January. The town has abundant precipitation, with an average annual total precipitation of 2053.1mm. However, the seasonal variation of rainfall is obvious. The dry season and the rainy season are distinct. The dry season is generally from December to April, and the rainfall in dry season is less than that of other seasons.

Boao Town has rich landscapes, including mountains, islands, rivers, estuary coasts, lagoons, etc.[14]. It not only has beautiful natural landscapes, but also has a strong cultural atmosphere. It is one of the top ten cultural towns in Hainan.

2.1.2. Data

Three remote sensing imageries from 2000 to 2020 are selected, two of which are Landsat5 TM imageries in 2000 and 2010, and the other one is Landsat8 OLI imageries in 2020. The acquired date of the imageries are all in June and July, with vigorous vegetation growth, less cloud cover and good quality. Their parameters are as follows:

Table 1. Remote sensing imageries.

Year	ID	Sensor	Time	Number	Cloud(%)
2000	LT51230472000209BJC00	TM	2000.07.27	123/47	3.55
2010	LT51230472010188BKT00	TM	2010.07.07	123/47	0.7
2020	LC81230472020152LGN00	OIL	2020.05.31	123/47	9.01

The elevation data used is SRTM elevation data with resolution of 30m, which can completely cover the study area. Using which, a DEM map of Boao town was built and shown in the Figure 2.

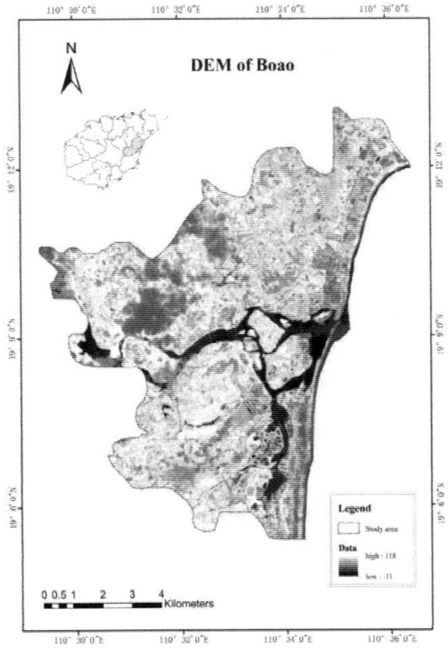

Figure 2. DEM map of Boao town.

2.2. Methods

2.2.1. Image pre-processing
Geographically register the administrative division maps of sub-townships in Hainan Province with remote sensing imageries, so that they can be displayed on top of each other, delimit the administrative boundary of Boao Town, and obtain the vector file of administrative scope of Boao Town.

The remote sensing image is pre-processed by the following steps: (1) geographic registration; (2) Radiation calibration; (3) Atmospheric correction; (4) Geometric correction; (5) Image clipping. Thus, the remote sensing imageries of Boao Town from 2000 to 2010 are obtained. Figure 3 shows the comparison between the original and the processed spectra.

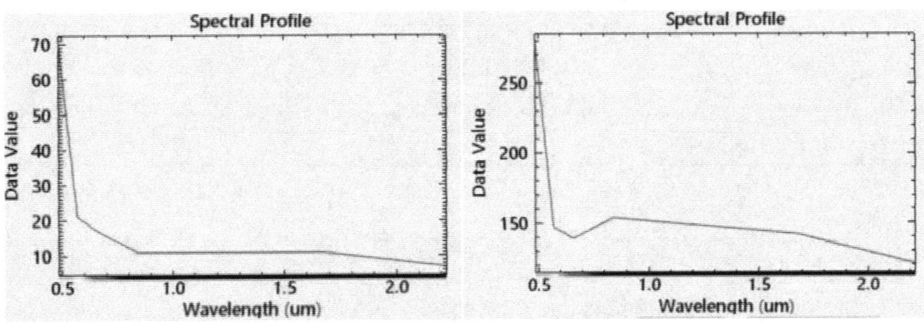

(a) (b)

Figure 3. Comparison between the original and the processed spectra

2.2.2. *Image classification*

Because the regional space of each study area is different, and the selected remote sensing data sources are different, the classification system of land use types is also different. According to the classification system of the National Land and Resources Division, considering the actual situation of Boao Town, the landscape types in the study area are divided into four types: forest land, cropland, water and built-up land.

As shown in Table 2 and 3, red, green and blue are the three primary colors, and it is their combinations that constructs this colorful world. At the same time, different bands contain different information, and the combination of multiple bands can provide us with abundant information, which helps us to improve the accuracy of visual interpretation.

Table 2. Landsat5 TM band combination.

Band combination	Significance
3 、2 、1 Red、Green、Blue	Natural true color
4、 3 、2 NIR、Red、Green	Standard false color, the vegetation shows red, the higher the vegetation coverage, the darker the red, and the color difference between the water body and other ground objects is obvious, which is mostly used for vegetation classification and wateridentification.
4、 5 、3 NIR、SWIR、Red	False color synthesis can be used to identify water
5 、4 、3 SWIR、NIR、Red	False color synthesis can be used to identify built-up land

Table 3. Landsat8 OIL band combination

Band combination	Significance
4 、3 、2 Red、Green、Blue	Natural true color
5、 4 、3 NIR、Red、Green	Standard false color, vegetation displayed in red, can be used to identify forest land
7、 6 、4 SWIR2、SWIR1、Red	False color synthesis, used for urban monitoring, can be used to identify construction land
6 、5 、2 SWIR1、NIR、Blue	False color synthesis, the cropland covered with crops shows green, and the bare land without cover shows pink
5 、6、 4 NIR、SWIR1、Red	False color synthesis can effectively distinguish land from water

According to the previous experience, the effect of supervised classification is generally better than unsupervised classification. Therefore, we choose the SVM support vector machine classification method in supervised classification, then use high-scoring imageries and Google Earth historical imageries to assist judgment. Next, select purer pixels as training samples, and finally classify the preprocessed imageries for land use to obtain the land use classification results in 2000, 2010 and 2020 shown in the Figure 4. In order to evaluate the accuracy of classification results, the random scatter function of ArcGIS was used to create verification samples, and the confusion matrix between land classification results and independent verification samples was constructed.

The results show that the overall accuracy of classification in 2000, 2010 and 2020 are 92%, 90% and 89%, respectively, and the Kappa statistical coefficients are 0.89, 0.85 and 0.86, respectively. Which indicates that the classification accuracy can meet the research needs.

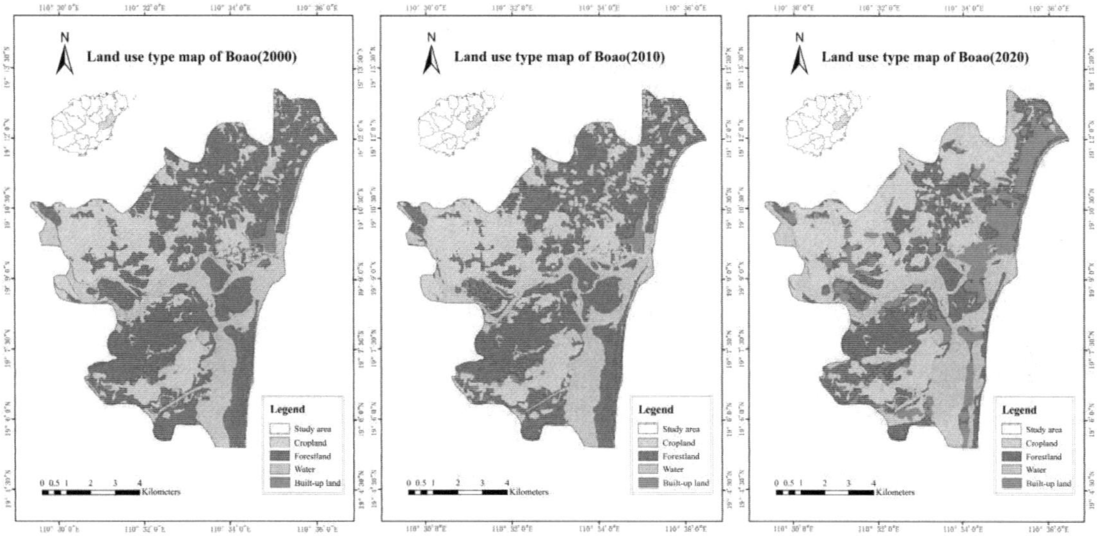

(a) (b) (c)

Figure 4. Land use type map of Boao（2000-2020）

2.2.3. *Landscape pattern index extraction*

Landscape pattern index can quantitatively describe the characteristics of landscape pattern, which can be divided into patch level, landscape type level and landscape level index from different scales. Different levels of index can reflect different landscape pattern characteristics. Landscape pattern index is helpful to understand the comprehensive information of ecological environment in a certain region, while landscape pattern index of different phases in the same region is helpful to understand the change of landscape pattern and the change trend of ecological environment in this region.

At present, there are many kinds of landscape pattern indexes, but there are interrelated relationships among many indexes. Therefore, when doing landscape pattern analysis, it is necessary to reasonably select various landscape pattern indexes to avoid redundant data, and to choose more basic indexes as much as possible.

The land classification data of Boao Town in the three phases (2000, 2010, and 2020) were resampled to the same resolution (30m*30m) by ArcGIS10.2, and exported to Fragstats4.2 to obtain the landscape index.

In this paper, landscape index at landscape scale level is selected from four aspects: (1) Shape index, which represents the geometric shape of landscape pattern; (2) Fragmentation index, which represents the fragmentation degree of landscape pattern; (3) Convergence index, representing the spatial distribution of landscape pattern; (4) Diversity index, representing the richness of landscape pattern types. Finally, as shown in Table 4, five class indexes, such as landscape shape index-LSI, patch number-NP, patch density-PD, patch COHESION and landscape separation-DIVISION are

IOP Conf. Series: Earth and Environmental Science 1004 (2022) 012024 doi:10.1088/1755-1315/1004/1/012024

selected. Six landscape indicators, including patch number-NP, patch density-PD, CONTAG index, landscape separation-DIVISION, Shannon diversity index-SHDI, Shannon uniformity index-SHEI, are selected to analyze the dynamic change characteristics of landscape pattern in Boao town.

Table 4. Selection of landscape pattern index.

Indicators	index	scale	Formula	Value range
Shape	Landscape Shape Index	Class	$LSI = \frac{e_i}{\min e_i}$ It is an index to describe the shape of landscape patches. The larger the LSI value, the more complex the shape, and the smaller the value, the more regular it is	≥1
Crushing	Number of Patches	Class /Land scape	$NP_i = N_i$ Indicates the total number of landscape patches	≥0
	Patch Density	Class /Land scape	$PD = \frac{\sum N_i}{A}$ Is the number of patches per unit area of landscape	≥0
Convergence and divergence	Landscape Division Index	Class /Land scape	$DIVISION = \left[1 - \sum_{i=1}^{m}\sum_{j=1}^{n}\left(\frac{a_{ij}}{A}\right)^2\right]$ Indicates the degree of separation between different patches in a certain landscape type	(0, 100]
	Patch Cohesion Index	Class	$COHESION = \left[1 - \frac{\sum_{j=1}^{n} P_{ij}}{\sum_{j=1}^{n} P_{ij}\sqrt{a_{ij}}}\right]\left[1 - \frac{1}{\sqrt{A}}\right]^{-}$ Indicates the degree of bonding between plaques. The greater the COHESION value, the higher the connectivity between plaques	(0, 100]
	Contagion Index	Land scape	$CONTAG = 1 + \sum_{i=1}^{n}\sum_{j=1}^{m} P_{ij} \ln P_{ij} \frac{P_{ij} \ln P_{ij}}{2 \ln M}$ It indicates the gathering or spreading trend of different patch types in the landscape. When the CONTAG value is small, it indicates that there are many small patches in the landscape. When it tends to 100, it indicates that there are dominant patch types with high connectivity in the landscape	(0, 100]
Diversity	Shannon's Diversity Index	Land scape	$SHDI = -\sum_{i=1}^{m} P_i \ln P_i$ Indicates the richness of each patch type in the landscape. When SHDI value is 0, it indicates that only one plaque type exists. The larger the value, the more	≥0

		patch types in the landscape or the balanced distribution of each patch type	
Shannon's Evenness Index	Land scape	$SHEI=\frac{-\sum_{i=1}^{m}P_i\ln P_i}{\ln m}$ When the SHEI value is 0, it indicates that the landscape is composed of only one patch type and there is no diversity. When the value is 1, it indicates that all plaque types are evenly distributed, and there is no dominant plaque type	[0, 1]

2.2.4. Calculation of NDVI and vegetation coverage

Normalized Vegetation Index NDVI is an important index that can be used to detect the growth state of vegetation and the coverage of regional vegetation, because it can eliminate most irradiance changes related to instrument calibration, sun angle, terrain, cloud, shadow and atmospheric conditions, and enhance the response ability to vegetation, so it has been widely used [15].

The NDVI calculation formula is:

$$NDVI=\frac{NIR-R}{NIR+R}$$

In which, NIR represents the near infrared band and R represents the red band. The range of NDVI is [-1,1], and the area with NDVI<0 is covered by water (cloud, snow, water, etc.), which has a large reflectivity to visible light; The area with NDVI=0 has no vegetation coverage at all, which may be bare rock or land, and the near infrared band value is equal to the red band value; The area with NDVI>0 is covered by vegetation, and the greater the NDVI value, the more the corresponding vegetation coverage.

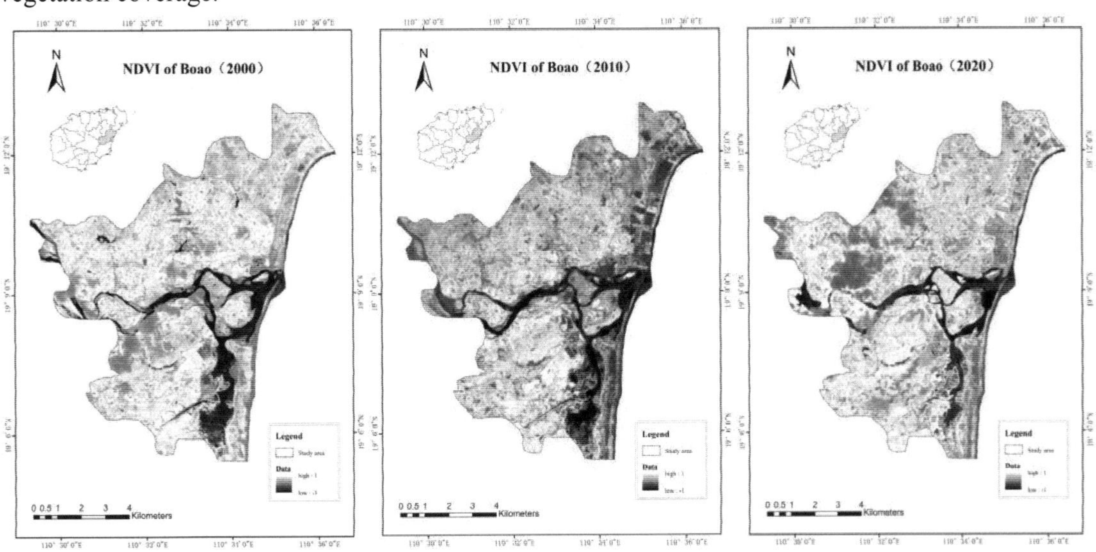

(a) (b) (c)

Figure 5. NDVI map of Boao（2000-2020）

After NDVI calculation is completed in the Figure 5, the corresponding vegetation coverage can be calculated according to NDVI value. Vegetation coverage refers to the percentage of the vertical projection area of vegetation (including leaves, stems and branches) on the ground in the total area of the statistical area. For each imagery, it can be roughly divided into three categories: waters, vegetation and buildings, and then the vegetation coverage is calculated by using the mixed pixel decomposition method, whose formula is shown as follows.

$$F_V = \frac{NDVI - NDVI_{soil}}{NDVI_{veg} - NDVI_{soil}}$$

In which, $NDVI_{soil}$ is the NDVI value of bare soil pixel and $NDVI_{veg}$ is the NDVI value of vegetation pixel. According to the previous researches, taking 0.70 and 0.05 as the thresholds. That is, when the $NDVI_{veg}$ of a pixel is greater than 0.70, F_V = 1; When it is less than 0.05, F_V =0. Vegetation coverage map in Boao is shown in Figure 6.

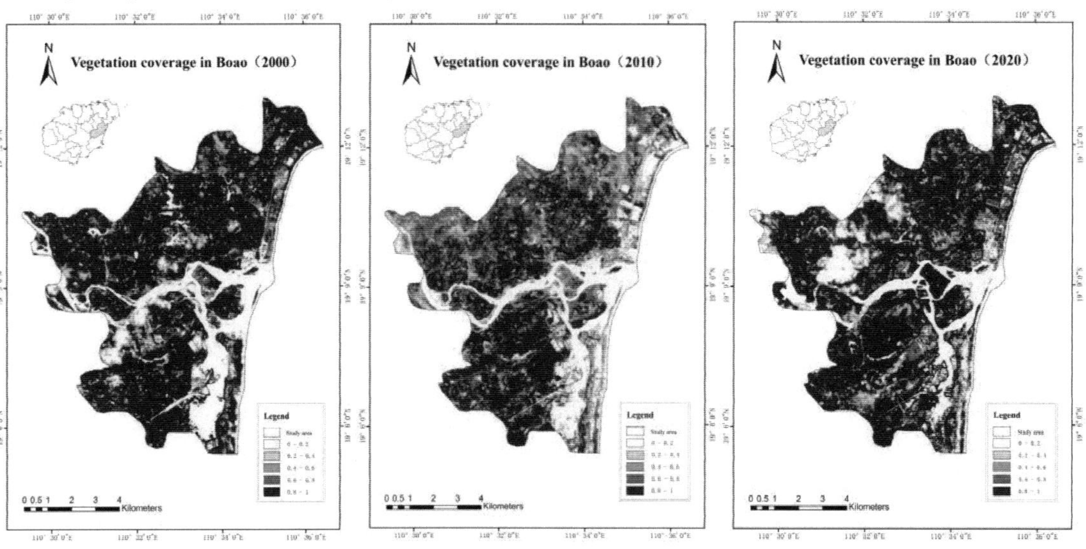

(a) (b) (c)

Figure 6. Vegetation coverage in Boao（2000-2020）

2.2.5. Driving forces analysis method

There are many natural factors affecting the landscape pattern changes in the study area, such as topographic factors, climatic factors and so on. Climate changes, such as precipitation, sunshine and temperature, will directly lead to the spatial heterogeneity of water, light and heat, thus affecting the changes of landscape pattern. Elevation, slope, aspect and other factors in terrain are often closely related to human production and construction activities, which have a profound impact on landscape pattern.

The function of principal component analysis is to divide the original multiple variables into a few comprehensive indicators through transformation on various variables. These indicators can not only fully reflect the information of more variables, but also weaken the mutual interference among the variables [16].

Climate, population, economy and technology are selected as the driving forces indicators to make principal component analysis and the results are shown in Table 5.

Table 5. Selection of driving factors.

Driving factors	Indicators
Climate	X1: annual average temperature (℃)
	X2: annual total sunshine hours (h)
	X3: annual total precipitation (mm)
Population	X4: total number of households (households)
	X5: total population (person)
	X6: agricultural population (person)
Economics	X7: GDP (ten thousand yuan)
	X8: primary industry (10,000 yuan)
	X9: secondary industry (ten thousand yuan)
	X10: third industry (ten thousand yuan)
	X11: per capita disposable income of farmers (yuan)
Technology	X12: total power of agricultural machinery (10,000 kilowatts)
	X13: Fertilizer application amount (10,000 tons)

Table 6 shows the correlation coefficient matrix of each selected index. From which, it can be seen that X1 has a highly positive correlation with X2, but its correlation with other indexes is not high. Because sunshine hours (X2) will affect air temperature (X1) to a certain extent, generally speaking, the higher the sunshine hours, the higher the corresponding air temperature will be.

The correlation coefficients of X4 with X5, X7, X8, X9, X10, X11, X12 and X13 are all more than 0.8, among which, its correlations with X5, X9 and X12 are more than 0.9, and the correlation with X12 is even more than 0.98. It shows that the total number of households has a high correlation with GDP, primary industry, third industry, per capita disposable income of farmers, fertilizer application amount, and a strong correlation with the total population, secondary industry value and total power of agricultural machinery, especially with the total power of agricultural machinery.

The correlations between X5 and X7, X8, X9, X10, X11 and X12 are more than 0.94, and the correlation between X5 and X9 is 0.99. It shows that the total population has a strong correlation with GDP, primary industry value, secondary industry value, third industry value, per capita disposable income of farmers and total power of agricultural machinery, among which, there is a highly positive correlation with the secondary industry.

The correlations between X6, X12 and X13 are higher than 0.7, which indicates that there is a certain positive correlation between agricultural population and fertilizer application amount and total power of agricultural machinery.

The correlations between X7 and X8, X9, X10 and X11 are more than 0.98, among which its correlations with X8, X10 and X11 are more than 0.995, and its correlation with X10 is 0.999. It shows that GDP has a strong correlation with primary industry value, secondary industry value, third industry value and per capita disposable income of farmers, especially with third industry value.

X8' s correlations with X9, X10 and X11 are higher than 0.98, and its correlations with X10 and X11 are 0.99, which indicates that the primary industry value has a high correlation with secondary industry value, third industry value and per capita disposable income of farmers, and has a significant positive correlation with third industry value and per capita disposable income of farmers.

Correlations between X9 and X10, X11 and X12 are more than 0.9, and correlations between X9 and X10, X11 are more than 0.98, which indicates that the value of secondary industry has a strong correlation with the value of third industry, farmers' per capita disposable income and fertilizer application amount, and there is a significant positive correlation with farmers' per capita disposable income and fertilizer application amount.

The correlation between X10 and X11 is 0.997, which indicates that there is a significant high positive correlation between the secondary industry value and the third industry value. The correlation between X12 and X13 is 0.879, which indicates that there is a strong correlation between fertilizer application amount and total power of agricultural machinery.

Table 6. Correlation coefficient matrix

	X1	X2	X3	X4	X5	X6	X7	X8	X9	X10	X11	X12	X13
X1	1	0.688	-0.064	0.248	0.268	-0.194	0.324	0.314	0.299	0.331	0.334	0.245	0.236
X2	0.688	1	-0.278	0.634	0.603	0.203	0.573	0.561	0.586	0.572	0.581	0.613	0.456
X3	-0.064	-0.278	1	-0.003	-0.056	0.206	-0.139	-0.12	-0.111	-0.152	-0.132	-0.106	-0.116
X4	0.248	0.634	-0.003	1	0.964	0.74	0.883	0.874	0.937	0.87	0.891	0.98	0.839
X5	0.268	0.603	-0.056	0.964	1	0.674	0.973	0.968	0.99	0.966	0.974	0.943	0.789
X6	-0.194	0.203	0.206	0.74	0.674	1	0.541	0.549	0.622	0.518	0.539	0.738	0.706
X7	0.324	0.573	-0.139	0.883	0.973	0.541	1	0.995	0.988	0.999	0.997	0.868	0.719
X8	0.314	0.561	-0.12	0.874	0.968	0.549	0.995	1	0.983	0.99	0.99	0.857	0.73
X9	0.299	0.586	-0.111	0.937	0.99	0.622	0.988	0.983	1	0.982	0.988	0.926	0.784
X10	0.331	0.572	-0.152	0.87	0.966	0.518	0.999	0.99	0.982	1	0.997	0.855	0.697
X11	0.334	0.581	-0.132	0.891	0.974	0.539	0.997	0.99	0.988	0.997	1	0.875	0.712
X12	0.245	0.613	-0.106	0.98	0.943	0.738	0.868	0.857	0.926	0.855	0.875	1	0.879
X13	0.236	0.456	-0.116	0.839	0.789	0.706	0.719	0.73	0.784	0.697	0.712	0.879	1

From the above analysis, it can be seen that some of the selected indexes have strong correlation with each other, so it is necessary to carry out principal component analysis.

3. Analysis of Ecological Environment Change

3.1. *Analysis of Land Use Change*

3.1.1. *Characteristics of quantity change*

The land use classification imageries of Boao Town in different years obtained by SVM are described and analyzed from qualitative and quantitative aspects.

Firstly, according to the land use classification imageries of Boao Town in 2000, 2010 and 2020, the area distribution of various land use types in Boao Town in each year is counted, and the basic information of land use in Boao Town in three different periods is obtained, as shown in Table 7.

Table 7. The area of land use types in Boao town（2000-2020）

Land use type	2000		2010		2020	
	Area /km^2	Proportion/%	Area /km^2	Proportion/%	Area /km^2	Proportion/%
Cropland	23.3	27.57	27.04	31.99	28.58	33.81
Forestland	46.42	54.92	43.37	51.31	25.65	30.35
Water	13.71	16.22	12.85	15.20	17.37	20.55
Built-up land	1.09	1.29	1.26	1.49	12.92	15.29

It can be seen that in the land use type information of Boao Town in each year, the area distribution of the four land use types is in the order of forestland > cropland > water > built-up land.

In 2000, forestland occupied the main position, with a total area of 46.42km^2, accounting for 54.92% of the total area of Boao Town, followed by cropland of 23.3km^2, and accounting for 27.57% of the total area of Boao Town. These two land use types accounted for 82.49% of the total area of Boao Town, while the area of built-up land was 1.09km^2, accounting for only 1.29% of the total area. Which indicates that in the early stage, there is little difference between the scale and structure of land use in 2010 and that in 2000. Cropland and built-up land increased slightly, while forest land and water body decreased slightly, which indicates that Boao Town developed slowly in the past 10 years. In 2020, forest land and cropland still occupy 64.26% of the total area, which is the main land. However, forest land is reduced to 25.65km^2, cropland is slightly increased, and built-up land is increased to 12.92km^2, accounting for 15.29% of the total area, which reflects the rapid development of Boao Town in the past 10 years.

From Table 8, we can see that from 2000 to 2020, the change rate of built-up land in Boao Town was the largest, and the scale continued to expand. From 2000 to 2020, the built-up land area increased by 11.66 km^2, and the change rate reached 1085.32%. Changes of forestland and water are also relatively high, and the scale of forestland decreased. From 2000 to 2020, the forestland area decreased by 20.77 km^2, with a change rate of-44.74%. The scale of waterand cropland is expanding, with areasof 3.66 km^2 and 5.28 km^2 respectively, and the change rates of 26.7% and 22.66%respectively.

Table 8. Area change of land use types in Boao town.

Year	2000-2010		2010-2020		2000-2020	
Change Type	Variation /km^2	Change rate/%	Variation /km^2	Change rate/%	Variation /km^2	Change rate/%
Cropland	3.74	16.05	1.54	5.70	5.28	22.66
Forestland	-3.05	-6.57	-17.72	-40.86	-20.77	-44.74
Water	-0.86	-6.27	4.52	35.18	3.66	26.70
Built-up land	0.17	15.60	11.66	925.40	11.83	1085.32

From Table 8, we can see that from the data of each period, there are differences in the changes of each land use type during different periods, and the period with the largest change of cropland is from 2000 to 2020, during which, the growth rate of cropland area is 22.66%; The period of the smallest change of cropland is from 2010 to 2020, during which, the cropland area in Boao Town increased by 1.54 km^2, with a growth rate of only 5.7%. The maximum change of forest land is from 2000 to 2020, with the change rate of -44.47%; the minimum change range of forest land was from 2000 to 2010, with a change rate of -6.57%. The maximum change in water is from 2010 to 2020, with a change rate of 35.18%; the minimum water change is from 2000 to 2010, with a decrease of 6.27%. The maximum built-up land change is from 2000 to 2020, with a rate of 1085.32%. The minimum change of built-up land is from 2000 to 2010, with a rate of 15.6%.

3.1.2. Dynamic attitude of land use changes

Single land use dynamic attitude is an index describing the change speed of a certain type of land use in the study area, and its formula is

$$K_i = \frac{A_b - A_a}{A_a} \times \frac{1}{T} \times 100\%$$

In which, A_a is the area of land use type at the beginning and A_b is the area of the end , and T is the time.

The results can be shown in Table 9.

Table 9. Dynamic degree of single land use in Boao town.

Type	2000-2010	2010-2020	2000-2020
Cropland	1.61%	0.57%	1.13%
Forestland	-0.66%	-4.09%	-2.24%
Water	-0.63%	3.52%	1.33%
Built-up land	1.56%	92.54%	54.27%

During 2000-2020, the maximum change type was cropland, which was 54.27%, while the smallest change type was built-up land, which was only 1.13%, indicating that the change speed of cropland was the fastest and the change speed of built-up land was the smallest. At the same time, it can be noted that compared with 2000-2010, the change of cropland in 2010-2020 was great, and the change of forest land and waters were also great, which showed that the change speed of cropland, forest land and water in the last 10 years was obviously faster than that in the previous 10 years.

Comprehensive land use dynamic attitude is an index describing the change speed of the overall land use in the study area, and its formula is:

$$S = \sum_{i=1}^{n} \frac{|\Delta A_i|}{nA} \times \frac{1}{T} \times 100\%$$

ΔA_i is the total area of land use type I converted to other land use types, and A is the total land use area at the initial stage of research, and T is the time.

The results can be shown in table 10.

Table 10. Dynamic degree of comprehensive land use in Boao town.

Time	2000-2010	2010-2020	2000-2020
Comprehensive dynamic attitude	1.20%	2.98%	1.15%

It can be seen from the table that the comprehensive dynamic attitude of land use in 2000-2020 is the smallest and the comprehensive dynamic attitude of land use in 2010-2020 is the largest, indicating that the land use in 2000-2020 was under tremendous changes in the last 10 years.

3.1.3. Transfer matrix analysis

Use the spatial statistical analysis function of ArcGIS to compare the land use data of Boao Town from 2000 to 2020, and generate the land use confusion matrix as shown in Table 11.

Table 11. Land use transfer matrix of Boao town（2000-2010）

Type	Cropland	Forestland	Water	Built-up land	Roll in	Total for 2010
Cropland	22.56	4.03	0.31	0.04	4.38	26.95
Forestland	0.13	40.6	2.6	0.08	2.81	43.42
Water	0.42	1.65	10.76	0	2.07	12.84
Built-up land	0.1	0.2	0	0.94	0.3	1.24
Roll out	0.65	5.88	2.91	0.12	9.56	
Total out 2000	23.21	46.49	13.68	1.07		84.45

From 2000 to 2010, the main transfer characteristics of land types are the transformation from forestland to cropland, and a small amount of water and forest land transform each other.

According to the characteristics of cropland transfer, the transferred area of cropland in Boao Town from 2000 to 2010 was about 7 times of the transferred area, and the increase of cropland area mainly came from forest land, with 22.56 km^2 of unchanged cropland, accounting for 97.2% of the total area of original cropland.

During 2000-2010, the area transferred out of forest land was twice that that of transferred in, and the decrement of forestland area was mainly transferred to cropland, with an area of 4.03 km^2, accounting for 8.67% of the total area of original forest land.

During 2000-2010, the amount of water transferred in and out was similar, most of which came from forestland, and most of which also transferred out into forestland, so the overall area changed little.

As shown in Table 12, the main characteristics of land type transfer from 2010 to 2020 were the transformation from forestland to built-up land, cropland and water body, and the transformation from cropland to built-up land and forestland.

Table 12. Land use transfer matrix of Boao town（2010-2020）

Type	Cropland	Forestland	Water	Built-up land	Roll in	Total in 2020
Cropland	19.71	7.37	1.38	0.09	8.84	28.55
Forestland	2.47	22.09	0.93	0.16	3.56	25.65
Water	1.7	5.68	9.95	0.01	7.39	17.34
Built-up land	3.06	8.25	0.56	0.99	11.87	12.86
Roll out	7.23	21.3	2.87	0.26	31.66	
Total out 2010	26.94	43.39	12.82	1.25		84.4

According to the characteristics of forestland transfer, from 2010 to 2020, the transferred area of forestland in Boao Town was about 6 times of the transferred area. The unchanged forestland was 22.09 km^2, accounting for 50.91% of the total area of original forestland, and the changed forest land was 21.3 km^2, among which the transferred areas for built-up land, cropland and water were 8.25 km^2, 7.37 km^2 and 5.68 km^2 respectively.

From 2010 to 2020, there was little difference between the transferred-out amount and the transferred-in amount of cropland. The transferred-out amount of cropland was mainly transferred to built-up land and forest land, which were 3.06 km^2 and 2.47 km^2 respectively, and the transferred-in amount mainly came from forestland, accounting for 83.37% of the total transferred-in amount.

As shown in Table 13, from 2000 to 2020, the main characteristics of land type transfer were the transformation from forestland to built-up land, cropland and water body.

According to the transferred-out area of forest land, from 2000 to 2020, the transferred-out area of forest land in Boao Town was 14 times of the transferred-in area, and the unchanged forestland was 24.14km^2, accounting for 51.94% of the total area of original forestland, and the forestland area of 22.34km^2 changed, among which the transferred-out areas for built-up land and cropland were 9.06km^2 and 8.72km^2, respectively, which were the main land use types of forestland.

Table 13. Land use transfer matrix of Boao town from（2000-2020）

Type	Cropland	Forestland	Water	Built-up land	Roll in	Total in 2020
Cropland	18.59	8.72	1.21	0.04	9.97	28.56
Forestland	1.07	24.14	0.40	0.05	1.52	25.66
Water	1.20	4.56	11.60	0.00	5.76	17.36
Built-up land	2.34	9.06	0.48	0.99	11.88	12.87
Roll out	4.61	22.34	2.09	0.09	29.13	
Total out 2000	23.2	46.48	13.69	1.08		84.45

3.2. Landscape pattern changes

3.2.1. Class level

The landscape indexes of each year are shown in Table 14.

Table 14. Landscape pattern indexes of class level.

Type	Year	LSI	NP (pieces)	PD (PCs /ha)	COHESION (%)	DIVISION (%)
Cropland	2000	16.61	165	1.95	97.96	0.98
	2010	15.05	127	1.50	98.14	0.98
	2020	13.90	119	1.41	98.46	0.98
Forestland	2000	15.29	96	1.14	99.33	0.91
	2010	15.57	139.00	1.64	99.27	0.93
	2020	18.21	155	1.83	98.21	0.99
Water	2000	10.55	33	0.39	99.04	0.98
	2010	12.04	53.00	0.63	99.11	0.98
	2020	11.51	67	0.76	98.76	0.98
Built-up land	2000	5.46	31	0.37	89.62	1.00
	2010	5.57	30.00	0.35	89.74	1.00
	2020	10.16	73	0.86	96.36	1.00

From 2000 to 2020, the LSI of forestland and built-up land in the study area showed a continuous growth trend. The landscape shape index of built-up land increased by 50%, and the change range of forestland was not as large as that of built-up land. This indicated that with the influence of the increment of human activity intensity in the study area, its fragmentation degree also increased. The LSI of cropland and water decreased with time, which indicated that the patches were influenced by human activities.

In total, cropland and forestland have higher NP and PD in each year, which indicates that the fragmentation of these two landscape types is higher, and the fragmentation of forestland landscape increases from 2000 to 2020 due to human interference. From 2000 to 2020, the PD of built-up land increased, which reflected that the degree of human interference on built-up land increased.

During 2000-2020, the DIVISION of each landscape type was small, which indicated that each landscape type was distributed in pieces. COHESION of built-up land increased obviously during 2010-2020, which indicated the trend of gradual agglomeration of built-up land.

3.2.2. landscape level

The landscape indexes of Boao Town in different years are shown in Table 15.

Table 15. Landscape pattern indexes of landscape level.

Index / Year	NP (pieces)	PD (PCs /ha)	CONTAG (%)	DIVISION (%)	SHDI	SHEI
2000	325	3.85	51.43	0.87	1.04	0.75
2010	349	4.13	50.45	0.89	1.06	0.76
2020	414	4.90	45.45	0.94	1.38	0.86

From 2000 to 2020, the patch number (NP) and plate density (PD) showed a gradual upward trend, indicating that the fragmentation degree of landscape pattern continued to increase from 2000 to 2020, and was affected by human activities.

The decreasing trend of CONTAG indicates that the patches of various landscape types do not form a good connection, and the number of small patches in the landscape is gradually increasing, and the connectivity of the landscape becomes worse. The increment of DIVISION also shows that the landscape types tended to be obviously divided in the past 20 years.

Shannon diversity index (SHDI) and Shannon evenness index (SHEI) increased slightly by 0.02 and 0.01 from 2000 to 2010, and increased by 0.32 and 0.1 from 2010 to 2020. This indicates that the diversity of patch types in Boao Town has increased in recent 20 years, and the distribution of each patch type is more uniform, and the landscape pattern tends to be diversified, separated and broken.

3.3. Changes in vegetation coverage

The vegetation coverage data of Boao Town from 2000 to 2020 were extracted. In order to analyze the vegetation change characteristics of Boao Town, the vegetation coverage was divided into five grades: low vegetation coverage ($F_V < 0.2$), medium-low vegetation coverage ($0.2 < F_V < 0.4$), medium vegetation coverage ($0.4 < F_V < 0.6$) and medium-high vegetation coverage ($F_V > 0.8$).According to above standards, the vegetation coverage data of Boao Town in different years are reclassified in ArcGIS and the area is counted by region. The results are shown in Table 16.

Table 16. Dynamic degree of comprehensive land use in Boao town（km^2）

Grade	2000	2010	2020
Low	14.23	13.03	12.34
Medium-low	4.88	8.35	6.45
Medium	6.55	12.56	7.65
Medium-high	10.05	21.38	10.38
High	48.78	29.17	47.67

According to statistics, the average vegetation coverage in Boao Town was 70.68% in 2000, 62.49% in 2010 and 71.65% in 2020. From 2000 to 2010, the vegetation coverage decreased by 8.19%, which showed that there was a certain degree of vegetation degradation in Boao Town during this period. In 2010-2020, the vegetation coverage increased by 9.16%, and the vegetation was restored to some extent.

From 2000 to 2020, the area of high vegetation coverage in Boao Town decreased first and then increased, while the area of medium, high and low vegetation coverage increased first and then decreased and the area of low vegetation coverage showed a continuous downward trend. Among them, in the first 10 years, the area covered by high, medium, medium and low vegetation changed greatly, all of which were close to the change rate of 50%. In the last 10 years, the areas with high, medium and high vegetation coverage changed greatly, and the area with high vegetation coverage increased by 63.42%.

4. Driving forces analysis

Table 17 shows the eigenvalue and variance contribution rate of each principal component, in which the contribution rate of principal component 1 is 67.879%, the contribution rate of principal component 2 is 15.566%, and the cumulative contribution rate of both exceeds 83.4%, which can cover most information of original variables.

Table 17. Eigenvalue and variance contribution rate.

Component	Initial eigenvalue			ExtractionSums of SquaredLoadings			RotationSums of SquaredLoadings		
	Total	% of variance	Cumulative%	Total	% of variance	Cumulative %	Total	% of variance	Cumulative %
1	9.216	70.89	70.89	9.216	70.89	70.89	8.824	67.879	67.879
2	1.632	12.555	83.445	1.632	12.555	83.445	2.024	15.566	83.445
3	0.963	7.41	90.855						
4	0.684	5.265	96.12						
5	0.305	2.344	98.464						
6	0.12	0.922	99.386						
7	0.058	0.449	99.836						
8	0.009	0.069	99.904						
9	0.006	0.045	99.949						
10	0.004	0.03	99.98						
11	0.003	0.02	100						
12	4.36E-05	0	100						
13	4.26E-17	3.28E-16	100						

The factor load matrix can be obtained after principal component analysis. For more intuitive analysis, the maximum variance method is used to rotate the factor load matrix, and the rotated factor load matrix is obtained in table 18.

Table 18. Factor and rotational factor load matrices.

Factor load matrix			The rotated factor load matrix		
	one	2		one	2
X5	0.992	0.073	X5	0.982	0.154
X9	0.99	0.022	X9	0.969	0.204
X11	0.97	-0.047	X4	0.966	0.091
X7	0.969	-0.042	X12	0.955	0.109
X8	0.965	-0.024	X7	0.934	0.261
X4	0.962	0.131	X11	0.934	0.266
X10	0.961	-0.062	X8	0.934	0.242
X12	0.955	0.111	X10	0.922	0.279
X13	0.831	0.154	X13	0.844	0.038
X6	0.658	0.624	X6	0.782	-0.458
X2	0.653	-0.574	X1	0.161	0.822
X1	0.343	-0.764	X2	0.505	0.707
X3	-0.117	0.513	X3	0.003	-0.526

From Table 18, we can see that the first principal component is mainly X5, X9, X4, X12, X7, X11, X8, X10, X13 and X6, and the second principal components are mainly X1, X2 and X3. That is to say, first principal component mainly represents population factors, economic factors and technical factors such as total households, total population, agricultural population, GDP, primary industry, secondary industry, third industry, per capita disposable income of farmers, total power of agricultural machinery, and fertilizer application amount; The second principal component mainly represents climatic factors such as annual average temperature, annual total sunshine hours and annual total precipitation.

Therefore, the main driving forces for the change of ecological environment in Boao Town is human factors, followed by climate and natural factors such as temperature and sunshine. The next is economic factors such as GDP, primary industry, secondary industry, third industry and per capita disposable income of farmers.

5. Conclusions

This paper analyzes the dynamic changes of ecological environment in Boao town in recent 20 years (land use, vegetation coverage and landscape pattern), and discusses the driving forces of ecological environment changes in Boao town from human and natural factors. The main conclusions are as follows:

(1) The main types of land use in Boao Town are forestland and cropland. In 2000, the distribution of land use in Boao Town was as follows: forestland accounted for 54.92%, cropland accounted for 27.57%, water accounted for 16.22%, and built-up land accounted for 1.29%. In 2010, the distribution of land use in Boao Town was: forest land accounted for 51.31%, cropland accounted for 31.99%, water accounted for 15.2%, and built-up land accounted for 1.49%; In 2020, the distribution of land use in Boao Town was as follows: forest land accounted for 30.35%, cropland accounted for 33.81%, water accounted for 20.55%, and built-up land accounted for 15.29%. The results showed that the city development of Boao Town was obvious in the past 20 years, and the built-up land area was greatly increased.

(2) From 2000 to 2020, the biggest change of land use attitude in Boao Town was built-up land, while the smallest change was cropland, which showed that the built-up land expanded rapidly in Boao Town in recent 20 years, while the change speed of cropland was slow, and the change speed of forest land and water was consistent. The dynamic attitudes of comprehensive land use in 2000, 2010 and 2020 were 1.20%(2000-2010), 2.98%(2010-2020) and 1.15%(2000-2020), respectively. This shows that the land use change rate of Boao Town in the last 10 years was greater than that of Boao Town in the first 10 years, then greater than that of Boao Town in the last 20 years. The land use change rate of Boao Town is gradually increasing with the increase of years.

(3) In 2000, the distribution of vegetation coverage in Boao Town was as follows: high vegetation coverage (58.07%) > low vegetation coverage (16.94%) > medium high vegetation coverage (11.96%) > medium vegetation coverage (5.81%) > medium low vegetation coverage (5.81%). The distribution of vegetation coverage in Boao Town in 2010 was as follows: high vegetation coverage area (34.73%) > medium-high vegetation coverage area (25.45%) > low vegetation coverage area (15.51%) > medium vegetation coverage area (14.95%) > medium-low vegetation coverage area (9.94%). The distribution of vegetation coverage in Boao Town in 2020 was as follows: high vegetation coverage area (56.75%) > low vegetation coverage area (14.69%) > medium high vegetation coverage area (12.36%) > medium vegetation coverage area (9.11%) > medium low vegetation coverage area (7.68%). The average vegetation coverage was 70.68% in 2000, 62.49% in 2010 and 71.65% in 2020. This shows that the vegetation coverage in Boao Town has a trend of decreasing first and then increasing in the past 20 years.

(4) The change trend of landscape pattern index at the type level from 2000 to 2020 was: the landscape shape index and landscape fragmentation of cropland gradually decreased, the cohesion

slightly increased and the change was not obvious, and the separation degree had no change; The landscape shape index of forest land continued to rise, and the upward trend in the last 10 years was more obvious than that in the previous 10 years. The landscape fragmentation continued to rise, the cohesion slightly decreased and the change was not obvious, but the separation gradually increased; The landscape shape index of water showed an overall upward trend of rising first and then falling: the landscape fragmentation continued to rise, the cohesion slightly increased and then decreased slightly, and the overall change was not significant, and the separation degree did not change. The landscape shape index and cohesion of built-up land showed an upward trend, which was obvious in the past10 years. The landscape fragmentation first decreased a little and then increased a lot, but the separation degree remained unchanged. This shows that in the past 20 years, human activities had obviously affected the landscape pattern changes of various land types, cropland tends to be simplified in shape rules and spatial agglomeration. The shape of forest land and built-up land tended to be complicated, and the fragmentation and discreteness increased.

(5) The trend of landscape pattern change index at landscape level from 2000 to 2020 was as follows: landscape fragmentation continued to increase, separation level continued to rise, diversity and evenness continued to increase, and the changes of all kinds of indexes were stronger in the last 10 years. That is, the landscape pattern changed obviously from 2010 to 2020. The change of landscape pattern index in the past 20 years also fully reflects the interference of human activities on the local landscape pattern, which makes the landscape pattern be discrete, richer and more diverse in spatial structure.

(6) Based on natural and human factors, this paper analyzes the driving forces that lead to the change of ecological environment in Boao Town. Natural factors include precipitation, sunshine, temperature and other climatic factors, while human factors include population, economy, technology and policy. The results of principal component analysis showed that human factors were the leading factors making ecological environment change in Boao Town, and economic factors such as GDP, third output value and per capita disposable income of farmers were the main driving forces of landscape pattern changes in Boao Town.

6. References

[1] Zhao J 2009 Ecological environment evolution and driving force analysis of Sanjiangyuan based on RS and GIS technology(Jilin University)

[2] Avashia V, Garg A, Dholakia H 2021 Understanding temperature related health risk in context of urban land use changes *Landscape and Urban Planning* p 212

[3] Maheng D, Pathirana A, Zevenbergen C 2021 A Preliminary Study on the Impact of Landscape Pattern Changes Due to Urbanization: Case Study of Jakarta, Indonesia *Land* **10** p 218

[4] Lasode M K, Esobi I C, Anyanwu C I, Lasode D O 2020 2013 Assessing Urban Land Use Change in New Braunfels *Journal of Geoscience and Environment Protection***08** p 232-243

[5] Lopes T R, Zolin C A, Mingoti R, Vendrusculo L G, Almeida F T, Souza A P, Oliveira R F, Paulino J, Uliana E M 2021 Hydrological regime, water availability and land use/land cover change impact on the water balance in a large agriculture basin in the Southern Brazilian Amazon *Journal of South American Earth Sciences* p 108

[6] Kowe P, Mutanga O, Odindi J, Dube T 2021 Effect of landscape pattern and spatial configuration of vegetation patches on urban warming and cooling in Harare metropolitan city, Zimbabwe *GIScience & Remote Sensing* **58** p 261-280
[7] Marina P I, Josep M E, Javier G, Joan P 2020 Changes in forest landscape patterns resulting from recent afforestation in Europe (1990 - 2012): defragmentation of pre-existing forest versus new patch proliferation *Annals of Forest Science: A journal of the French National Institute for Agriculture, Food and Environment (INRAE)* **77** p 971-980
[8] Sharareh M, Carlos S C 2020 Statistical index (SI) as spatial modeling to evaluate the potential for greenbelt development in medium-sized Iranian cities *Spatial Information Research* **28** p 537-546
[9] Roger F A, Darrell E N, Steven K, Thomas R M J, Kristi L S 2012 The Driving Forces Of Land Change In The Northern Piedmont Of The United States *Geographical Review* **102** p 53-75
[10] Guilherme T N P L, Vivian C S H, Lidia S B, Rozely F S 2016 Identifying driving forces of landscape changes: Historical relationships and the availability of ecosystem services in the Atlantic forest *Ecosystem Services* **22** p 11-17
[11] Jitendra T Quantifying 2018 The Importance of Driving Forces of Landscape Change Using Remote Sensing: A Case Study *International Journal of Engineering and Future Technology* **15** p 15
[12] Zbelo T, Jan N, Tesfaalem G, Kelemework T, Amanuel Z, Jean P, Seppe D, Veerle V E 2019 Transhumance as a driving force of landscape change in the marginal grabens of northern Ethiopia *Singapore Journal of Tropical Geography* **40** p 20
[13] Difanty A, Supriatna S 2021 Spatial modeling for prediction agricultural land-use change in Jampang Kulon, Sukabumi Regency *IOP Conference Series: Earth and Environmental Science* **623** p 012-084
[14] Yang W 2014 Study on land use change in typical coastal zone: taking Sanya and Boao as examples(Nanjing University)
[15] Liu J 2008 Study on agricultural drought evaluation in three gorges reservoir area based on remote sensing technology(Southwest University)
[16] Lin H, Li N, Huang B D, Teng Y M 2020 Water quality evaluation of Nanliu river based on principal component analysis *Guangdong chemical industry* **47** p 144-146+148.

Acknowledgement
This work was financially supported by Hainan Provincial Department of Science and Technology under Grant No. ZDKJ2019006 and Hundred person project of Hainan province under Grant No.[2018]31.

Research on Disaster Acquisition Method Based on Point and Surface Combination of UAV Technology

Xuehua ZHANG, Duo DENG

National Earthquake Response Support Service, Beijing, China

E-mail: zhangxhua_nerss@163.com

Abstract. UAV (Unmanned Aerial Vehicle) remote sensing technology plays an increasingly important role in disaster acquisition. At present, after the disaster occurs, the orthophoto images of the UAV in the disaster area can be quickly obtained, but the bidimensionality of the orthophoto image limits the accuracy of disaster information extraction; it is required to improve the timeliness of the large-area 3D model established using the oblique photography technology in the disaster area. Combining the advantages of orthophoto and 3D models, this paper proposes a method for rapid acquisition of disasters based on point-surface combination. The Tibet Milin M6.9 earthquake is used as an example to analyze the advantages and disadvantages of this method in practical applications.

1. Introduction

After the disaster occurs, it is important to obtain disaster information quickly so that decision makers can reasonably dispatch rescue equipment and other forces according to the extent of damage and resources available in different areas, and minimize the damage and loss of people and property in the disaster area. In particular, the occurrence of major natural disaster often causes problems such as communication disruption and traffic damage in the disaster area, leading to a failure in timely conveying the situation of the disaster area, which brings certain difficulties to disaster assessment, command and rescue, and other work (Zheng, *et al.*2010, Wang, *et al.*2015). After years of development and practice, UAV remote sensing technology plays an important role in disaster information acquisition and monitoring, emergency rescue, disaster assessment, and post-disaster reconstruction in the face of major disasters such as earthquakes, geologic hazards, and freezing rains and snow (Shuai, *et al.*2018, Liang, *et al.*2013, He, *et al.*2017, Lei, *et al.*2011).

UAV remote sensing technology is used to obtain the spatial remote sensing information of ground features by carrying optical cameras, multispectral and other payloads on UAV flight platforms. UAV has been increasingly used in aerial photography, reconnaissance and monitoring because of its low cost, easy take-off and landing, and beyond-visual-range (BVR) autonomous flight. Meanwhile, when encountering interrupted communication and blocked traffic in the disaster, UAV can quickly cross mountains and rivers and penetrate deeply into the disaster area to obtain relevant information, providing accurate information for rescue and relief in a timely manner (Lu, *et al.*2011).

At present, the main way of using UAV remote sensing technology to timely acquire disaster information is to obtain wide-range, high-resolution orthophoto remote sensing images of the disaster area using the fixed-wing UAV mounted optical cameras, so as to quickly grasp the overall disaster situation. However, as these data are 2D images, the amount of information is limited and cannot express the disaster information of the disaster-affected area in an all-round way, which cannot meet the needs of accurate disaster assessment and other aspects. The development of oblique photography by UAV has broken through the limitations in orthographic aerial survey by 2D images, namely to obtain multi-angle image information of vertical and inclined targets simultaneously by mounting multiple cameras in different angles on the same flying platform. A high-precision and omni-directional three-dimensional model can be established for targets through data processing of professional software. Combined with measuring tools, the length, width, area and volume of damages to the ground objects can be measured, so as to give an accurate assessment of the disaster situation from a quantitative perspective (Gerke, *et al.*2011). However, the current oblique photography data processing method has the disadvantages of immature algorithm, slow processing speed and long-time consumption. Even using medium- and high-end computer hardware equipment, it takes 80–100 h/km2 to establish the 3D fine model of disaster areas, which is far from meeting the demand for rapid emergency response in large areas (Huang, *et al.*2017). Therefore, it is important to find a suitable method that can satisfy both demands for large-area disaster information acquisition and accurate disaster assessment.

Depending on the orthophoto aerial survey that can quickly acquire large area disaster information, and the oblique photography that can display disaster information in an all-round and quantitative manner, this paper proposes a rapid disaster information acquisition method based on point-surface combination, aiming to quickly and comprehensively display the situation in the disaster area and provide emergency rescue, disaster assessment, and other services.

2. Methodology

To meet the emergency demand for rapid response and the purpose of accurate disaster assessment, emergency aerial surveys by orthophoto were first carried out in key disaster-affected areas, UAV oblique photography technology was used to take oblique photography of the key and typical damage points found during the aerial survey and after data processing, and the orthophoto aerial survey data was quickly stitched and processed at the disaster site to obtain the Digital Orthophoto Map (DOM) of large disaster areas. Meanwhile, the 3D models with different degrees of refinement were processed and constructed taking into account the number of oblique shot points, the area, the processing equipment on site and other factors. The technology roadmap is described as follows:

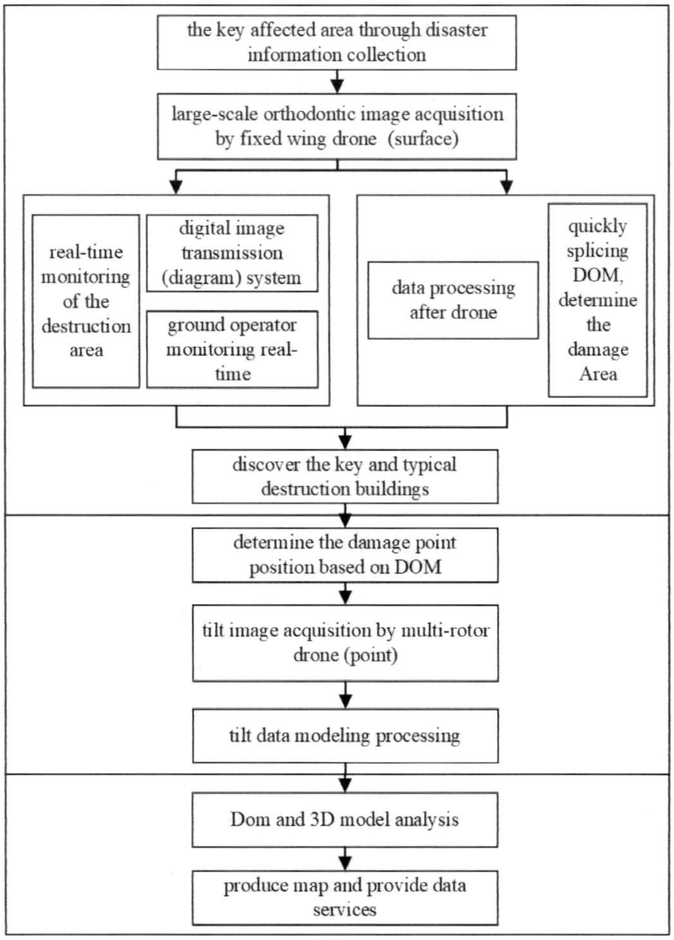

Figure 1. Technology roadmap

The fixed-wing UAV or multi-rotor UAV is used to fly at high altitudes to quickly acquire orthophoto data of the disaster area. The UAV remote sensing system equipped with a digital image transmission system can perform a visual interpretation during the UAV flight and record the current flight path points of UAV to determine the location and orientation of key and typical damage points. The UAV remote sensing system without a digital transmission system can conduct the DOM fast stitching processing of remote sensing data after the UAV has landed, and then carry out the visual interpretation or automatic computer interpretation of DOM to determine the location of key and typical damage points.

After determining the location of the key and typical damage points, the multi-rotor UAV is used to acquire oblique photography data of the damage points. The analysis of the data post-processing process shows that the core task of data acquisition is to obtain all-round image information of the target, so in addition to using a professional multi-lens oblique photography system, a consumer-grade single-lens UAV can also be used to take low-altitude flight shots of the target point in accordance with the requirements of oblique photography technology (Gu, *et al.*2017, Li, *et al.* 2016).

In the data acquisition phase, "surface" refers to the orthophoto image of a large disaster area acquired using a fixed-wing or high-altitude multi-rotor UAV, while "point" refers to the oblique image of a key and typical damage point acquired using multi-rotor oblique photography technology. Through integrating the respective advantages of the fixed-wing UAV or high-altitude multi-rotor UAV orthophotography system and the multi-rotor oblique photography system, the acquisition of disaster image data based on the point-surface combination can not only obtain wide-range rough disaster information, but also show the details of small-scale disaster damage and help perform accurate measurement, thus achieving the complementation between the information of different levels and heights, provding wide-range disaster depictions, and characterizing the reproduction of small-range typical damage, so as to meet the demands for emergency rescue, disaster assessment and other work.

3. Data processing technology

3.1 Orthphoto aerial survey data processing

Automatic stitching of UAV data is the process of automatically stitching the image sequences with a certain degree of overlap into a ground panoramic image with a large field of view, which has been developed maturely in foreign countries and formed a relatively complete theory (Li, *et al.*2012). At present, there is a lot of software for fast stitching orthophoto in the market, and most of the oblique image processing software has the function of fast stitching orthophoto, such as Photoscan, Pix4Dmapper, PixelGrid (an integrated mapping system for high-resolution remote sensing image data developed by Chinese Academy of Surveying & Mapping), MapMatrix and so on.

3.2 Oblique photography data processing

In terms of oblique image processing, apart from pre-processing such as geometric correction and image enhancement for the original image data, the main processing steps also include aerial triangulation, multi-view image dense matching and 3D modelling, and etc. The resulting remote sensing images not only provide a true and comprehensive picture of the ground conditions, but also accurate geographical information.

At present, there are mature image post-processing technologies and intelligent image processing software systems for oblique photography data. Common software includes Pictometry oblique image processing software, Smart3D (Bentley), Pixel Factory (Infoterra), LPS Workstation(Leica), DP-Modeler (TJH) and other oblique photography software (Yang, *et al.*2016). The data post-processing software not only integrates the data processing functions of oblique photography technology, but also has spatial analysis and application functions, which can accurately determine the 3D spatial coordinates of each point on the image, measure the length and volume, etc., and then explore more real environmental information. It plays an increasingly important role in disaster emergency, recovery and reconstruction.

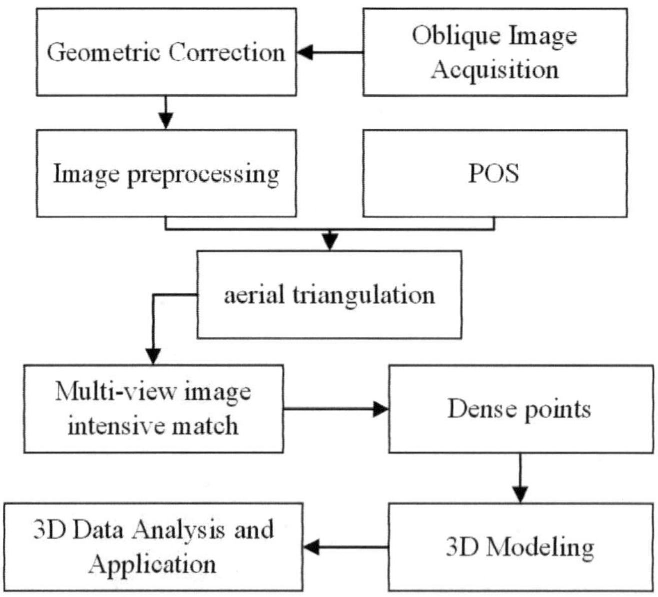

Figure 2. Oblique images processing flow

IOP Conf. Series: Earth and Environmental Science 1004 (2022) 012025 doi:10.1088/1755-1315/1004/1/012025

4. Applications analysis

As measured by the China Earthquake Networks Center, on November 18, 2017 at 06:34 Beijing time, a M6.9 earthquake occurred in Milin County, Nyingchi Prefecture, Tibet, with the epicenter located at 29.75°N, 95.02°E, a hypocentral depth of 10km and a maximum earthquake intensity of VIII (8 degrees). The epicenter was located in the uninhabited area of a high mountain valley. The village and town nearest to the epicenter location was Jiala Village, Pai Town, Milin County, with a distance of approximately 12 km, which was sparsely populated and had no casualties (Figure 3).

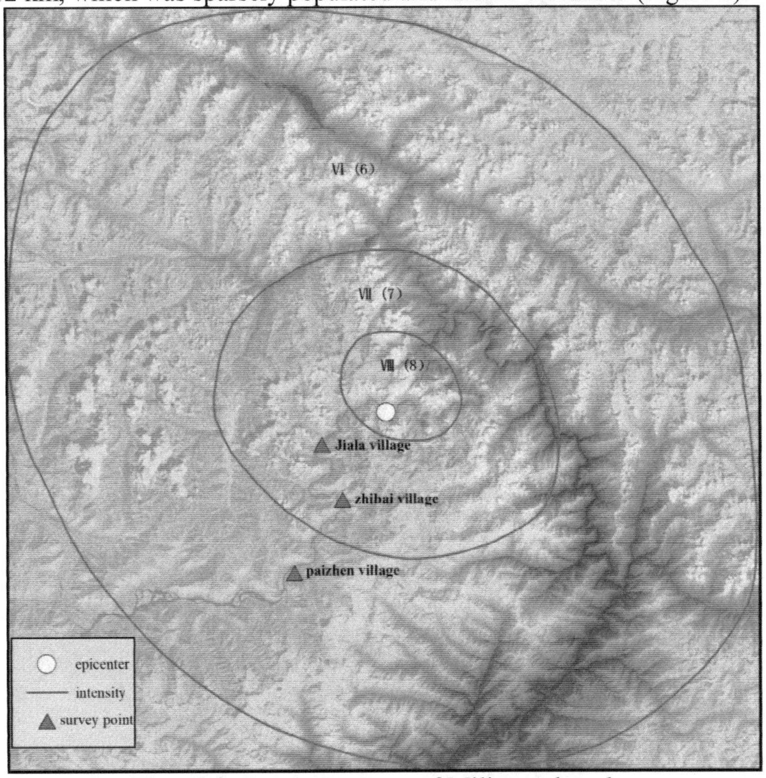

Figure 3. Landscape map of Milin earthquake

The author, as a member of the on-site team of the China Earthquake Administration, carried out an earthquake damage survey in Pai Town, Zhibai Village and Jiala Village in Milin County. It is found that there is no obvious damage to buildings in Pai Town on the UAV images, the houses and population in Jiala Village are less and concentrated. The earthquake damage survey was completed through visits. In Zhibai Village, the buildings are scattered and some buildings have serious damage. Therefore, this paper takes Zhibai Village as an example for illustration.

As Zhibai Village was approximately 16.5km away from the epicenter, the earthquake did not cause any obstruction to the roads to Zhibai Village, so the investigators were able to enter the village directly. During the survey of this village, orthophotography of the entire village area was conducted using the multi-rotor UAV with automatic route planning function. During the shooting, the real-time visual interpretation of the returned images revealed the presence of a rubble and stick pile under the eaves of a building located southeast of the take-off point, which was preliminarily concluded to be a vandalized structure (Figure 4). An investigation was then performed using another multi-rotor UAV, and oblique aerial photography was carried out to produce a refined 3D model on site, which was combined with ground survey data to make thematic maps for the command.

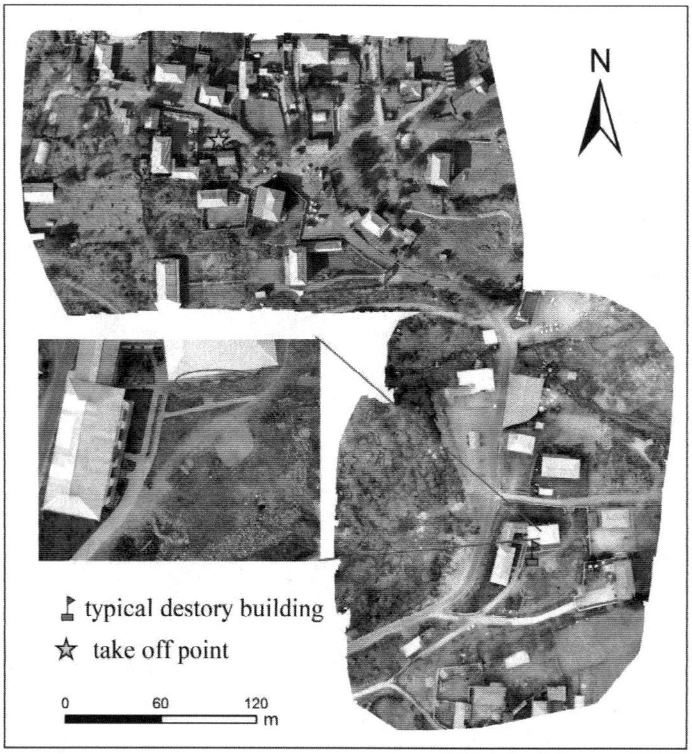

Figure 4. Orthphoto and aerial location map of Zhibai village

5. Conclusion

Combining the respective advantages of orthophoto and oblique image, the UAV damage survey method based on point-surface combination can achieve wide-range disaster depictions and small-area disaster reproduction, and fully take into account the wide range and processing speed of orthophoto and the detailed display characteristics of oblique image, thus providing a reliable basis for rapid and accurate assessment of the disaster. Due to the relatively minor damage caused by the Milin earthquake, the advantages of this method have not been well illustrated, but the usefulness of this method for accurate disaster assessment has been demonstrated. In subsequent potential disaster investigations, the author will continue to use the UAV disaster acquisition method based on point-surface combination to obtain disaster data to serve all levels of command.

Reference

[1] Zheng T Y, Li Y, Hou J S, et al. 2010. Review on earthquake disaster loss in Chinese mainland in 2008. *Journal of Catastrophology.* Vol **25**, No 2, pp 112-118.

[2] Wang X Q, Dou A X, Ding X, et al. 2015. Advance on the RS-based emergency seismic intensity assessment. *Journal of Geo-information Science.* Vol **17**, No 12, pp 1356-1544.

[3] Shuai X H, Liu Q, Zhen M, et al. 2018. Application of the UVA oblique photography technique in the field of Ludian earthquake, Yunnan. *Technology for Earthquake Disaster Prevention.* Vol **13**, No 1, pp 158-167.

[4] Liang J T, Cheng Y L, Wang J, et al. 2013. Variations of Galactic. *The Chinese Journal of Geological Hazard and Control.* Vol **24**, No 3, pp 54-61.

[5] He J, Tang C, Wang S Y, et al. 2017. Application of UAV images to geological disaster investigation. *Engineering of Surveying and Mapping.* Vol **26**, No 5, pp 40-45.

[6] Lei T J, Gong A D, Li C C, et al.2011. Application of UAV remote sensing system monitoring in the low-temperature and frozen ice-snow disaster. *Journal of Anhui Agri.Sci.* Vol **39**, No 4, pp 2417-2419,2423.
[7] Lu B D, Meng A W, Lu M, et al. 2011. Application and exploration of unmanned aerial vehicle in major natural disasters. *Journal of Catastrophology*. Vol **25**, No 2, pp 122-126.
[8] Gerke M, Kerle N. 2011. Automatic structural seismic damage assessment with airborne oblique pictometry imagery. *Photogrammetric Engineering & Remote Sensing*. Vol **77**, No 9, pp 885-898.
[9] Huang M E, Nan S, Zhou X H, et al. 2017. The application on emergency surveying and mapping service based on the vertical and oblique image fusion in comprehensive and individual way. *Urban Geotechnical Investigation & Surveying*. No **2**, pp 89-92.
[10] Gu G J, Zhang K P, Liu Z C, et al. 2017. Introduction to the standard of technology of tilt photogrammetry for UAV. *Bulletin of Surveying and Mapping*. **S1**, pp 210-213.
[11] Li X, Quan C W, Chen L. 2016. Research on the method of consumption-type UAV oblique photography data collection. Surveying Mapping. Vol **39**, No 6, pp 263-265,280.
[12] Li M, Liu H, Zhu X Y. 2012. Approach to fast mosaic UAV images for disaster emergency. *Journal of Catastrophology*. Vol **27**, No 3, pp 139-144.
[13] Yang G D, Wang M S. 2016. The tilt photographic measuration technique and expectation. *Geomatics & Spatial Information Technology*. Vol **39**, No 1, pp 13-15.

Acknowledgments
The authors would like to thank the anonymous reviewers. This work was supported by the National Key R&D program of China (2017YFB0504104).

AUTHOR INDEX